Efficient Chemical Technologies and Adsorbents for Environmental Pollution Removal and Wastes Recycling II

Efficient Chemical Technologies and Adsorbents for Environmental Pollution Removal and Wastes Recycling II

Guest Editors

Monika Wawrzkiewicz
Anna Wołowicz

MDPI

Basel • Beijing • Wuhan • Barcelona • Belgrade • Novi Sad • Cluj • Manchester

Guest Editors

Monika Wawrzkiewicz
Department of Inorganic Chemistry
Maria Curie-Sklodowska University
Lublin
Poland

Anna Wołowicz
Department of Inorganic Chemistry
Maria Curie-Sklodowska University
Lublin
Poland

Editorial Office

MDPI AG
Grosspeteranlage 5
4052 Basel, Switzerland

This is a reprint of the Special Issue, published open access by the journal *Molecules* (ISSN 1420-3049), freely accessible at: https://www.mdpi.com/journal/molecules/special_issues/2L728C0Q79.

For citation purposes, cite each article independently as indicated on the article page online and as indicated below:

Lastname, A.A.; Lastname, B.B. Article Title. *Journal Name* **Year**, *Volume Number*, Page Range.

ISBN 978-3-7258-7745-4 (Hbk)
ISBN 978-3-7258-7746-1 (PDF)
https://doi.org/10.3390/books978-3-7258-7746-1

Contents

About the Editors

Monika Wawrzkiewicz

Since 2005, Monika Wawrzkiewicz has worked as a Professor in the Department of Inorganic Chemistry, Faculty of Chemistry, Maria Curie-Sklodowska University, in Lublin (MCSU). She obtained her Ph.D. in Chemistry (2005) and then Habilitation (2015) at MCSU. She is a lecturer in inorganic chemistry for students of the Faculty of Chemistry at MCSU, and her research interests include the sorption and separation of inorganic and organic compounds, with particular emphasis on dyes and heavy metal ions on various sorbents, synthesis, and characterization of hybrid adsorbents and composites, as well as environmental protection. She is the author of more than 64 international peer-reviewed papers and book chapters and is a member of the Polish Chemical Society and the European Chemical Society.

Anna Wołowicz

Anna Wołowicz is an Assistant Professor in the Department of Chemistry at the Faculty of Chemistry of Maria Curie-Sklodowska University in Lublin (MCSU). She obtained her Ph.D. in Chemistry (2010) and then Habilitation (2024) at MCSU. Her research interests in the field of exact and natural sciences, within the discipline of chemical sciences, focus on environmental chemistry, particularly the sorption and separation of toxic contaminants, especially heavy metal ions, from water and wastewater. Her research includes the use of commercial sorbents and the design and modification of inorganic, polymeric, composite and hybrid adsorbents for contaminant removal. She has published over 50 scientific articles in JCR journals (45) (citations 1355/1179, H-index 21 according to Scopus) and non-JCR journals (7), 2 monographs, about 60 chapters in monographs, and 120 conference presentations. She is a member of the Polish Chemical Society and the European Chemical Society.

Preface

This preprint has been produced in response to the growing need to disseminate reliable scientific knowledge on innovative methods of removing pollutants from the environment. Its aim is to present the current state of knowledge and research on adsorption materials and their practical applications in the remediation of water, wastewater, and soil. The collected papers demonstrate the diversity of research approaches and the interdisciplinary nature of contemporary environmental research. This preprint is intended for scientists and professionals seeking proven, innovative environmental protection solutions.

Adsorption has been an effective method of removing pollutants from water and industrial wastewater for many years. Recently, however, it has also gained importance as a means of recovering valuable (strategic) or scarce (critical) substances. Interest in adsorption processes occurring in the aquatic environment is steadily growing for several important reasons. These include increasingly stringent legal requirements regarding the quality of drinking and irrigation water, limits on wastewater discharge, the identification of new and emerging contaminants and their potential negative impact on human health and ecosystems, and the need to further improve wastewater treatment technologies to enable reuse. Further impetus for research in this area comes from promoting ecological solutions, shifting towards a circular economy, and the growing importance of managing natural resources and energy efficiently. This Special Issue, titled "Advanced Materials in Catalysis and Adsorption", contains nine scientific articles, including one review article, published by scientists from ten countries.

The review by Santoso et al. indicates that metal phenolic networks (MPNs) based on gallic acid (GA-MPNs) are a promising class of functional materials with significant potential for adsorption, surface modification, and applications in fields like the environment, energy, and biomedicine, due to their eco-friendliness, versatile adhesion, and tunable coordination chemistry. The review also explores synthesis strategies and ways to enhance the performance of GA-MPN adsorbents, including stand-alone MPNs, MPN-mediated mesoporous materials, MPN-MOF composites, and MPN-coated substrates. The authors point out that the further development of these materials requires advances in controlled synthesis and structural modification, as well as expanding their applications to areas such as catalysis and energy storage. In the context of sustainable development, a particularly promising area of research is the use of phenolic ligands derived from biomass, which could significantly improve the cost-effectiveness of practical GA-MPN solutions.

The remaining eight experimental studies focus on the practical application of innovative sorbents and biological methods for removing organic contaminants such as bisphenol A (Zaborowska et al., 2024), phenol (Khumalo et al., 2023) and synthetic dyes, e.g., Reactive Red 120 (Kuśmierek et al., 2024), Reactive Black 5 (Hussain et al., 2024), and Rhodamine B (Elamin et al., 2023), as well as inorganic heavy metal ions including Cu(II) (Kuśmierek et al., 2024), Pb(II) (Vázquez-Sánchez et al., 2023), and Cr(VI) (Khan et al., 2024). Modified mineral materials such as functionalized halloysite (Kuśmierek et al., 2024), acrylonitrile-grafted cellulose (Khan et al., 2024), nanocomposites (Elamin et al., 2023) and plant- or waste-derived biosorbents (Vázquez-Sánchez et al., 2023; Li et al., 2023) demonstrated high adsorption efficiency. For instance, using unmodified halloysite (NM-H) or halloysite modified with (3-aminopropyl)triethoxysilane (H-APTES) or sulfuric acid (H-SA) increased the adsorption efficiency of Cu(II) ions and Reactive Red 120 dye (Kuśmierek et al., 2024), in a similar way to that of chemically modified cellulose derived from sugarcane bagasse for removing Cr(VI) from wastewater (Khan et al., 2024). Some studies extend classical adsorption processes to include catalytic ozonation (Hussain et al., 2024) and the bioconversion of organic waste (Li et al., 2023). Catalytic ozonation using iron-loaded dead leaf ash (Sapindus Mucorossi) has enabled the near-complete degradation of the textile dye Reactive Red 120 (Hussain et al., 2024). An important aspect of the research was to assess the impact of contaminants and sorbents on ecosystem functioning, including soil enzymatic activity and plant growth (Zaborowska et al., 2024), as well as adsorption mechanisms and sorption optimization. Additionally, the use of Black Soldier Fly Larvae has been demonstrated as an efficient biological strategy for the treatment and bioconversion of mixed organic wastes, reducing pollution while producing valuable biomass (Li et al., 2023).

The results confirm the significant potential of low-cost, sustainable environmental remediation technologies with high practical applicability.

Monika Wawrzkiewicz and Anna Wołowicz
Guest Editors

molecules

MDPI

Article

Evaluation of the Effectiveness of Innovative Sorbents in Restoring Enzymatic Activity of Soil Contaminated with Bisphenol A (BPA)

Magdalena Zaborowska , Jadwiga Wyszkowska *, Agata Borowik and Jan Kucharski *

Department of Soil Science and Microbiology, University of Warmia and Mazury in Olsztyn, Plac Łódzki 3, 10-727 Olsztyn, Poland; m.zaborowska@uwm.edu.pl (M.Z.); agata.borowik@uwm.edu.pl (A.B.)
* Correspondence: jadwiga.wyszkowska@uwm.edu.pl (J.W.); jan.kucharski@uwm.edu.pl (J.K.)

Abstract: As part of the multifaceted strategies developed to shape the common environmental policy, considerable attention is now being paid to assessing the degree of environmental degradation in soil under xenobiotic pressure. Bisphenol A (BPA) has only been marginally investigated in this ecosystem context. Therefore, research was carried out to determine the biochemical properties of soils contaminated with BPA at two levels of contamination: 500 mg and 1000 mg BPA kg^{-1} d.m. of soil. Reliable biochemical indicators of soil changes, whose activity was determined in the pot experiment conducted, were used: dehydrogenases, catalase, urease, acid phosphatase, alkaline phosphatase, arylsulfatase, and β-glucosidase. Using the definition of soil health as the ability to promote plant growth, the influence of BPA on the growth and development of *Zea mays*, a plant used for energy production, was also tested. As well as the biomass of aerial parts and roots, the leaf greenness index (SPAD) of *Zea mays* was also assessed. A key aspect of the research was to identify those of the six remediating substances—molecular sieve, zeolite, sepiolite, starch, grass compost, and fermented bark—whose use could become common practice in both environmental protection and agriculture. Exposure to BPA revealed the highest sensitivity of dehydrogenases, urease, and acid phosphatase and the lowest sensitivity of alkaline phosphatase and catalase to this phenolic compound. The enzyme response generated a reduction in the biochemical fertility index (BA_{21}) of 64% (500 mg BPA) and 70% (1000 mg BPA kg^{-1} d.m. of soil). The toxicity of BPA led to a drastic reduction in root biomass and consequently in the aerial parts of *Zea mays*. Compost and molecular sieve proved to be the most effective in mitigating the negative effect of the xenobiotic on the parameters discussed. The results obtained are the first research step in the search for further substances with bioremediation potential against both soil and plants under BPA pressure.

Keywords: BPA; remediating substances; adsorbents; *Zea mays*; enzymes activity

1. Introduction

Modern agricultural practices have played a significant role in increasing productivity in the 20th century. However, they have also led to environmental degradation in many parts of the world, as agriculture occupies about 38% of the world's land area, totaling about 5 billion hectares [1]. Therefore, one of the greatest challenges is to meet the growing global demand for food and energy crops while taking into account the concept of a 'circular economy' [2]. This has led to an 'ecological shift' that is currently practiced in 187 countries and covers 72.3 million hectares of agricultural land [3]. These efforts are compounded and aligned with global industrialization, which, through industrial expansion, has contributed to a 65% increase in environmental pollution levels over the last 20 years [4]. Global market statistics indicate that plastics, including the production and use of plastic additives such as bisphenols, are most responsible for these trends [5,6].

It is not without reason that BPA has been commercially produced in the United States since 1957 [7]. Its potential lies in retarding the oxidative degradation of plastics exposed to

UV radiation [8]. Desired are also its properties, such as high strength, heat resistance, and chemical stability [9], hence the impressive list of applications for BPA. Among the most important are the production of polycarbonates and associated electrical and electronic devices as well as building materials [10], epoxy resins, PVC, and polyesters [11]. Growing urbanization and changing lifestyle patterns contribute to the demand for plastic products, thereby driving the development of the BPA market. This trend reflects the expected compound annual growth rate of the global market for plastic additives, which is predicted to be 5.7% for the period 2021–2028 [12]. BPA production is also forecast to grow by 6% per year until 2027 [13].

Stringent government regulations on reclassifying BPA as a xenobiotic are emerging as a result of numerous scientific reports on its significant distribution in all ecosystems and, more alarmingly, its toxicity to organisms [14–16]. These regulations aim to limit the market expansion of BPA. One of the more significant regulations that came into force in 2023 is the acceptable daily intake of BPA at the level of 0.2 ng kg^{-1} body weight [17]. BPA has also been included in the Human Biomonitoring for Europe database, which defines the level of toxicity of this phenolic compound for humans [18], as well as in the Community Rolling Action Plan, a list of chemical compounds that are being evaluated for the risk they pose not only to human health but also to the environment [19]. These decisions are well founded, as the adverse effects of prolonged exposure to BPA have been linked to endocrine disruption [20], brain, and nervous system damage, and fetal development [21,22]. It is also a phenolic compound responsible for the induction of carcinogenic changes [23], insulin resistance, type 2 diabetes [24], and male infertility [25].

The dispersion of BPA in environmental media primarily refers to air, water, and soil. In the air, the main source of BPA is the thermal treatment of waste, mainly e-waste [26], as well as emissions resulting from its storage and transformation. In India, these practices have contributed to an increase in airborne BPA concentrations up to 4550 pg m^{-3} of air [27]. Currently, scientific discourse also considers the role of microplastics, which are thought to act as vectors for the transport of bisphenols in the air due to their adsorption by hydrogen and halogen bonds [28]. Leakages from landfills and discharges from wastewater treatment plants are the main sources of BPA pollution in aquatic environments. This issue is particularly prominent in rivers in Asia. In India, BPA levels of up to 14,800 ng dm^{-3} have been detected in river water [29]. In Europe, the highest levels of BPA in surface water have been found in Spain (28–560 ng dm^{-3} of water) [30].

Soil is exposed to BPA contamination from several undisputed sources. These include e-waste landfills [31], herbicide leaching [32], and treated wastewater used for irrigation [33]. Biosolids are also currently receiving considerable attention [34]. In Australia, their agricultural application increased to 73% in 2021, equivalent to 350,000 tons of biosolids [35]. In the European Union and the United States, it oscillates around 35% and 55%, respectively [36,37]. Staples et al. [38] documented the highest concentrations of BPA in European sediments (95,000 µg kg^{-1} d.m. of sediments) and much lower levels in North America (14,200 µg kg^{-1} d.m. of sediments).

The growing awareness of the persistent environmental pollution caused by xenobiotics has led to a wide range of global solutions to this problem. In order to maintain the philosophy of sustainable development, sorbents are increasingly being used for the remediation of contaminants in soils. This practice is supported not only by their affordable price but also by their market availability and, importantly, their cation exchange capacity [39]. Similarly, high-quality compost is receiving considerable attention from researchers, as it promotes the proliferation of microorganisms, thereby increasing the pool of enzymes [40–42]. Equally noteworthy as a soil fertilizing substance is coniferous bark, which constitutes over half of the 22 Mt of industrial bark produced annually in the European Union, particularly in Sweden and Finland [43]. Therefore, valuable applications for it are being sought, including in the production of ethanol, methane, and tannins, and as a substance with bioremediation properties [44,45].

An important component of the scientific debate on soil response to the pressure of organic pollutants is the analysis and consideration of soil enzyme responses. This variable, which describes soil quality, is recognized as a critical parameter and thus particularly reliable in assessing the extent of disturbance to its equilibrium resulting from the interaction of soil enzymes with phenolic compounds, including BPA [46,47]. These disturbances are also reflected in the response of cultivated plants. BPA is rapidly metabolized by a wide range of plant species, and the extent of disruption to their growth and development is closely related to the dose of this xenobiotic [48].

The current research objective was formulated taking into account observed trends related to the quest for answers to the question: (1) what is the potential scale of the inhibitory effect of BPA on soil condition? (2) It aimed to determine the response of seven soil enzymes to soil contamination with BPA, (3) expanded with the response of maize to the applied phenolic compound, and (4) to diagnose the bioremediation potential of (i) molecular sieve, (ii) zeolite SO1, (iii) sepiolite, (iv) starch, (v) grass compost, and (vi) fermented bark. Importantly, there have been no previous studies verifying the effectiveness of these sorbents against BPA pressure. The hypothesis was posited that this phenolic compound is responsible for significant soil equilibrium disturbance, reflected in both the reaction of soil enzymes and *Zea mays*. It was also assumed that the applied remediation substances would demonstrate varied potential in improving soil conditions, giving the research strictly practical significance.

2. Results

2.1. Enzymatic Activity of Soil

The results obtained from the biochemical analyses indicate that bisphenol A (BPA) is a potent inhibitor of soil enzymatic activity, with different responses observed between individual enzymes to this xenobiotic. This is demonstrated by the mean values obtained and their corresponding homogeneous groups in each treatment (Table 1). Considering the sensitivity of the seven enzymes analyzed to both 500 mg and 1000 mg BPA kg^{-1} of soil, as determined based on the mean values of these parameters, the following ranking can be proposed: dehydrogenases (Deh) > urease (Ure) > acid phosphatase (AcP) > β-glucosidase Glu > arylsulfatase (Aryl) > catalase (Cat) > alkaline phosphatase (AlP) (500 mg BPA) and Ure > Deh > AcP > Aryl > Glu > Cat > AlP (1000 mg BPA). The extent of BPA inhibition is also illustrated by the mean values of the soil fertility index (BA_{21}), derived from the sum of all enzyme activities. It was observed that the application of 500 mg BPA kg^{-1} of soil resulted in a 64% lower BA_{21} value, and 1000 mg BPA kg^{-1} of soil resulted in a 70% decrease compared to the control. However, it should be emphasized that the activities of Deh, Ure, AcP, Aryl, and Glu were inhibited by BPA regardless of the soil contamination level with this phenolic compound. Conversely, an opposite trend was observed for AlP, and in the case of soil contaminated with 1000 mg BPA kg^{-1}, Cat exhibited a different response. It was therefore essential to study the response of each enzyme separately in order to accurately describe the extent of the disturbance of the soil equilibrium assessed by its biochemical properties.

Table 1. Enzyme activity in 1 kg d.m. of soil per 1 h.

Object	0_BPA	500_BPA	1000_BPA	Average
		Dehydrogenases, µmol TFF		
Control (Ct)	10.636 ^{b}	1.955 $^{f–h}$	1.167 ^{h}	4.586 ^{C}
Molecular sieve (M)	13.484 ^{a}	6.509 ^{c}	7.051 ^{c}	9.015 ^{A}
Zeolite (Z)	10.079 ^{b}	1.038 ^{h}	1.379 ^{gh}	4.165 ^{D}
Sepiolite (Sep)	10.754 ^{b}	3.366 ^{de}	2.285 $^{e–g}$	5.468 ^{B}
Starch (St)	10.355 ^{b}	2.674 ^{ef}	2.413 $^{e–g}$	5.147 ^{B}
Compost (Cp)	10.754 ^{b}	4.160 ^{d}	1.094 ^{h}	5.336 ^{B}
Bark (Bk)	10.012 ^{b}	1.863 $^{f–h}$	1.358 ^{gh}	4.412 ^{C}
Average	10.868 ^{I}	3.081 ^{II}	2.392 ^{III}	

Table 1. *Cont.*

Object	0_BPA	500_BPA	1000_BPA	Average
		Catalase, mol O_2		
Control (Ct)	0.243 ^{g}	0.238 ^{g}	0.274 ^{f}	0.252 ^{D}
Molecular sieve (M)	0.305 ^{de}	0.343 ^{c}	0.335 ^{c}	0.328 ^{B}
Zeolite (Z)	0.313 ^{d}	0.159 ^{i}	0.179 ^{h}	0.217 ^{E}
Sepiolite (Sep)	0.334 ^{c}	0.180 ^{h}	0.272 ^{f}	0.262 ^{CD}
Starch (St)	0.340 ^{c}	0.239 ^{g}	0.237 ^{g}	0.272 ^{C}
Compost (Cp)	0.387 ^{a}	0.287 ^{ef}	0.368 ^{b}	0.347 ^{A}
Bark (Bk)	0.287 ^{ef}	0.339 ^{c}	0.366 ^{b}	0.331 ^{B}
Average	0.316 ^{I}	0.255 ^{III}	0.290 ^{II}	
		Urease, mmol N-NH_4		
Control (Ct)	0.345 ^{g}	0.156 ^{i}	0.020 ^{j}	0.174 ^{F}
Molecular sieve (M)	1.512 ^{b}	0.430 ^{f}	0.252 ^{h}	0.731 ^{B}
Zeolite (Z)	1.211 ^{d}	0.028 ^{j}	0.026 ^{j}	0.422 ^{E}
Sepiolite (Sep)	1.329 ^{c}	0.030 ^{j}	0.114 ^{i}	0.491 ^{D}
Starch (St)	1.685 ^{a}	0.351 ^{g}	0.311 ^{g}	0.783 ^{A}
Compost (Cp)	1.206 ^{d}	0.225 ^{h}	0.246 ^{h}	0.559 ^{C}
Bark (Bk)	1.132 ^{e}	0.252 ^{h}	0.132 $_{i}$	0.505 ^{D}
Average	1.203 ^{I}	0.210 ^{II}	0.157 ^{III}	
		Acid phosphatase, mmol PN		
Control (Ct)	4.058 ^{c}	2.456 ^{fg}	2.364 ^{gh}	2.959 ^{C}
Molecular sieve (M)	2.319 ^{gh}	1.915 ^{i}	1.411 ^{j}	1.882 ^{E}
Zeolite (Z)	1.499 ^{j}	2.663 ^{ef}	2.321 ^{gh}	2.161 ^{D}
Sepiolite (Sep)	2.773 ^{e}	2.118 ^{hi}	1.554 ^{j}	2.148 ^{D}
Starch (St)	5.990^{a}	2.674 ^{ef}	2.250 ^{gh}	3.638 ^{B}
Compost (Cp)	5.249 ^{b}	3.494 ^{d}	3.511 ^{d}	4.085 ^{A}
Bark (Bk)	4.267 ^{c}	1.031 ^{k}	0.943 ^{k}	2.081 ^{D}
Average	3.737 ^{I}	2.336 ^{II}	2.051 ^{III}	
		Alkaline phosphatase, mmol PN		
Control (Ct)	0.330 ^{m}	0.558 $^{j–l}$	0.544 $^{j–l}$	0.477 ^{F}
Molecular sieve (M)	0.525 $^{j–l}$	0.605 $^{h–j}$	0.487 ^{kl}	0.539 ^{E}
Zeolite (Z)	0.460 ^{l}	0.518 $^{j–l}$	0.579 $^{i–k}$	0.519 ^{E}
Sepiolite (Sep)	0.751 ^{fg}	0.803 ^{ef}	0.877 ^{de}	0.810 ^{C}
Starch (St)	0.738 ^{fg}	0.556 $^{j–l}$	0.686 ^{gh}	0.660 ^{D}
Compost (Cp)	0.978 ^{cd}	0.667 $^{g–i}$	0.984 ^{c}	0.876 ^{B}
Bark (Bk)	0.719 ^{fg}	1.458 ^{a}	1.229 ^{b}	1.135 ^{A}
Average	0.643 ^{II}	0.738 ^{I}	0.769 ^{I}	
		Arylsulfatase, mmol PN		
Control (Ct)	0.176 $^{b–d}$	0.141 $^{e–g}$	0.121 ^{gh}	0.146 ^{C}
Molecular sieve (M)	0.177 ^{bc}	0.156 $^{c–f}$	0.179 ^{bc}	0.171 ^{B}
Zeolite (Z)	0.197 ^{b}	0.097 ^{h}	0.093 ^{h}	0.129 ^{D}
Sepiolite (Sep)	0.171 $^{b–d}$	0.233 ^{a}	0.169 $^{b–e}$	0.185 ^{A}
Starch (St)	0.228 ^{a}	0.148 $^{d–g}$	0.135 $^{f–g}$	0.170 ^{B}
Compost (Cp)	0.192 ^{b}	0.141 $^{e–g}$	0.148 $^{d–g}$	0.158 ^{C}
Bark (Bk)	0.163 $^{c–f}$	0.162 $^{c–f}$	0.156 $^{c–f}$	0.160 ^{BC}
Average	0.186 ^{I}	0.154 ^{II}	0.143 ^{III}	
		β-glucosidase, mmol PN		
Control (Ct)	0.472 ^{de}	0.377 $^{h–k}$	0.357 $^{i–k}$	0.402 ^{C}
Molecular sieve (M)	0.435 $^{e–h}$	0.345 ^{jk}	0.260 ^{l}	0.346 ^{D}
Zeolite (Z)	0.417 $^{e–i}$	0.467 $^{d–f}$	0.331 ^{k}	0.405 ^{C}
Sepiolite (Sep)	0.425 $^{e–h}$	0.381 $^{g–k}$	0.350 ^{jk}	0.384 ^{C}
Starch (St)	0.808 ^{a}	0.344 ^{jk}	0.405 $^{f–j}$	0.519 ^{B}
Compost (Cp)	0.568 ^{c}	0.447 $^{e–g}$	0.553 ^{c}	0.522 ^{B}
Bark (Bk)	0.569 ^{c}	0.732 ^{b}	0.519 ^{cd}	0.606 ^{A}
Average	0.528	0.442 ^{II}	0.396 ^{III}	

Table 1. *Cont.*

Object	0_BPA	500_BPA	1000_BPA	Average
		BA_{21}		
Control (Ct)	16.260 [c]	5.880 [g–i]	4.848 [jk]	8.996 [E]
Molecular sieve (M)	18.757 [b]	10.303 [e]	9.977 [e]	13.012 [A]
Zeolite (Z)	14.175 [d]	4.970 [jk]	4.909 [jk]	8.018 [F]
Sepiolite (Sep)	16.536 [c]	7.111 [f]	5.620 [i–k]	9.756 [D]
Starch (St)	20.145 [a]	6.986 [fg]	6.437 [f–i]	11.189 [C]
Compost (Cp)	19.333 [ab]	9.422 [e]	6.903 [f–h]	11.884 [B]
Bark (Bk)	17.148 [c]	5.836 [h–j]	4.703 [k]	9.229 [F]
Average	17.478 [I]	7.216 [II]	6.200 [III]	

BPA—bisphenol A; 0_BPA, 500_BPA, 1000_BPA—doses of BPA kg^{-1} d.m. of soil. Homogeneous groups were determined individually for each enzyme and BA_{21} (represented by letters [a–m]); averaged values, regardless of the level of soil pollution with BPA, are marked with letters [A–F], while the mean values, regardless of the type of remediation substance, are indicated by numbers [I–III]. significant at $p = 0.05$, $n = 63$.

In uncontaminated soil, the activity of dehydrogenases was measured at 10.636 µmol TFF kg^{-1} of soil h^{-1} (Table 1). The application of 500 mg of the phenolic compound resulted in an 82% inhibition of the activity of these oxidoreductases, while 1000 mg resulted in 89% inhibition compared to the control. The values of the bisphenol influence factor (IF_{BPA}) for the indicated treatments highlighted the trends observed for this parameter, assuming negative values of −0.816 and −0.890, respectively (Figure 1).

Studies indicate that catalase is less sensitive to bisphenol exposure than dehydrogenases. This is evidenced by the activity of this enzyme (Table 1), which increased by 13% at higher levels of BPA soil contamination (IF_{BPA} = 0.13) (Figure 1).

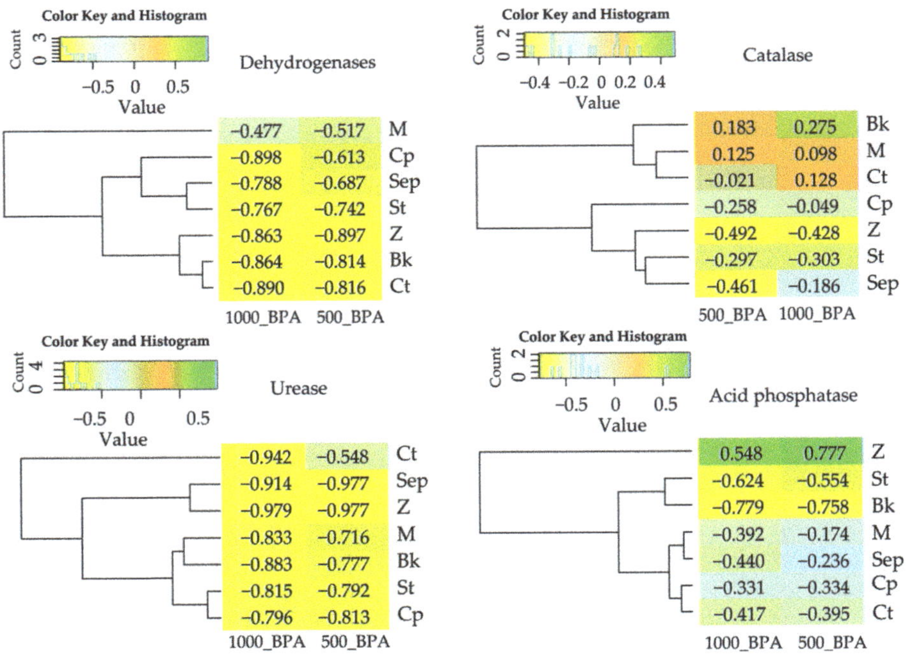

Figure 1. *Cont.*

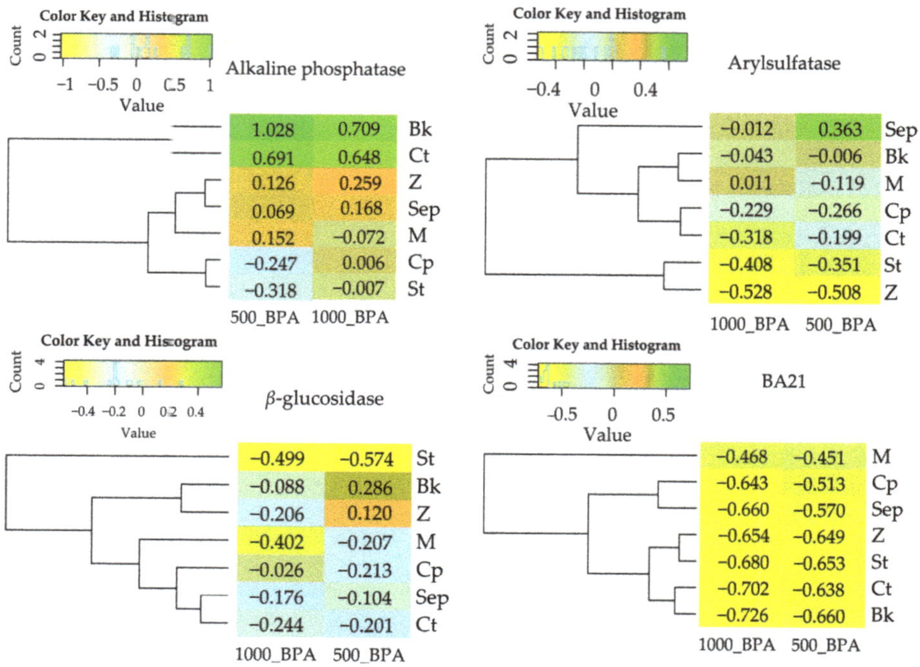

Figure 1. Index of the influence of BPA (IF_{BPA}) on the activity of soil enzymes. Ct—soil without remediating substances, M—molecular sieve, Z—zeolite, Sep—sepiolite, St—Starch, Bk—pre-fermented bark, Cp—grass compost, 500_BPA, 1000_BPA—doses of BPA kg^{-1} d.m. of soil. The abbreviation IF_{BPA} is explained in the Section 4.

Within the hydrolase class, urease was the most sensitive enzyme to the influence of BPA. Importantly, in contrast to dehydrogenases, the effects of inhibition at 500 mg of this phenolic compound were not as drastic as at 1000 mg BPA kg^{-1} d.m. of soil. The pressure of individual doses of the xenobiotic resulted in a 55% and 94% inhibition of Ure activity, respectively, compared to the control objects. The trends observed corresponded to the negative values obtained for IF_{BPA} (−0.548 and −0.942) (Figure 1). The responses of acid phosphatase and alkaline phosphatase to BPA soil contamination were different. Importantly, the activity of AcP was observed to be 11-fold higher than that of AlP in soil not contaminated with the phenolic compound. However, the sensitivity of AcP to BPA action was disproportionately higher than that of AlP. Remarkably, regardless of the level of soil contamination with the xenobiotic, AcP activity was inhibited in the range of 40–42%, whereas AlP activity increased by 69% in response to 500 mg BPA and by 65% after the application of 1000 mg of the phenolic compound to the soil compared to the control.

The response of arylsulfatase and β-glucosidase to soil contamination with 500 mg of BPA was similar. In the sensitivity ranking of enzymes to this phenolic compound proposed above, Aryl and Glu ranked fourth and fifth, respectively. An escalation of Aryl inhibition was observed in soil exposed to 1000 mg BPA kg^{-1} d.m. of soil. This resulted in a 32% inhibition of enzyme activity and a negative IF_{BPA} index value of −0.318.

In pursuit of one of the defined research objectives, the potential of seven sorbents to restore the biochemical equilibrium of the soil was determined (Table 1, Figure 2). Mitigation of the negative effects of BPA on Deh was effective with the application of a molecular sieve (M). The adsorbent not only increased enzyme activity by 27% in uncontaminated soil but also induced a 233% increase in activity in soil exposed to 500 mg and a remarkable 504% increase in soil exposed to 1000 mg BPA kg^{-1} d.m. of soil. Positive influence indices of amendments on Deh activity also indicate the beneficial effects of grass compost (Cp) in soil exposed to 500 mg BPA (IF_{RS} = 1.128) and starch (St) in soil exposed to 1000 mg BPA kg^{-1} of soil (IF_{RS} = 1.068) (Figure 2). Although Cat proved to be a more resistant enzyme to the inhibitory effect of the phenolic compound, the applied sorbents did not play a significant role in stimulating its activity. The obtained positive IF_{RS} values for M, Co, and B in BPA-contaminated objects ranged only between 0.206 and 0.441. The effectiveness of the remediation substances, assessed on the basis of the urease reaction, confirmed the bioremediation potential of St, M, and Co, especially in soil contaminated with the highest dose of BPA. The urease activity increased from 0.020 mmol $N\text{-}NH_4$ kg^{-1} of soil h^{-1} (Ct) to 0.351 (St), 0.252 (M), and 0.246 (Cp) mmol $N\text{-}NH_4$ kg^{-1} of soil h^{-1} (Table 1), and the IF_{SR} values were 14.550 (St), 11.600 (M), and 11.300 (Cp), respectively (Figure 2). It is also worth noting that all applied substances had a positive effect on Ure activity. Based on the verification of changes in AcP and AlP activity, it was found that grass compost effectively mitigated the inhibitory effect of BPA on these enzymes. However, pre-fermented bark generated the highest increase in AlP activity, by 161% in soil contaminated with 500 mg BPA and by 126% after application of 1000 mg BPA kg^{-1} d.m. of soil.

The values of the influence indices (IF_{RS}) on Aryl activity assigned to the objects with the compilation of BPA and individual sorbents indicated their relatively small positive effect on the soil condition, except for zeolite. Conversely, the assessment of their effectiveness through the prism of Glu reaction allowed for the identification of the bioremediation potential of pre-fermented bark. This was reflected both in the obtained average enzyme activity values, the assigned homogeneous groups, and the corresponding IF_{RS} values (0.942 and 0.454) in BPA-contaminated soil.

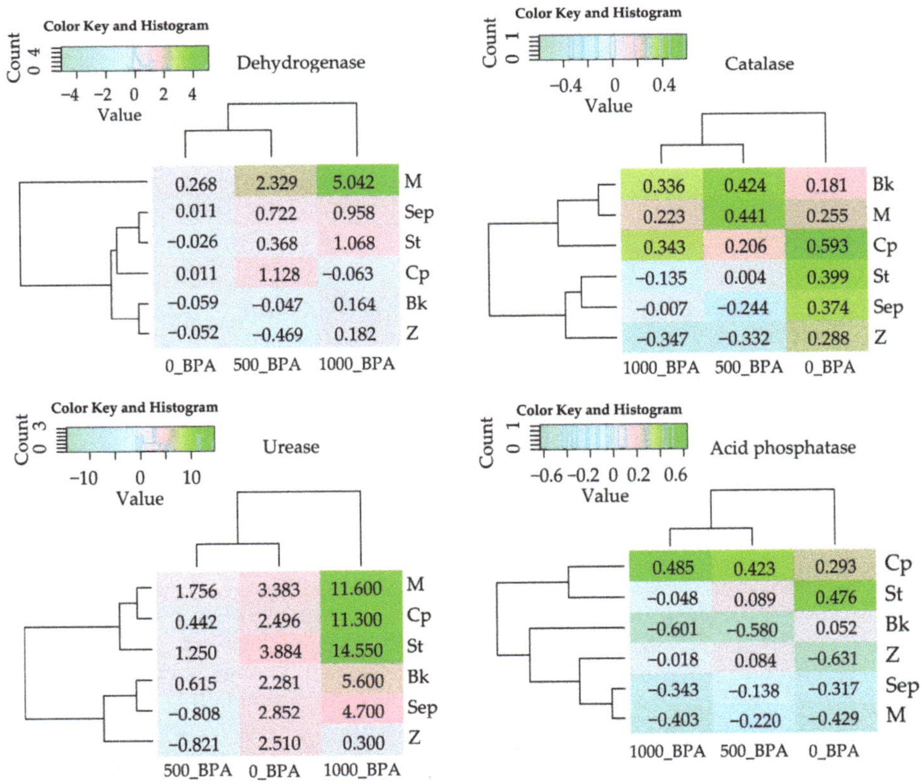

Figure 2. *Cont.*

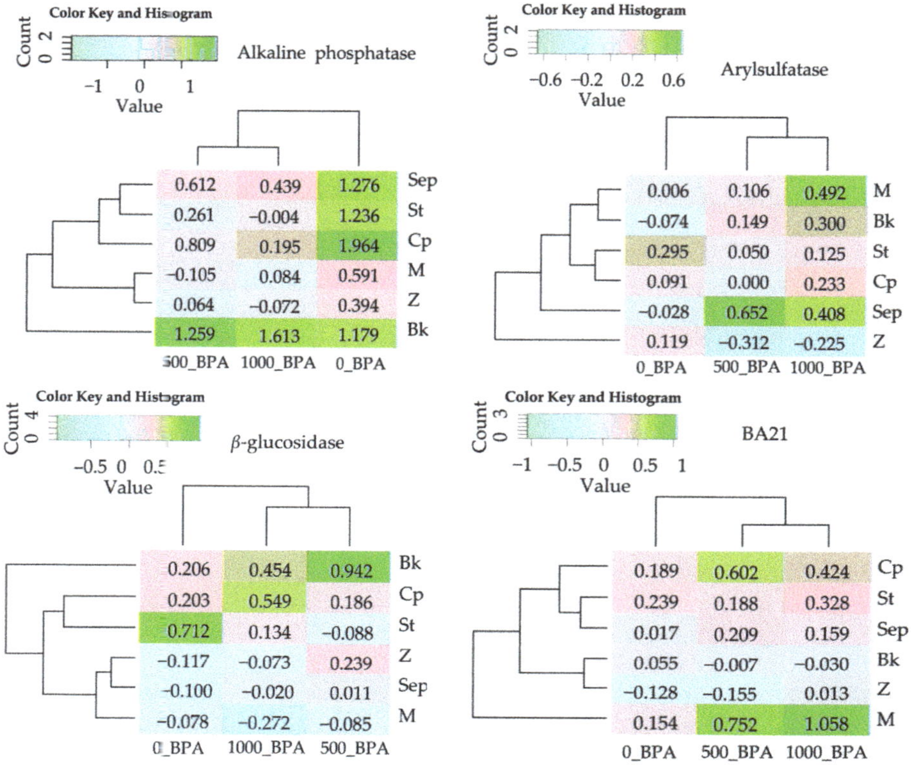

Figure 2. Index of the influence of remediation substances (IF_{RS}) on the activity of soil enzymes. The abbreviations M, Z, Sep, St, Bk, and Cp are explained in Figure 1. The abbreviation IF_{Rs} is explained in the Section 4

2.2. *Zea mays*

Given that soil condition defines its productivity, the response of *Zea mays* to BPA was monitored (Table 2). Considering the changes in the biomass of aerial parts and roots of the plants, a much stronger negative effect of higher doses of the xenobiotic on the growth and development of the cultivated plant was observed. The greater toxicity of BPA to its roots was also proven. The application of 500 mg BPA kg^{-1} soil resulted in a 60% inhibition of root yield, while the yield of aerial parts was reduced by 44% compared to the control. Conversely, in the group of objects subjected to the pressure of 1000 mg BPA kg^{-1} soil, the trend observed was a reduction in the yield of aerial parts of *Zea mays* similar to the yield of its roots. It oscillated at levels of 71% and 84%, respectively.

The response of *Zea mays* to soil contamination with this phenolic compound, analyzed on the basis of mean yield values and their corresponding homogeneous groups, was reflected in the negative values obtained for the index of bisphenol A influence (IF_{BPA}) in soil subjected to increasing levels of the xenobiotic (500 mg and 1000 mg). These values were significantly lower, particularly evident in the case of root biomass, with values of −0.596 and −0.836, respectively (Figure 3).

Soil contamination with BPA significantly disrupted the growth and development of *Zea mays* to such an extent that not all remediation substances applied were able to at least partially mitigate its adverse effects (Table 2, Figure 4). In the BPA-uncontaminated objects, the yield of aerial parts of *Zea mays* increased significantly after the application of compost, which was not observed in the case of plant roots. However, it is worth noting that in soil exposed to 500 mg kg^{-1} of soil, the bioremediation function was fulfilled by the molecular sieve (M) and compost (Cp). Their potential was further confirmed by the positive values of sorbent influence indices (IF_{RS}) on the aerial parts of the plant, which were 0.191 and 0.260, respectively. In soil with a compilation of 1000 mg BPA and sorbents, only M and Cp were identified as being equally effective in promoting the growth and development of *Zea mays*. These compounds contributed to an increase in aerial part yield by 90% (M) and 33% (Cp) compared to the control, and in parallel objects, promoted root biomass growth of 42% (M) and 64% (Cp) (Figure 4). The observed trends were confirmed by positive IF_{RS} values ranging from 0.326 (Cp, aerial parts of plants) to 0.896 (M, aerial parts of plants), and from 0.423 (M, roots of plants) to 0.637 (Cp, roots of plants).

Table 2. The yield of *Zea mays* (d.m. g pot^{-1}).

Object	0_BPA	500_BPA	1000_BPA	Average
		Aerial parts		
Control (Ct)	46.475 ^{bc}	26.130 ^{g}	13.570 ^{l}	28.725 ^{B}
Molecular sieve (M)	48.129 ^{bc}	31.109 ^{f}	25.732 ^{g}	34.990 ^{A}
Zeolite (Z)	45.988 ^{c}	20.702 ^{h}	15.896 ^{jk}	27.529 ^{BC}
Sepiolite (Sep)	44.333 ^{d}	24.965 ^{g}	15.868 ^{ij}	27.066 ^{B}
Starch (St)	35.693 ^{d}	23.165 ^{h}	15.246 ^{jk}	24.559 ^{C}
Compost (Cp)	52.997 ^{a}	32.924 ^{f}	17.999 ^{kl}	34.640 ^{A}
Bark (Bk)	48.352 ^{b}	20.419 ^{i}	9.400 ^{m}	26.057 ^{BC}
Average	45.995 ^{I}	25.632 ^{II}	16.244 ^{III}	
		Roots		
Control (Ct)	11.813 ^{a}	4.771 ^{d}	1.937 ^{fg}	6.174 ^{AB}
Molecular sieve (M)	11.581 ^{a}	3.909 ^{de}	2.757 ^{ef}	6.082 ^{AB}
Zeolite (Z)	11.795 ^{a}	4.520 ^{d}	2.409 ^{fg}	6.241 ^{A}
Sepiolite (Sep)	8.700 ^{b}	4.864 ^{d}	2.382 ^{fg}	5.315 ^{B}
Starch (St)	11.281 ^{a}	5.054 ^{d}	2.950 ^{ef}	6.428 ^{A}
Compost (Cp)	11.218 ^{a}	6.368 ^{c}	3.171 ^{ef}	6.919 ^{A}
Bark (Bk)	12.166 ^{a}	5.150 ^{cd}	1.137 ^{g}	6.151 ^{AB}
Average	11.222 ^{I}	4.984 ^{II}	2.392 ^{III}	

BPA—bisphenol A; 0_BPA, 500_BPA, 1000_BPA—doses of BPA kg^{-1} d.m. of soil. Homogeneous groups were determined individually for each part of *Zea mays* (represented by letters $^{a–m}$); averaged values, regardless of the level of soil contamination with BPA, are marked with letters $^{A–C}$, while the mean values, regardless of the type of remediation substance, are indicated by numbers $^{I–III}$. significant at p = 0.05, n = 63.

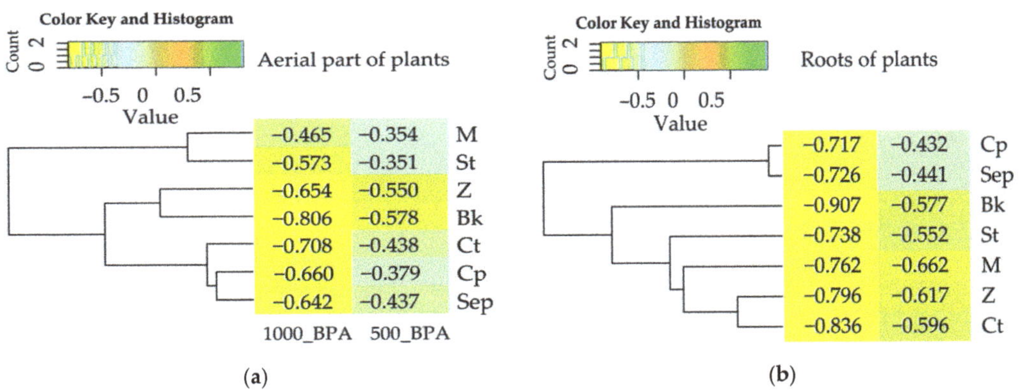

Figure 3. Index of the influence of BPA (IF_{BPA}) on the yield of *Zea mays* (**a**) aerial parts of plants, (**b**) roots of plants. The abbreviations Ct, M, Z, Sep, St, Bk, and Cp are explained in Figure 1 and the abbreviation IF_{BPA} is explained in the Section 4.

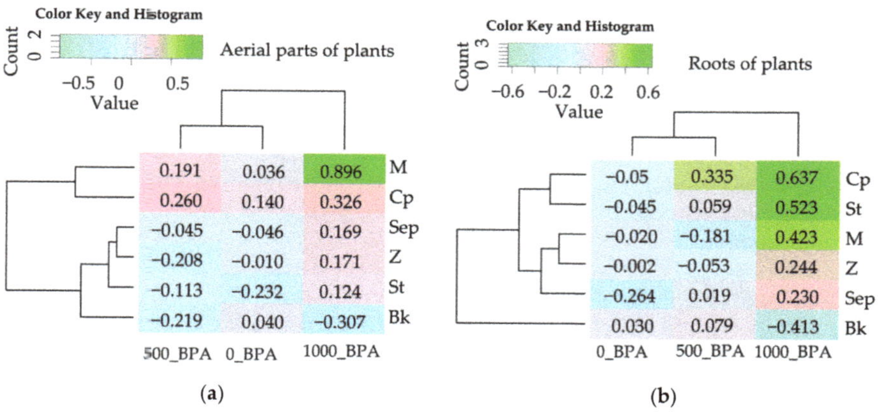

Figure 4. Index of the influence of remediation substances (IF_{RS}) on the yield of *Zea mays* (**a**) aerial parts of plants, (**b**) roots of plants. The abbreviations M, Z, Sep, St, Bk, and Cp are explained in Figure 1 and the abbreviation IF_{Rs} is explained in the Section 4.

A validating parameter for the relationships described was the estimated ratio of aerial parts biomass to root biomass (PR) of *Zea mays* (Figure 5). Its value was significantly positively correlated with increasing soil contamination with BPA. The combination of the phenolic compound and the molecular sieve generated the highest PR values regardless of the magnitude of the xenobiotic level.

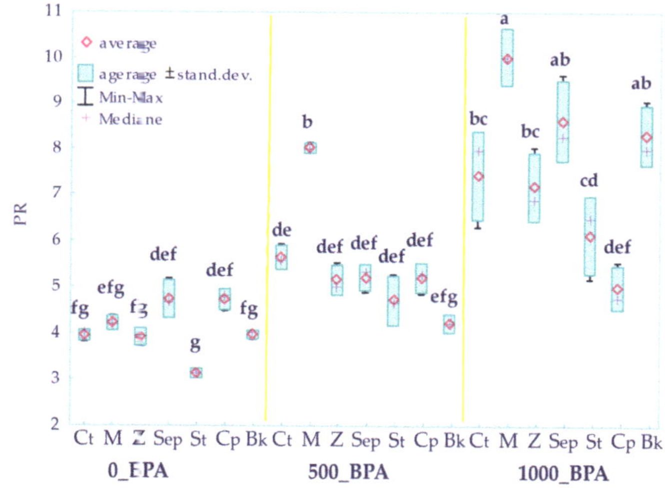

Figure 5. The ratio of the mass of the aerial parts of plants to the roots (PR) in the soil contaminated with BPA. The abbreviations are explained in Figure 1. Homogeneous groups were determined for the PR ratio assigned to all objects sown with *Zea mays* (represented by letters a–g), significant at $p = 0.05$, $n = 63$.

In the study conducted, the relative chlorophyll content was assessed using the SPAD leaf greenness index (Table 3). Increasing levels of soil contamination with BPA stimulated the synthesis of the photosynthetic pigment. The application of 500 mg BPA led to an increase in its content by 6%, and 1000 mg kg^{-1} of soil by 15%. The highest index values were recorded in soil subjected to the combinations of BPA and Cp, BPA and S, as well as in control objects enriched with Cp and Sep.

Table 3. The mean values of SPAD on the 50th day of growth of *Zea mays*.

Object	0_BPA	500_BPA	1000_BPA	Average
Control (Ct)	35.835^{ij}	$37.946^{f–i}$	$41.066^{c–e}$	39.618^{ABC}
Molecular sieve (M)	$36.277^{h–j}$	$39.610^{d–g}$	$38.960^{e–h}$	38.842^{C}
Zeolite (Z)	34.803^{j}	$38.952^{e–h}$	$41.351^{c–e}$	39.957^{ABC}
Sepiolite (Sep)	$38.025^{f–i}$	42.337^{cd}	$41.600^{c–e}$	41.407^{AB}
Starch (St)	21.226^{l}	35.797^{ij}	$39.884^{d–f}$	36.676^{BC}
Compost (Cp)	$36.843^{g–j}$	47.983^{a}	43.646^{bc}	44.095^{A}
Bark (Bk)	30.439^{k}	45.873^{ab}	40.655^{de}	40.969^{BC}
Average	33.350^{II}	41.214^{I}	41.023^{I}	

BPA—bisphenol A; 0_BPA, 500_BPA, 1000_BPA—doses of BPA kg^{-1} d.m. of soil. Homogeneous groups were determined for all objects (represented by letters $^{a–l}$); averaged values, regardless of the level of soil pollution with BPA, are marked with letters $^{A–C}$, while the mean values, regardless of the type of remediation substance, are indicated by numbers ^{I,II}, significant at $p = 0.05$, $n = 63$.

The phenolic compound introduced into the soil slightly moderated the nitrogen and organic carbon content, thereby affecting the soil C:N ratio (Figure 6a). The application of 500 mg BPA resulted in a 19% increase in nitrogen and a 5% increase in carbon ultimately leading to a decrease in the C:N value. Among the six remediating substances applied, Cp was responsible for a spectacular increase in soil organic carbon. It was twice as high in objects with soil exposed to 1000 mg BPA kg^{-1} of soil and in objects not contaminated with this xenobiotic compared to the control soil. Pre-fermented bark (Bk) was the second most effective in maintaining the trend described. Therefore, after its supplementation into the soil, the highest C:N ratio values were recorded in two sets of objects: Bk_0 = 7.561 and Bk_500 = 7.573, indicating a 23% increase in this parameter in each of them compared to the control. Regardless of the scale of soil contamination with BPA, molecular sieve and sepiolite induced an average increase in soil pH of one pH unit (Figure 6b). Pre-fermented bark also increased soil pH values, but only in objects with the combined effect of this remediation substance and the phenolic compound.

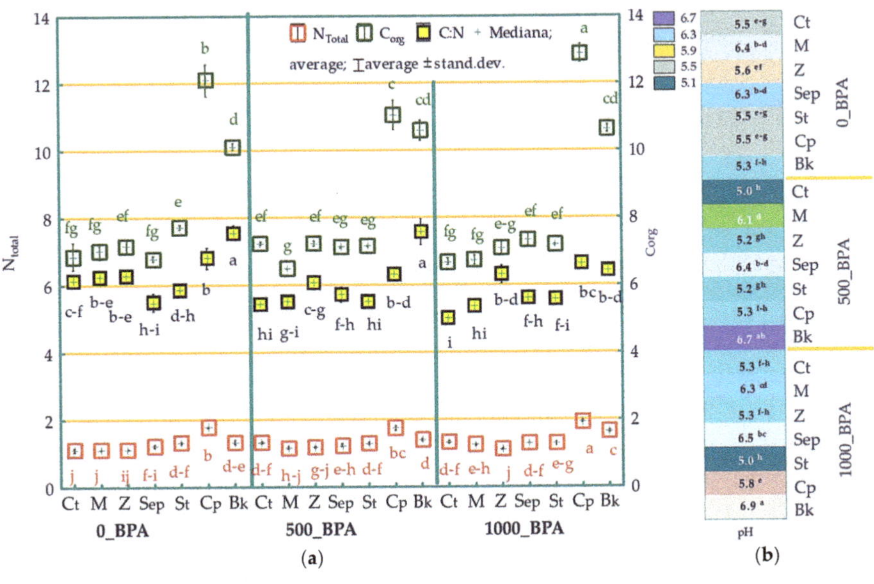

Figure 6. Total nitrogen content, the organic carbon content in g kg^{-1} d.m. of soil, and C:N ratio (**a**) and soil pH (**b**). Homogeneous groups were determined individually for N, C, C:N, and pH (represented by letters a–j (N_{total}), a–g (C_{org}), a–i (C:N), and a–h (pH). Abbreviations are explained in Figure 1, significant at $p = 0.05$, $n = 63$.

2.3. *The Relationships between the Examined Properties: Percentage of Variability of the Analyzed Variable (η^2) and PCA*

All the dependent variables characterized in the studies were moderated to varying degrees by the independent variables defined in the experiment (Figure 7). Based on the obtained η^2 values, it was estimated that biotic stress induced by soil contamination with BPA affected the response of *Zea mays* more than soil enzymes or changes in total nitrogen (N_{total}), organic carbon (C_{org}), and soil pH. Under increasing BPA pressure, the yield of the aerial parts of the cultivated plant and the roots underwent changes at levels of 84% and 93%, respectively. Analyzing the different enzyme responses to BPA in the soil revealed the following sequence: Ure (78%) > Deh (79%) > AcP (33%) > Aryl (27%) > Glu (17%) > Cat (15%) > AlP (4%). The remediation substances applied proved to be a much more significant factor in determining the biochemical activity of the soil. Considering the extent of their intervention, enzymes were ranked in the following order: AlP (70%) > Cat (49%) > Glu (44%) > AcP (38%) > Aryl (26%) > Deh (18%) > Ure (12%). Sorbent supplementation proved to be less significant for the growth and development of *Zea mays*. However, it strongly influenced the content of C_{org} (96%), N_{total} (88%), and soil pH (71%).

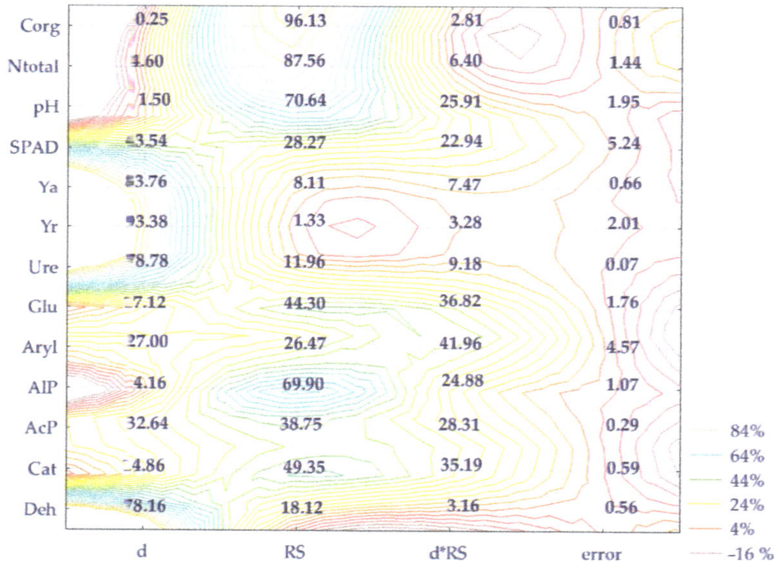

Figure 7. The share of independent variables (η^2) in the change of soil enzyme activity, a yield of aerial parts and roots of *Zea mays*, SPAD, soil pH, total nitrogen, organic carbon, in %. d—dose of BPA kg^{-1} d.m of soil, RS—remediation substance, Y_a—aerial part of plants, Y_r—roots of plants, N_{total}—total nitrogen, C_{org}—organic carbon, Deh—dehydrogenases, Cat—catalase, Ure—urease, AcP—acid phosphatase, AlP—alkaline phosphatase, Aryl -arylsulfatase, Glu—β-glucosidase.

The association of thirteen research parameters allowed the delineation of the relationships between them (Figure 8). These were explored using multivariate PCA. The PCA highlighted the extent of BPA interference and the bioremediation potential of the applied remediation substances. 39.33% of the total data variance, explained by the first principal component variable, was related to five soil enzymes: Deh, Ure, AcP, Aryl, Glu, and the SPAD index. The second variable, describing 26.81% of the variable variance (PCA2), illustrated the response of the Cat enzyme and AlP, parameters characterizing the response of *Zea mays*: Y_a, Y_r, as well as changes in soil N_{total}, C_{org}, and pH. The distribution of all cases demonstrated that BPA in the soil had an inhibitory effect on the six enzymes mentioned above. Their displacement further highlighted the beneficial effect of the sorbents applied to the soil on the biomass of *Zea mays* aerial parts. It also indicates a positive correlation between AlP activity and soil C_{org} and N_{total} contents, while emphasizing the importance of soil pH in this context.

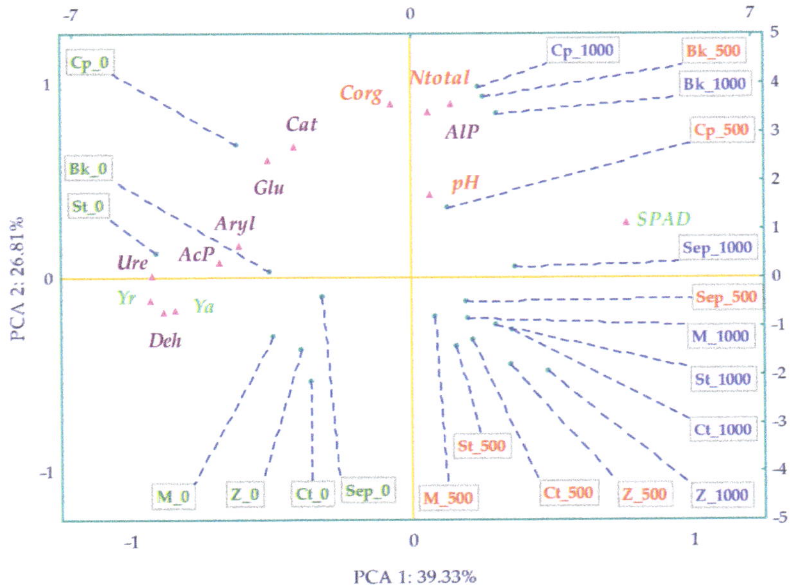

Figure 8. Interdependencies among all parameters characterized in the research—PCA method. Abbreviations are explained in Figures 1 and 7.

3. Discussion

3.1. Soil Enzymes

The selection of soil enzymes as reliable indicators of soil condition was dictated by the fact that they play a key role in catalyzing the decomposition of organic matter by reducing the activation energy associated with these reactions [49]. The defined activity of seven soil enzymes in our study and the biochemical fertility index BA_{21} estimated from them accurately delineated the extent of the disturbance in soil equilibrium resulting from increasing bisphenol A (BPA) pressure on these parameters. Each of the soil enzymes analyzed shaped its value to a different extent. Deh, Ure, AcP, Aryl, and Glu were sensitive to BPA soil contamination.

Particularly sensitive to the tested xenobiotic were dehydrogenases, enzymes involved in the anabolic and catabolic pathways of living microorganisms, located in their polysomes and cytoplasm [50]. Scientific reports [51,52], which do not unequivocally indicate the toxicity of the tested phenolic compound to Deh, prompt a discussion on the reactions of these enzymes. In the study by Zaborowska et al. [51], Deh activity was found to be three times higher in soil contaminated with 100 mg BPA kg^{-1} d.m. of soil compared to control objects. It could be presumed to be the correct trend, considering that these are enzymes involved in the dehydrogenation of phenolic compounds. The hydrogen obtained in the bioutilization process of the organic substrate is transferred to the microbiological respiratory chain, ultimately benefiting the biosynthesis process and cellular metabolism [53]. Dehydrogenases also participate in the final degradation of acetophenone to benzaldehyde and benzoic acid [52]. However, to specify, the mechanism of stimulation was not necessarily based on an increase in the activity of Deh exposed to BPA, but rather on the accumulation of the enzyme by microorganisms, as a response to oxidative stress [54]. Moreover, in other studies, the application of 800 mg BPA kg^{-1} d.m. of soil [55] and 1000 mg BPA kg^{-1} d.m.

of soil [56] resulted in inhibition of Deh activity by 21% and 50%, respectively, compared to the control The most credible explanation for these correlations is the demonstrated toxicity of BPA biodegradation intermediates, mainly hydroquinone [57].

The inhibitory effect of BPA on urease exposed in our studies is confirmed by the results obtained by other researchers [34] and our previous reports [55,56]. They showed an inhibitory effect of the phenolic compound on urease at the level of 22% under the pressure of 800 mg BPA [55] and 13% below 100 mg BPA kg^{-1} dry soil [51] compared to uncontaminated soil. The positioning of the hydroxyl group and other important substituents, including N1 and N2-diaryl derivatives in the phenyl ring, played a significant role, as suggested by Mustafa et al. [58] and Perveen et al. [59]. More controversy was raised by the effect of BPA on phosphatase activity, particularly in inducing an increase in alkaline phosphatase (AlP) activity. Similar trends were observed in our previous studies [56], although the stimulation strength at 1000 mg BPA kg^{-1} d.m. of soil was higher than in the current studies. In contrast, exposure to 100 [51] and 800 [55] mg BPA kg^{-1} d.m. of soil increased AlP activity by 23% and 8%, respectively. Interestingly, similar to urease, the activation of this enzyme was due to the presence of hydroxyl and carboxyl groups configured in the phenyl ring, which enhances AlP adsorption to soil colloids with a positive effect [60].

It should also be emphasized that the soil pH is a major determinant of biochemical activity [61]. In particular, this parameter notably moderates not only enzyme sorption but also proteolysis and enzyme inactivation [62]. Moeskops et al. [63] postulate that an increase in pH affects the destabilization of ionic and hydrogen bonds in the active center of dehydrogenases. Conversely, Kappaun et al. [64] report that the optimal pH for urease is 7–8. Importantly, the synthesis of this enzyme by microorganisms may be their response to stress, including excessively low soil pH.

Of the seven sorbents verified for their bioremediation potential, two were selected: the molecular sieve and grass compost, which is particularly important for improving the biochemical properties of the soil. The compost revealed effectiveness against Deh, Cat, Ure, AcP, and AlP. The research results obtained are in line with the attributed function of this organic substance as a moderator in the C, N, and P cycles, closely associated with the stimulation of dehydrogenases, urease, and phosphatases [42,64]. Additionally, compost is recognized as a source of electrons essential for the catabolic neutralization or elimination of organic pollutants, including BPA, conducted through redox reactions [42]. The molecular sieve demonstrated efficacy against Deh, Cat, and Ure. Its capabilities are attributed to the high number of nanopores providing adsorption of pollutants, associated with the silicate hydroxyl group, specifically ion exchange on its surface [65]. Pre-fermented bark also attenuated the inhibitory effect of BPA on Glu activity and enhanced AlP activity. Its beneficial effect is probably due to the richness and diversity of polyphenolic structures, including phenolic acids, stilbenes, flavonoids, and glucosides [44].

3.2. *Zea mays*

The demonstrated sensitivity of *Zea mays* to increasing soil contamination with bisphenol (BPA) undoubtedly requires discussion and justification, taking into account the disturbance of root growth and development as well as the aerial parts of the plant. A much higher toxicity of BPA to *Zea mays* roots was observed in our research, similar to previous studies by other authors [47,66]. They also explain many of the mechanisms activated in response to this xenobiotic. One of the fundamental mechanisms is the reduction in the activity of nitrate reductase, which is responsible for the reduction of NO_3^- to NO_2^- in the cytosol of maize roots. Importantly, in response to BPA pressure, the activity of key enzymes involved in ammonia assimilation also decreases, occurring through the GS/GOGAT (glutamine synthetase/glutamine oxoglutarate aminotransferase) pathway and GDH—glutamate dehydrogenase [66].

The research results obtained can be justified by referring to the findings of Bahmani et al. [67], which indicate that BPA inhibits root growth by redistributing, or more precisely accumulating, auxin in the root meristem in the elongation zone. This process, in turn, is associated with increased expression of the PIN1 and PIN4 genes and decreased expression of the EXPA8 and 10 genes.

Another significant phenomenon that supports the greater toxicity of BPA to plant roots than to their aerial parts is the documented reduction in abscisic acid (ABA) content and increase in gibberellic acid (GA) and ethylene in roots, as reported by Li et al. [68]. The greater sensitivity of *Zea mays* roots compared to aerial parts to the applied xenobiotic in our studies may also be attributed to the increase in mitochondrial reactive oxygen species (ROS), including H_2O_2 and O_2, as indicated by Xiao et al. [69], which are associated with cell death in the apical meristem of roots. According to the researchers, such drastic effects of BPA action are also evidenced by the increase in malondialdehyde (MDA) levels, a flagship marker of oxidative stress that reflects the degree of cell damage [70]. Growth and development disturbances in *Zea mays* could also result from a decrease in glutathione content in maize root cells, induced by an increase in glutathione peroxidase (GPX) and ascorbate peroxidase (APX) activity [71]. It is worth noting that Zhao et al. [66] demonstrated that exposure to 50 mg BPA kg^{-1} of soil reduced proline and protein content in maize seedling roots by 12% and 29%, respectively.

Based on the results of the research carried out, a significant trend observed was the induction of negative changes by BPA in the aerial parts of *Zea mays*, although this was not as pronounced as in the case of the roots. Undoubtedly, the hydrophobicity of BPA ($\log_{Kow}$ BPA = 3.40) plays an important role in the conditions obtained, favoring the weak migration of this phenolic compound, which consequently corresponds to lower bioconcentration factors of BPA in stems and leaves compared to plant roots [72]. One of the reasons for this is that the hydroxyl group of BPA interacts with binary ions in the soil, with which it forms conjugated estrogens, which in turn interfere with the mobility of this xenobiotic in the plant [73].

To mitigate excessive damage, including oxidative stress, plants activate defense systems against the adverse effects of bisphenols. As contested by Tossounian et al. [74], glutathione S-transferases (GSTs) play a very important role in the detoxification of exogenous and endogenous compounds in plants. The marker gene commonly used to assess plant sensitivity to stress induced by BPA is the enzyme ATGSTT8, assigned to the phi class of GSTs. An adaptive response to BPA toxicity also involves an increased accumulation of amino acids in the plant, such as arginine, serine, and tyrosine [75]. One mechanism that has been well-described by researchers [76,77] is the regulation of detoxification gene expression. This mechanism is particularly significant in *Zea mays*, as it corresponds to the activation of up to ten genes, including lignin, flavonoid, and phenylpropanoid genes. They are thought to have the function of initiating the phenylpropanoid pathway, providing a milder response of maize to BPA pressure [77].

An important research step was to determine the chlorophyll content in *Zea mays* leaves exposed to BPA, expressed as SPAD index values. The research results obtained were quite controversial, as many researchers [78–80] pointed to a disruption of chlorophyll synthesis in plant leaves, mediated by photosystem I (PSI) and photosystem II (PSII), mainly argued by the limitation of the function of the stomatal apparatus due to its reduced size and oxidative damage to photosynthetic pigments. However, in the current scientific discussion, it is necessary to take into account the fact that plants respond differently to BPA due to their species diversity and directly related to it, the different rates of BPA metabolism [81,82]. This hypothesis is confirmed by our previous research findings [56] in which exposure to 1000 BPA kg^{-1} d.m. of soil did not disrupt chlorophyll synthesis in the leaves of both *Zea mays* and *Brassica naupus*. The importance of this postulate may also be strengthened by the aforementioned hydrophobicity of the characterized phenolic compound, which determines its low mobility in the plant.

The starting point for analyzing the potential of bioremediating substances to mitigate the toxic effects of BPA on the growth and development of *Zea mays* was the fact that carbon assimilation by the plant results in the activation of reaction centers in the photosynthetic system, and that BPA inhibits this process [83]. The hypothesis put forward by the researchers explains the high SPAD values obtained, and thus the higher yield of *Zea mays* in plots with a combination of BPA and compost, as well as BPA and pre-fermented bark in our research. Both compounds significantly increased soil organic carbon content. In turn, the values of this parameter were correlated with the values of the C:N ratio.

4. Materials and Methods

4.1. Materials

4.1.1. Soil

The soil for the study was collected from the morainic region of the undulating ground of the Olsztyn Lakeland, covering an area of 1845 km^2. It was an agriculturally utilized area located at the geographical coordinates: NE Poland, 53.713° N, 20.432° E, in the north-eastern part of Lakeland. The research was conducted on the soil of the Eutric Cambisol type (FAO 2006), sampled from a cultivated field at a depth of 0–20 cm. Before setting up the experiment in the vegetation hall of the Didactic-Experimental Centre of the University of Warmia and Mazury in Olsztyn, analyses were performed to characterize the basic properties of the tested soil using standard analytical methods. The particle size distribution was determined using the Malvern Mastersizer 3000 laser diffraction analyzer (Malvern, Worcestershire, UK) [84,85], revealing it to be loamy sand composed of sand (37.14%), silt (54.71%), and clay (8.15%). C_{org} and N_{total} contents were also determined and were 6.28 g and 1.16 g kg^{-1} of soil dry matter (d.m.), respectively, with a C:N ratio of 5.42. Hydrolytic acidity (HAC) was determined using the Kappen method ($-$17.25 mM($^+$) kg^{-1} d.m. of soil), along with exchangeable base cations (EBC)—196.00 mM($^+$) kg^{-1} d.m. of soil [86]. Based on the obtained values of HAC and EBC, two other soil properties were calculated: cation exchange capacity (CEC) (213.23 mM($^+$) kg^{-1} d.m. of soil), and alkaline cation saturation (ACS) (91.91%). The studied soil exhibited a neutral reaction (pH 6.7 in 1 mol KCl dm^{-3}). pH determination was carried out using a pH meter HI 2221 (Hanna Instruments, Washington, DC, USA) [87]. The characterization of the soil was expanded to include its biochemical properties (Table 4).

Table 4. Some biochemical properties of the soil used in the experiment.

* Type of Enzyme	Enzymatic Activity per 1 kg d.m. h^{-1}	Unit	Methodical References
Deh	3.716	µmol TPF	[88]
Cat	0.210	mol O_2	[89]
Ure	0.266	mmol N-NH_4	[90]
AcP	3.207	mmol PN	
AlP	0.712	mmol PN	
Aryl	0.132	mmol PN	
Glu	0.381	mmol PN	

* Ure—urease, AcP—acid phosphatase, AlP—alkaline phosphatase, Aryl—arylsulfatase, Glu—β-glucosidase, TPF—triphenyl formazan, PN—4-nitrophenol.

4.1.2. BPA

According to the Sigma Aldrich (St. Louis, MO, USA) Material Safety Data Sheet, bisphenol A (BPA) CAS: 80-05-7 is a crystalline white substance with a purity of $\geq$98.0% (HPLC). Its synonyms include 4,4′-isopropylidenediphenol and 2,2-bis(4-hydroxyphenyl)-propane). The most important physicochemical properties of BPA, which influence the interactions of this phenolic compound in soil, are the bioconcentration factor (BCF), vapor pressure (V_P), and water solubility (S_W) [91]. The fate of BPA in soil is significantly influenced by its hydrophobic properties [92]. These properties are also defined by the soil adsorption coefficient (logK_{OC}) (Table 5) [26].

Table 5. Selected physicochemical properties of BPA [89].

Acronym	Molecular Weight $g\ mol^{-1}$	Total Formula	Structural Formula	BCF	$logK_{OC}$	S_W $mg\ dm^{-3}$	V_P (Pa)
BPA	228.29	$C_{15}H_{16}O_2$	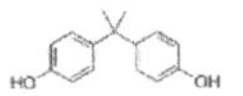	71.85	4.88	120	5.6×10^{-6}

4.1.3. Characteristics of Remediating Substances

In the experiment, the potentials of three mineral sorbents and three organic nutrient substances were identified. The group of mineral sorbents includes:

Molecular sieve (M). The average micropore diameter of the utilized aluminosilicate was 0.3 mm, with a pH of 8.5. According to the product specification, the maximum volatile content in the product at 950 °C was 2.5%. The commercial name of the product is Silosiv A3 (Company Sylosiv, Columbia, MD, USA).

Zeolite Bio.Zeo.S.01 (Z). The main compounds composing the zeolite used in the study are SiO_2 and Al_2O_3, constituting 70.6% and 12.32%, respectively. The total content of Fe, Ti, Mn, Ca, Mg, K, and Na was 8.64%. The mineral applied is a source of clinoptilolite, which accounts for 60% of the zeolite. Clinoptilolite is characterized by a high cation exchange capacity and is used in zeoponic substrates for plant cultivation and as a fertilizer. The manufacturer of zeolite BIO.Zeo.S.1 is the company Bio-Drain (Rzeszów, Poland). It is an inorganic compound with a three-dimensional crystalline structure.

Sepiolite (Sep). A clay mineral ($Mg_4[Si_6O_{15}(OH)_2]6H_2O$) produced by the company Sepiolsa Minersa Group (Guadalajara, Spain). According to the manufacturer, the product contained 70% sepiolite, with a pH_{KCl} = 7.1. Four key characteristics describing sepiolite include fibrous morphology, atypical pore structure, high adsorption capacity, and specific surface area. A comprehensive characterization of molecular sieve, sepiolite, and zeolite is presented in the research by Strachel et al. [93] and Boros et al. [94].

The three organic nutrient substances used are starch, grass compost, and pre-fermented bark.

Starch (St). According to the Sigma Aldrich data sheet, the molar mass of soluble starch ($C_6H_{10}O_5$) is 162.1 g mol^{-1} and the water solubility is 50 g dm^{-3} (90 °C), with a pH range of 6.0–7.5.

Compost (Cp). The grass compost used in the experiment was characterized by the following parameters: N_{total} (20.18), C_{org} (146.61), soil organic matter (SOM, 252.76 g kg^{-1} d.m.), P (3.41), K (9.25), and Mg (5.69, mg kg^{-1} d.m.). The pH_{KCl} of the compost was 6.1. A detailed description of this nutrient can be found in the study by Wyszkowska et al. [95].

Pre-fermented bark (Bk). The supplier of pre-fermented bark from coniferous trees was the company "Athena Bio-Produkty" Sp. z o.o. (Golczewo, Poland). According to the manufacturer's description, the product's fraction size ranged from 20–50 mm, organic matter content ≥ 50%, dry matter content ≥ 30%, and $pH_{H2O} \leq 6.0$.

4.1.4. Characteristics of Zea Mays

Zea mays was first classified as a domestic crop in Mexico around 7000 BCE [96]. According to the FAO Report, the global maize harvest in 2023 increased by 1.2% (33.3×10^9) compared to 2022, with global production of this plant exceeding 1×10^9 annually (1147.7×10^9) [97]. Maximizing the production and productivity of *Zea mays* meets the nutritional needs of the growing world population [98]. It is cultivated in more than 170 countries worldwide, covering approximately 193.7 million hectares. Maize has also attained the status of a global industrial crop, being used in 83% of the feed, starch, and biofuel industries [97]. In the experiment, the soil was sown with the maize of the DS1897B variety.

4.2. Methods

4.2.1. Design and Procedure for Conducting a Greenhouse Experiment Using *Zea mays*

After conducting research aimed at determining the physicochemical and biochemical properties of soil sieved through a 0.5 cm mesh sieve, a pot experiment was set up in a greenhouse with four replicates. The experimental factors were: (1) the level of soil contamination with BPA: 0, 500, 1000 mg BPA kg^{-1} d.m. of soil—3 objects—and (2) the type of remediation substance: control (Ct), molecular sieve (M), zeolite SO1 (Z), sepiolite (Sep), starch (St), compost from grass (Cp), pre-fermented bark (Bk)—7 objects—giving a total of 21 objects. As the experiment was conducted in four replicates, a total of 286 kg of soil packed at 3.4 kg in each of the 84 pots was used. Remediation substances were added to the individual treatments at a rate of 20 g kg^{-1} d.m. of soil. As the soil was sown with *Zea mays* to assess the sensitivity of this crop to the increasing pressure from the phenolic compound, pre-sowing fertilization with N and P (at 150 mg kg^{-1}), K (at 50 mg kg^{-1}), and Mg (at 20 mg kg^{-1}) was necessary to provide optimum nutrient levels for maize. After thorough mixing with the soil, the xenobiotic (BPA), mineral fertilizers, and remediation substances, the soil material was packed into the pots, arranged in planned configurations, and the soil moisture level was adjusted to 60% of the maximum soil moisture content in each pot. Subsequently, eight maize (*Zea mays* L.) seeds were sown in each pot. Five days after sowing, when the coleoptile emerged on the soil surface, four plants were left in each pot. The experiment was terminated after 55 days by harvesting *Zea mays* and collecting soil samples for biochemical analyses as well as the determination of C_{org} and N_{total} content

4.2.2. Measurement of SPAD and *Zea mays* Biomass

According to the BBCH scale used in the European Union to identify the phenological stages of crops, plant harvesting was performed at the initial stage of the emergence of the panicle (BBCH 51). Just before harvesting, the average SPAD value was determined based on eight readings on the 5th leaf of each plant. The SPAD 502 Chlorophyll Meter 2900P (KONICA MINOLTA, Inc., Chiyoda, Japan) was used for measurements. On the day of harvest, the yield of aerial and root parts of *Zea mays* was estimated. The dry weight of maize was determined after 5 days of drying at 60 °C.

4.2.3. Determination of Soil Enzyme Activities

To diagnose the condition of soil subjected to BPA pressure, the activity of seven soil enzymes was determined: dehydrogenases (Deh) [88] and catalase (Cat) [89], urease (Ure), acid phosphatase (AcP), alkaline phosphatase (AlP), β-glucosidase (Glu), and arylsulfatase (Aryl) [90]. The selection of these soil enzymes was dictated by the aim of obtaining holistic data on the extent of disruption of the biochemical transformations of carbon (Deh, Cat, Glu), nitrogen (Ure), phosphorus (AcP, AlP), and sulfur (Aryl), in which individual enzymes are involved. The significant parameters of individual biochemical analyses were as follows:

Substrates: Deh—2,3,5–Triphenyl tetrazolium chloride (TTC); Cat—H_2O_2—aqueous solution; Ure—urea—aqueous solution; AcP and AlP—Disodium 4-nitrophenyl phosphate hexahydrate (PNP); Glu—4-nitrophenyl-β-d-glucopyranoside (PNG); Aryl—Potassium-4-nitrophenylsulfate (PNS),

Products: Deh—triphenyl formazan (TFF); Cat—O_2; Ure—N-NH_4; AcP, AlP, Glu, and Aryl-4-nitrophenol (PN). The activity of Deh, Ure, Glu, Aryl, AcP, and AlP was determined, in triplicate, using a Perkin-Elmer Lambda 25 spectrophotometer (Waltham, MA, USA). A detailed description of the methods used to determine the activity of all enzymes has been provided in previous studies [99].

4.2.4. Methodology for Calculations and Statistical Data Analysis

The scale of the inhibitory interaction of BPA was verified using the bisphenol influence factor (IF_{BPA}) The potential of the applied mineral sorbents and organic nutrient substances to mitigate the hypothetical negative effect of BPA on enzyme activity, as well as on the growth and development of *Zea mays*, was traced through the prism of the influence factor of the remediation substance (IF_{Rs}) on the studied parameters. The response of the seven enzymes was verified and described based on the BPA influence index (IF_{BPA}) and the index of bioremediating substances (IF_{RS}) on the parameters tested. The formulas used to determine the index values are:

$$IF_{BPA} = \frac{A_{BPA}}{A_C} - 1 \quad (1)$$

where:

IF_{BPA}—index of influence of soil contamination with BPA

IF_{Rs}—index of influence of bioremediating substances

IF_{BPA_Rs}—<1—inhibition; >1—stimulation of individual enzyme activity and growth and development of *Zea mays*

A_{BPA_Rs}—activity of individual enzymes and yield of *Zea mays* in soil contaminated with 500 and 1000 mg BPA kg^{-1} d.m. of soil

A_C—activity of individual enzymes and yield of *Zea mays* in soil not contaminated with BPA

$$IF_{Rs} = \frac{A_{Rs}}{A_C} - 1 \quad (2)$$

where

IF_{Rs}—index of influence of bioremediating substances

IF_{Rs}—<1—inhibition; >1—stimulation of individual enzyme activity and growth and development of *Zea mays*

A_{Rs}—activity of individual enzymes and yield of *Zea mays* in soil contaminated with 500 and 1000 mg BPA kg^{-1} d.m. of soil

A_C—activity of individual enzymes and yield of *Zea mays* in soil not contaminated with BPA

Based on the sum of the activities of seven enzymes assigned to each treatment, the biochemical soil fertility index (BA_{21}) was also calculated, the formula for which was proposed in the publication by Wyszkowska et al. [100]. Additionally, the ratio of aerial parts biomass to root biomass was calculated [56]. To verify the obtained data, an analysis of variance (ANOVA) was conducted to describe the percentage of variability of the analyzed variable (η^2). Principal component analysis (PCA) was used to trace the interdependencies between thirteen parameters, including seven enzymes (Deh, Cat, AcP, AlP, Ure, Glu, Aryl), nitrogen content (N_{total}), organic carbon content (C_{org}), the leaf greenness index (SPAD) of *Zea mays*, aerial parts biomass (Ya), root biomass (Yr), and soil pH. Tukey's test at $p = 0.05$ was used to determine homogeneous variances between the variables [101].

5. Conclusions

Bisphenol A proved to be a significant moderator of soil biochemical activity. Application of this phenolic compound to soil had a negative effect on the activity of all enzymes analyzed, except alkaline phosphatase and catalase. However, the crux of the response to the research question regarding the extent of BPA inhibition highlighted the particular sensitivity of dehydrogenases, urease, and acid phosphatase to this xenobiotic, regardless of the level of contamination applied. Soil contamination with 500 mg BPA kg^{-1} d.m. of soil resulted in an 82% inhibition of Deh activity, while 1000 mg kg^{-1} d.m. of soil generated an 89% inhibition compared to the control. In contrast, exposure to increasing doses of this xenobiotic resulted in 55% and 94% inhibition of Ure activity, respectively. As enzymes serve as early indicators of soil equilibrium disturbances, their response to BPA was also reflected in the growth and development of *Zea mays*. Toxicological stress caused a dramatic inhibition of the root system growth and consequently a reduction in the aerial parts biomass of the cultivated plant. Soil contamination at 500 and 1000 mg BPA kg^{-1} d.m. of soil reduced aerial part yield by 44% and 71% and Zea mays root yield by 60%

and 84%, respectively. Therefore, the use of remediation substances to verify their potential was highly justified. The application of compost from grass and molecular sieves can be recommended as an effective practice to be implemented within common agricultural practices. They not only neutralize the toxic effects of BPA on soil enzymes but also restore the soil's ability to promote *Zea mays*. The results obtained can be described as a matrix for further research steps in the search for substances that not only restore soil health but also reduce the sensitivity of *Zea mays*, a plant used for energy production, to the effects of BPA.

Author Contributions: Conceptualization and experimental design and methodology, M.Z., J.W., A.B. and J.K.; investigation, M.Z., J.W., A.B. and J.K.; statistical analyses, M.Z., J.W., A.B. and J.K.; visualization, J.W., A.B., M.Z. and J.K., writing—original draft preparation, M.Z., writing—review and editing, M.Z. J.W., A.B. and J.K.; supervision, J.K. All authors have read and agreed to the published version of the manuscript.

Funding: This research was funded by the University of Warmia and Mazury in Olsztyn, Faculty of Agriculture and Forestry, Department of Soil Science and Microbiology (grant No. 30.610.006-110), project financially supported by the Minister of Science under "the Regional Initiative of Excellence Program" and Research project (NGR/319/2022) "Rector's Scientific Grant—Edition II", University of Warmia and Mazury in Olsztyn.

Institutional Review Board Statement: Not applicable.

Informed Consent Statement: Not applicable.

Data Availability Statement: Data are available by contacting the authors.

Conflicts of Interest: The authors declare no conflicts of interest.

Abbreviations

BPA—bisphenol A, 0, 500, 1000—doses of BPA kg^{-1} d.m. of soil, Ct—soil without remediating substances, M—molecular sieve, Z—zeolite, Sep—sepiolite, St—starch, Cp—grass compost, Bk—fermented-bark, Deh—dehydrogenases, Cat—catalase, Ure—urease, AcP—acid phosphatase, AlP—alkaline phosphatase, Glu—β-glucosidase, Aryl—arylsulfatase, BA_{21}—the biochemical fertility index, IF_{BPA}—index of the influence of BPA, IF_{RS}—index of the influence of remediating substances, SPAD—the leaf greenness index, PR—the ratio of the mass of the aerial parts of plants to the roots, Y_a—aerial part of plants, Y_r—roots of plants, N_{total}—total nitrogen, C_{org}—organic carbon.

References

1. FOA. Sustainable Food And Agriculture [Online]. Food and Agriculture Organization of the United Nations. 2020. Available online: https://www.fao.org/sustainability/news/detail/en/c/1274219/ (accessed on 31 January 2024).
2. Mason, A.R.G.; Salomon, M.J.; Lowe, A.J.; Cavagnaro, T.R. Microbial Solutions to Soil Carbon Sequestration. *J. Clean. Prod.* **2023**, *417*, 137993. [CrossRef]
3. Willer, H.; Schlatter, B.; Trávníček, J. *The World of Organic Agriculture. Statistics and Emerging Trends 2023*; Willer, H., Schlatter, B., Trávníček, J., Eds.; Research Institute of Organic Agriculture FiBL: Frick, Switzerland; IFOAM—Organics International: Bonn, Germany, 2023; pp. 1–358.
4. Sagheer, U.; Al-Kindi, S.; Abohashem, S.; Phillips, C.T.; Rana, J.S.; Bhatnagar, A.; Gulati, M.; Rajagopalan, S.; Kalra, D.K. Environmental Pollution and Cardiovascular Disease: Part 2 of 2: Soil, Water, and Other Forms of Pollution. *JACC Adv.* **2024**, *3*, 100815. [CrossRef] [PubMed]
5. Maddela, N.R.; Ramakrishnan, B.; Kadiyala, T.; Venkateswarlu, K.; Megharaj, M. Do Microplastics and Nanoplastics Pose Risks to Biota in Agricultural Ecosystems? *Soil Syst.* **2023**, *7*, 19. [CrossRef]
6. Bodor, A.; Feigl, G.; Kolossa, B.; Mészáros, E.; Laczi, K.; Kovács, E.; Perei, K.; Rákhely, G. Soils in Distress: The Impacts and Ecological Risks of (Micro)Plastic Pollution in the Terrestrial Environment. *Ecotoxicol. Environ. Saf.* **2024**, *269*, 115807. [CrossRef] [PubMed]
7. Corrales, J.; Kristofco, L.A.; Steele, W.B.; Yates, B.S.; Breed, C.S.; Williams, E.S.; Brooks, B.W. Global Assessment of Bisphenol A in the Environment: Review and Analysis of Its Occurrence and Bioaccumulation. *Dose-Response* **2015**, *13*, 1559325815598308. [CrossRef] [PubMed]

8. Hahladakis, J.N.; Velis, C.A.; Weber, R.; Iacovidou, E.; Purnell, P. An Overview of Chemical Additives Present in Plastics: Migration, Release, Fate and Environmental Impact during Their Use, Disposal and Recycling. *J. Hazard. Mater.* **2018**, *344*, 179–199. [CrossRef]
9. ChemSpider. Search and Share Chemistry. 2020. Available online: https://www.chemspider.com/ (accessed on 31 January 2024).
10. Xue, J.; Liu, W.; Kannan, K. Bisphenols, Benzophenones, and Bisphenol A Diglycidyl Ethers in Textiles and Infant Clothing. *Environ. Sci. Technol.* **2017**, *51*, 5279–5286. [CrossRef] [PubMed]
11. Huang, Y.Q.; Wong, C.K.C.; Zheng, J.S.; Bouwman, H.; Barra, R.; Wahlström, B.; Neretin, L.; Wong, M.H. Bisphenol A (BPA) in China: A Review of Sources, Environmental Levels, and Potential Human Health Impacts. *Environ. Int.* **2012**, *42*, 91–99. [CrossRef] [PubMed]
12. Da Costa, J.P.; Avellan, A.; Mouneyrac, C.; Duarte, A.; Rocha-Santos, T. Plastic Additives and Microplastics as Emerging Contaminants: Mechanisms and Analytical Assessment. *TrAC Trends Anal. Chem.* **2023**, *158*, 116898. [CrossRef]
13. Bisphenol-A (BPA) Market—Growth. Trends. COVID-19 Impact. and Forecasts (2022–2027). 2022. Report ID 5318392:150. Available online: https://www.reportlinker.com/market-report/Chemicals/522572/Bisphenol-A (accessed on 31 January 2024).
14. Wang, L.; Zhang, Z.-F.; Liu, L.-Y.; Zhu, F.-J.; Ma, W.-L. National-Scale Monitoring of Historic Used Organochlorine Pesticides (OCPs) and Current Used Pesticides (CUPs) in Chinese Surface Soil: Old Topic and New Story. *J. Hazard. Mater.* **2023**, *443*, 130285. [CrossRef] [PubMed]
15. Liu, S.; Gao, A.; Ma, Y.; Ding, Z.; Wang, S.; Seif, M.; Xu, X. Nonspecific Immune, Histology and Accumulation of Marine Worm, *Urechis Unicinctus* in Response to Bisphenol A (BPA). *Ecotoxicol. Environ. Saf.* **2024**, *271*, 115993. [CrossRef] [PubMed]
16. Kearns, K.A.; Naeher, L.P.; McCracken, J.P.; Boyd Barr, D.; Saikawa, E.; Hengstermann, M.; Mollinedo, E.; Panuwet, P.; Yakimavets, V.; Lee, G.E.; et al. Estimating Personal Exposures to Household Air Pollution and Plastic Garbage Burning among Adolescent Girls in Jalapa, Guatemala. *Chemosphere* **2024**, *348*, 140705. [CrossRef] [PubMed]
17. EFSA. Re-Evaluation of the Risks to Public Health Related to the Presence of Bisphenol A (PBA) in Foodstuffs. *EFSA J.* **2023**, *21*, 6857. [CrossRef]
18. Vaccher, V.; Lopez, M.E.; Castaño, A.; Mol, H.; Haji-Abbas-Zarrabi, K.; Bury, D.; Koch, H.M.; Dvorakova, D.; Hajslova, J.; Nübler, S.; et al. European Interlaboratory Comparison Investigations (ICI) and External Quality Assurance Schemes (EQUAS) for the Analysis of Bisphenol A, S and F in Human Urine: Results from the HBM4EU Project. *Environ. Res.* **2022**, *210*, 112933. [CrossRef] [PubMed]
19. ECHA. European Chemicals Agency. 2020. Available online: https://echa.europa.eu/substanceinformation/-/substanceinfo/100.001.133 (accessed on 1 February 2024).
20. Roy, N.; Sinha, A.; Mukherjee, A. Insights into the Removal of Endocrine Disruptor, Bisphenol A (BPA) from Aqueous Solution Using Fc-rGO/nZVI Nanocomposite System. *Surf. Interfaces* **2024**, *44*, 103708. [CrossRef]
21. Castro, B.; Sánchez, P.; Torres, J.M.; Ortega, E. Bisphenol A, Bisphenol F and Bisphenol S Affect Differently 5α-Reductase Expression and Dopamine–Serotonin Systems in the Prefrontal Cortex of Juvenile Female Rats. *Environ. Res.* **2015**, *142*, 281–287. [CrossRef] [PubMed]
22. Sendra, M.; Cavia-Saiz, M.; Múñiz, P. Are the BPA Analogues an Alternative to Classical BPA? Comparison between 2D and Alternative 3D in Vitro Neuron Model to Assess Cytotoxic and Genotoxic Effects. *Toxicology* **2024**, *502*, 153715. [CrossRef] [PubMed]
23. Zhang, D.; Zhao, K.; Han, T.; Zhang, X.; Xu, X.; Liu, Z.; Ren, X.; Zhang, X.; Lu, Z.; Qin, C. Bisphenol A Promote the Cell Proliferation and Invasion Ability of Prostate Cancer Cells via Regulating the Androgen Receptor. *Ecotoxicol. Environ. Saf.* **2024**, *269*, 115818. [CrossRef]
24. Wade, M.; Delawder, V.; Reneau, P.; dos Santos, J.M. The Effect of BPA Exposure on Insulin Resistance and Type 2 Diabetes—The Impact of Muscle Contraction. *Med. Hypotheses* **2020**, *140*, 109675. [CrossRef] [PubMed]
25. Kiwitt-Cárdenas, J.; Arense-Gonzalo, J.J.; Adoamnei, E.; Sarabia-Cos, L.; Vela-Soria, F.; Fernández, M.F.; Gosálvez, J.; Mendiola, J.; Torres-Cantero, A.M. Urinary Concentrations of Bisphenol A, Parabens and Benzophenone-Type Ultra Violet Light Filters in Relation to Sperm DNA Fragmentation in Young Men: A Chemical Mixtures Approach. *Sci. Total Environ.* **2024**, *912*, 169314. [CrossRef] [PubMed]
26. Vasiljevic, T.; Harner, T. Bisphenol A and Its Analogues in Outdoor and Indoor Air: Properties, Sources and Global Levels. *Sci. Total Environ.* **2021**, *789*, 148013. [CrossRef] [PubMed]
27. Fu, P.; Kawamura, K. Ubiquity of Bisphenol A in the Atmosphere. *Environ. Pollut.* **2010**, *158*, 3138–3143. [CrossRef] [PubMed]
28. Evangeliou, N.; Grythe, H.; Klimont, Z.; Heyes, C.; Eckhardt, S.; Lopez-Aparicio, S.; Stohl, A. Atmospheric Transport Is a Major Pathway of Microplastics to Remote Regions. *Nat. Commun.* **2020**, *11*, 3381. [CrossRef] [PubMed]
29. Lalwani, D.; Ruan, Y.; Taniyasu, S.; Yamazaki, E.; Kumar, N.J.I.; Lam, P.K.S.; Wang, X.; Yamashita, N. Nationwide Distribution and Potential Risk of Bisphenol Analogues in Indian Waters. *Ecotoxicol. Environ. Saf.* **2020**, *200*, 110718. [CrossRef] [PubMed]
30. Esteban, S.; Gorga, M.; Petrovic, M.; González-Alonso, S.; Barceló, D.; Valcárcel, Y. Analysis and Occurrence of Endocrine-Disrupting Compounds and Estrogenic Activity in the Surface Waters of Central Spain. *Sci. Total Environ.* **2014**, *466–467*, 939–951. [CrossRef] [PubMed]
31. Huang, D.-Y.; Zhao, H.-Q.; Liu, C.-P.; Sun, C.-X. Characteristics, Sources, and Transport of Tetrabromobisphenol A and Bisphenol A in Soils from a Typical e-Waste Recycling Area in South China. *Environ. Sci. Pollut. Res.* **2014**, *21*, 5818–5826. [CrossRef] [PubMed]

32. Chen, M.; Xu, P.; Zeng, G.; Yang, C.; Huang, D.; Zhang, J. Bioremediation of Soils Contaminated with Polycyclic Aromatic Hydrocarbons, Petroleum, Pesticides, Chlorophenols and Heavy Metals by Composting: Applications, Microbes and Future Research Needs. *Biotechnol. Adv.* **2015**, *33*, 745–755. [CrossRef] [PubMed]
33. Bisphenol, S. National Toxicology Program. 2014. Available online: https://Ntpsearch.Niehs.Nih.Gov/?query=bisphenol+S&e=False&suffixes=false (accessed on 1 February 2024).
34. Pérez, R.A.; Albero, B.; Férriz, M.; Tadeo, J.L. Rapid Multiresidue Determination of Bisphenol Analogues in Soil with On-Line Derivatization. *Anal. Bioanal. Chem.* **2017**, *409*, 4571–4580. [CrossRef] [PubMed]
35. Vero, C. Biosolids Production and Use Survey. In *Australia and New Zealand Biosolids Partnership*; Pollution Solutions & Designs (PSD) Pty Ltd.: Northern Territor, Australia, 2021.
36. Hušek, M.; Moško, J.; Pohořelý, M. Sewage Sludge Treatment Methods and P-Recovery Possibilities: Current State-of-the-Art. *J. Environ. Manag.* **2022**, *315*, 115090. [CrossRef] [PubMed]
37. Lu, Q.; He, Z.L.; Stoffella, P.J. Land Application of Biosolids in the USA: A Review. *Appl. Environ. Soil Sci.* **2012**, *2012*, e201462. [CrossRef]
38. Staples, C.; van der Hoeven, N.; Clark, K.; Mihaich, E.; Woelz, J.; Hentges, S. Distributions of Concentrations of Bisphenol A in North American and European Surface Waters and Sediments Determined from 19 Years of Monitoring Data. *Chemosphere* **2018**, *201*, 448–458. [CrossRef] [PubMed]
39. Oustriere, N.; Marchand, L.; Lottier, N.; Motelica, M.; Mench, M. Long-Term Cu Stabilization and Biomass Yields of Giant Reed and Poplar after Adding a Biochar, Alone or with Iron Grit, into a Contaminated Soil from a Wood Preservation Site. *Sci. Total Environ.* **2017**, *579*, 620–627. [CrossRef] [PubMed]
40. Siedt, M.; Schäffer, A.; Smith, K.E.C.; Nabel, M.; Roß-Nickoll, M.; van Dongen, J.T. Comparing Straw, Compost, and Biochar Regarding Their Suitability as Agricultural Soil Amendments to Affect Soil Structure, Nutrient Leaching, Microbial Communities, and the Fate of Pesticides. *Sci. Total Environ.* **2021**, *751*, 141607. [CrossRef] [PubMed]
41. Kranz, C.N.; McLaughlin, R.A.; Johnson, A.; Miller, G.; Heitman, J.L. The Effects of Compost Incorporation on Soil Physical Properties in Urban Soils—A Concise Review. *J. Environ. Manag.* **2020**, *261*, 110209. [CrossRef] [PubMed]
42. Cao, X.; Williams, P.N.; Zhan, Y.; Coughlin, S.A.; McGrath, J.W.; Chin, J.P.; Xu, Y. Municipal Solid Waste Compost: Global Trends and Biogeochemical Cycling. *Soil Environ. Health* **2023**, *1*, 100038. [CrossRef]
43. Thorenz, A.; Wietschel, L.; Stindt, D.; Tuma, A. Assessment of Agroforestry Residue Potentials for the Bioeconomy in the European Union. *J. Clean. Prod.* **2018**, *176*, 348–359. [CrossRef] [PubMed]
44. Borrega, M.; Kalliola, A.; Määttänen, M.; Borisova, A.S.; Mikkelson, A.; Tamminen, T. Alkaline Extraction of Polyphenols for Valorization of Industrial Spruce Bark. *Bioresour. Technol. Rep.* **2022**, *19*, 101129. [CrossRef]
45. Şen, U.; Esteves, B.; Pereira, H. Pyrolysis and Extraction of Bark in a Biorefineries Context: A Critical Review. *Energies* **2023**, *16*, 4848. [CrossRef]
46. DeLuca, T.H.; Pingree, M.R.A.; Gao, S. Chapter 16—Assessing Soil Biological Health in Forest Soils. In *Developments in Soil Science*; Busse, M., Giardina, C.P., Morris, D.M., Page-Dumroese, D.S., Eds.; Global Change and Forest Soils; Elsevier: Amsterdam, The Netherlands, 2019; Volume 36, pp. 397–426.
47. Zaborowska, M.; Wyszkowska, J.; Borowik, A.; Kucharski, J. Bisphenol A—A Dangerous Pollutant Distorting the Biological Properties of Soil. *Int. J. Mol. Sci.* **2021**, *22*, 12753. [CrossRef] [PubMed]
48. Xiao, C.; Wang, L.; Zhou, Q.; Huang, X. Hazards of Bisphenol A (BPA) Exposure: A Systematic Review of Plant Toxicology Studies. *J. Hazard. Mater.* **2020**, *384*, 121488. [CrossRef]
49. Zuccarini, P.; Sardans, J.; Asensio, L.; Peñuelas, J. Altered Activities of Extracellular Soil Enzymes by the Interacting Global Environmental Changes. *Glob. Chang. Biol.* **2023**, *29*, 2067–2091. [CrossRef] [PubMed]
50. Wallenstein, M.D.; Burns, R.G. Ecology of Extracellular Enzyme Activities and Organic Matter Degradation in Soil: A Complex Community-Driven Process. In *Methods of Soil Enzymology*; Dick, R.P., Ed.; Soil Science Society of America: Madi-Son, WI, USA, 2011; pp. 35–55.
51. Zaborowska, M.; Wyszkowska, J.; Borowik, A. Soil Microbiome Response to Contamination with Bisphenol A, Bisphenol F and Bisphenol S. *Int. J. Mol. Sci.* **2020**, *21*, 3529. [CrossRef] [PubMed]
52. Daudzai, Z.; Treesubsuntorn, C.; Thiravetyan, P. Inoculated *Clitoria Ternatea* with *Bacillus Cereus* ERBP for Enhancing Gaseous Ethylbenzene Phytoremediation: Plant Metabolites and Expression of Ethylbenzene Degradation Genes. *Ecotoxicol. Environ. Saf.* **2018**, *164*, 50–60. [CrossRef] [PubMed]
53. Dong, Y.; Zhao, J.; Wang, L.; Wang, H.; Zou, X.; Zhang, J. Effect of Bisphenol A and Pentachlorophenol on Different Enzymes of Activated Sludge. *Sci. Total Environ.* **2019**, *671*, 1170–1178. [CrossRef]
54. Carvalho, M.B.; Tavares, S.; Medeiros, J.; Núñez, O.; Gallart-Ayala, H.; Leitão, M.C.; Galceran, M.T.; Hursthouse, A.; Silva Pereira, C. Degradation Pathway of Pentachlorophenol by *Mucor Plumbeus* Involves Phase II Conjugation and Oxidation–Reduction Reactions. *J. Hazard. Mater.* **2011**, *198*, 133–142. [CrossRef] [PubMed]
55. Zaborowska, M.; Wyszkowska, J.; Kucharski, J. Role of Chlorella Sp. and Rhamnolipid 90 in Maintaining Homeostasis in Soil Contaminated with Bisphenol A. *J. Soils Sediments* **2021**, *21*, 27–41. [CrossRef]
56. Zaborowska, M.; Wyszkowska, J.; Borowik, A.; Kucharski, J. Effect of Separate and Combined Toxicity of Bisphenol A and Zinc on the Soil Microbiome. *Int. J. Mol. Sci.* **2022**, *23*, 5937. [CrossRef] [PubMed]

57. Perotti, E.B.R. Impact of Hydroquinone Used as a Redox Effector Model on Potential Denitrification, Microbial Activity and Redox Condition of a Cultivable Soil. *Rev. Argent. Microbiol.* **2015**, *47*, 212–218. [CrossRef] [PubMed]
58. Mustafa, S.; Perveen, S.; Khan, A. Synthesis, Enzyme Inhibition and Anticancer Investigation of Unsymmetrical 1,3-Disubstituted Ureas. *J. Serbian Chem. Soc.* **2014**, *79*, 1–10. [CrossRef]
59. Perveen, S.; Khan, K.M.; Lodhi, M.A.; Choudhary, M.I.; Atta-ur-Rahman; Voelter, W. Urease and α-Chymotrypsin Inhibitory Effects of Selected Urea Derivatives. *Lett. Drug Des. Discov.* **2008**, *5*, 401–405. [CrossRef]
60. Khadem, A.; Raiesi, F. Response of Soil Alkaline Phosphatase to Biochar Amendments: Changes in Kinetic and Thermodynamic Characteristics. *Geoderma* **2019**, *337*, 44–54. [CrossRef]
61. Wang, G.; Post, W.M.; Mayes, M.A. Development of Microbial-Enzyme-Mediated Decomposition Model Parameters through Steady-State and Dynamic Analyses. *Ecol. Appl.* **2013**, *23*, 255–272. [CrossRef] [PubMed]
62. Menezes-Blackburn, D.; Giles, C.; Darch, T.; George, T.S.; Blackwell, M.; Stutter, M.; Shand, C.; Lumsdon, D.; Cooper, P.; Wendler, R.; et al. Opportunities for Mobilizing Recalcitrant Phosphorus from Agricultural Soils: A Review. *Plant Soil* **2018**, *427*, 5–16. [CrossRef] [PubMed]
63. Moeskops, B.; Sukristiyonubowo; Buchan, D.; Sleutel, S.; Herawaty, L.; Husen, E.; Saraswati, R.; Setyorini, D.; De Neve, S. Soil Microbial Communities and Activities under Intensive Organic and Conventional Vegetable Farming in West Java, Indonesia. *Appl. Soil Ecol.* **2010**, *45*, 112–120. [CrossRef]
64. Kappaun, K.; Piovesan, A.R.; Carlini, C.R.; Ligabue-Braun, R. Ureases: Historical Aspects, Catalytic, and Non-Catalytic Properties—A Review. *J. Adv. Res.* **2018**, *13*, 3–17. [CrossRef] [PubMed]
65. Rizwan, M.; Ali, S.; Rizvi, H.; Rinklebe, J.; Tsang, D.C.W.; Meers, E.; Ok, Y.S.; Ishaque, W. Phytomanagement of Heavy Metals in Contaminated Soils Using Sunflower: A Review. *Crit. Rev. Environ. Sci. Technol.* **2016**, *46*, 1498–1528. [CrossRef]
66. Zhao, C.; Shi, Y.; Xu, Y.; Lin, N.; Dong, H.; Bei, L. Effects of Bisphenol A on Antioxidation and Nitrogen Assimilation of Maize Seedlings Roots. *Ecotoxicol. Environ. Saf.* **2022**, *247*, 114255. [CrossRef] [PubMed]
67. Bahmani, R.; Kim, D.; Modareszadeh, M.; Thompson, A.J.; Park, J.H.; Yoo, H.H.; Hwang, S. The Mechanism of Root Growth Inhibition by the Endocrine Disruptor Bisphenol A (BPA). *Environ. Pollut.* **2020**, *257*, 113516. [CrossRef] [PubMed]
68. Li, X.; Wang, L.; Shen, F.; Zhou, Q.; Huang, X. Impacts of Exogenous Pollutant Bisphenol A on Characteristics of Soybeans. *Ecotoxicol. Environ. Saf.* **2018**, *157*, 463–471. [CrossRef] [PubMed]
69. Xiao, C.; Wang, L.; Hu, D.; Zhou, Q.; Huang, X. Effects of Exogenous Bisphenol A on the Function of Mitochondria in Root Cells of Soybean (*Glycine Max* L.) Seedlings. *Chemosphere* **2019**, *222*, 619–627. [CrossRef]
70. Rosalie, R.; Joas, J.; Deytieux-Belleau, C.; Vulcain, E.; Payet, B.; Dufossé, L.; Léchaudel, M. Antioxidant and Enzymatic Responses to Oxidative Stress Induced by Pre-Harvest Water Supply Reduction and Ripening on Mango (Mangifera Indica L. Cv. 'Cogshall') in Relation to Carotenoid Content. *J. Plant Physiol.* **2015**, *184*, 68–78. [CrossRef] [PubMed]
71. Saleem, M.; Fariduddin, Q.; Castroverde, C.D.M. Salicylic Acid: A Key Regulator of Redox Signalling and Plant Immunity. *Plant Physiol. Biochem.* **2021**, *168*, 381–397. [CrossRef] [PubMed]
72. Wang, S.; Liu, F.; Wu, W.; Hu, Y.; Liao, R.; Chen, G.; Wang, J.; Li, J. Migration and Health Risks of Nonylphenol and Bisphenol a in Soil-Winter Wheat Systems with Long-Term Reclaimed Water Irrigation. *Ecotoxicol. Environ. Saf.* **2018**, *158*, 28–36. [CrossRef] [PubMed]
73. Goeppert, N.; Dror, I.; Berkowitz, B. Fate and Transport of Free and Conjugated Estrogens during Soil Passage. *Environ. Pollut.* **2015**, *206*, 80–87. [CrossRef] [PubMed]
74. Tossounian, M.-A.; Wahni, K.; Van Molle, I.; Vertommen, D.; Astolfi Rosado, L.; Messens, J. Redox-Regulated Methionine Oxidation of Arabidopsis Thaliana Glutathione Transferase Phi9 Induces H-Site Flexibility. *Protein Sci.* **2019**, *28*, 56–67. [CrossRef] [PubMed]
75. Zanganeh, R.; Jamei, R.; Rahmani, F. Role of Salicylic Acid and Hydrogen Sulfide in Promoting Lead Stress Tolerance and Regulating Free Amino Acid Composition in *Zea mays* L. *Acta Physiol. Plant.* **2019**, *41*, 94. [CrossRef]
76. Zhang, J.; Wang, L.; Zhou, Q.; Huang, X. Reactive Oxygen Species Initiate a Protective Response in Plant Roots to Stress Induced by Environmental Bisphenol A. *Ecotoxicol. Environ. Saf.* **2018**, *154*, 197–205. [CrossRef] [PubMed]
77. Yang, F.; Li, W.; Jiang, N.; Yu, H.; Morohashi, K.; Ouma, W.Z.; Morales-Mantilla, D.E.; Gomez-Cano, F.A.; Mukundi, E.; Prada-Salcedo, L.D.; et al. A Maize Gene Regulatory Network for Phenolic Metabolism. *Mol. Plant* **2017**, *10*, 498–515. [CrossRef] [PubMed]
78. Bourgeade, P.; Aleya, E.; Alaoui-Sosse, L.; Herlem, G.; Alaoui-Sosse, B.; Bourioug, M. Growth, Pigment Changes, and Photosystem II Activity in the Aquatic Macrophyte Lemna Minor Exposed to Bisphenol A. *Environ. Sci. Pollut. Res.* **2021**, *28*, 68671–68678. [CrossRef] [PubMed]
79. Kim, D.; Kwak, J.I.; An, Y.-J. Effects of Bisphenol A in Soil on Growth, Photosynthesis Activity, and Genistein Levels in Crop Plants (*Vigna Radiata*). *Chemosphere* **2018**, *209*, 875–882. [CrossRef] [PubMed]
80. Sharma, A.; Kumar, V.; Shahzad, B.; Ramakrishnan, M.; Singh Sidhu, G.P.; Bali, A.S.; Handa, N.; Kapoor, D.; Yadav, P.; Khanna, K.; et al. Photosynthetic Response of Plants Under Different Abiotic Stresses: A Review. *J. Plant Growth Regul.* **2020**, *39*, 509–531. [CrossRef]
81. Jiao, L.; Ding, H.; Wang, L.; Zhou, Q.; Huang, X. Bisphenol A Effects on the Chlorophyll Contents in Soybean at Different Growth Stages. *Environ. Pollut.* **2017**, *223*, 426–434. [CrossRef] [PubMed]

82. Zhang, J.; Wang, L.; Li, M.; Jiao, L.; Zhou, Q.; Huang, X. Effects of Bisphenol A on Chlorophyll Fluorescence in Five Plants. *Environ. Sci. Pollut. Res.* **2015**, *22*, 17724–17732. [CrossRef] [PubMed]
83. Li, Y.-T.; Liang, Y.; Li, Y.-N.; Che, X.-K.; Zhao, S.-J.; Zhang, Z.-S.; Gao, H.-Y. Mechanisms by Which Bisphenol A Affect the Photosynthetic Apparatus in Cucumber (*Cucumis Sativus* L.) Leaves. *Sci. Rep.* **2018**, *8*, 4253. [CrossRef] [PubMed]
84. *PN-R-04032*; Soil and Mineral Materials—Sampling and Determination of Particle Size Distribution. Polish Committee for Standardization: Warsaw, Poland, 1998.
85. *ISO 11464*; Soil Quality—Pre-Treatment of Samples for Physico-Chemical Analysis. International Organization for Standardization: Geneva, Switzerland, 2006.
86. Klute, A. Methods of Soil Analysis. In *Agronomy Monograph 9*; American Society of Agronomy: Madison, WI, USA, 1996.
87. *ISO 10390*; In Soil Quality—Determination of PH. International Organization for Standardization: Geneva, Switzerland, 2005. Available online: https://www.iso.org/standard/75243.html (accessed on 10 July 2023).
88. Öhlinger, R. Dehydrogenase Activity with the Substrate TTC. In *Methods in Soil Biology*; Schinner, F., Ohlinger, R., Kandler, E., Margesin, R., Eds.; Springer: Berlin/Heidelberg, Germany, 1996; pp. 241–243.
89. Johnson, J.L.; Temple, K.L. Some Variables Affecting the Measurement of "Catalase Activity" in Soil. *Soil Sci. Soc. Am. J.* **1964**, *28*, 207–209. [CrossRef]
90. Alef, K.; Nannipieri, P. (Eds.) *Methods in Applied Soil Microbiology and Biochemistry*; Academic London: London, UK, 1988; pp. 316–365.
91. Hu, Y.; Zhu, Q.; Yan, X.; Liao, C.; Jiang, G. Occurrence, Fate and Risk Assessment of BPA and Its Substituents in Wastewater Treatment Plant: A Review. *Environ. Res.* **2019**, *178*, 108732. [CrossRef] [PubMed]
92. Wang, H.; Liu, Z.; Zhang, J.; Huang, R.; Yin, H.; Dang, Z.; Wu, P.; Liu, Y. Insights into Removal Mechanisms of Bisphenol A and Its Analogues in Municipal Wastewater Treatment Plants. *Sci. Total Environ.* **2019**, *692*, 107–116. [CrossRef] [PubMed]
93. Strachel, R.; Wyszkowska, J.; Baćmaga, M. An Evaluation of the Effectiveness of Sorbents in the Remediation of Soil Contaminated with Zinc. *Water Air Soil Pollut.* **2018**, *229*, 235. [CrossRef] [PubMed]
94. Boros-Lajszner, E.; Wyszkowska, J.; Kucharski, J. Use of a Zeolite and Molecular Sieve to Restore Homeostasis of Soil Contaminated with Cobalt. *Minerals* **2020**, *10*, 53. [CrossRef]
95. Wyszkowska, J.; Borowik, A.; Kucharski, J. The Role of Grass Compost and Zea Mays in Alleviating Toxic Effects of Tetracycline on the Soil Bacteria Community. *Int. J. Environ. Res. Public Health* **2022**, *19*, 7357. [CrossRef] [PubMed]
96. Kennett, D.J.; Prufer, K.M.; Culleton, B.J.; George, R.J.; Robinson, M.; Trask, W.R.; Buckley, G.M.; Moes, E.; Kate, E.J.; Harper, T.K.; et al. Early Isotopic Evidence for Maize as a Staple Grain in the Americas. *Sci. Adv.* **2020**, *6*, eaba3245. [CrossRef] [PubMed]
97. FAO (Food and Agricultural Organization). *FAOSTAT: Production: Crops and Livestock Products*; FAO: Rome, Italy, 2023. Available online: https://www.fao.org/3/cc9205en/cc9205en.pdf (accessed on 17 February 2024).
98. Basnet, B.; Khanal, S. Quantitative Trait Loci and Candidate Genes for Iron and Zinc Bio-Fortification in Genetically Diverse Germplasm of Maize (*Zea Mays* L): A Systematic Review. *Heliyon* **2022**, *8*. [CrossRef] [PubMed]
99. Borowik, A.; Wyszkowska, J.; Wyszkowski, M. Resistance of Aerobic Microorganisms and Soil Enzyme Response to Soil Contamination with Ekodiesel Ultra Fuel. *Environ. Sci. Pollut. Res.* **2017**, *24*, 24346–24363. [CrossRef]
100. Wyszkowska, J.; Kucharski, J.; Kucharski, M.; Borowik, A. Applicability of Biochemical Indices to Quality Assessment of Soil Polluted with Heavy Metals. *J. Elem.* **2014**, *18*, 733–756. [CrossRef]
101. Tibco Software Inc. *Data Analysis Software System*; Statistica, Version 13; Tibco Software Inc.: Palo Alto, CA, USA, 2021. Available online: http://statistica.io (accessed on 10 July 2023).

molecules MDPI

Article

Modified Halloysite as an Adsorbent for the Removal of Cu(II) Ions and Reactive Red 120 Dye from Aqueous Solutions

Krzysztof Kuśmierek [1,*], Andrzej Świątkowski [1], Ewa Wierzbicka [2] and Izabella Legocka [2]

[1] Institute of Chemistry, Military University of Technology, 00-908 Warsaw, Poland; a.swiatkowski@wp.pl
[2] Department of Polymer Technology and Processing, Łukasiewicz-Industrial Chemistry Institute, 01-793 Warsaw, Poland; e.wierzbicka@outlook.com (E.W.); izabella.legocka@ichp.lukasiewicz.gov.pl (I.L.)
* Correspondence: krzysztof.kusmierek@wat.edu.pl

Abstract: The adsorption of copper ions and Reactive Red 120 azo dye (RR-120) as models of water pollutants on unmodified halloysite (H-NM), as well as halloysites modified with sulfuric acid (H-SA) and (3-aminopropyl)triethoxysilane (H-APTES), was investigated. The results showed that adsorption of both the adsorbates was pH-dependent and increased with the increase in halloysite dosage. The adsorption kinetics were evaluated and the results demonstrated that the adsorption followed the pseudo-second-order model. The adsorption isotherms of Cu(II) ions and RR-120 dye on the halloysites were described satisfactorily by the Langmuir model. The maximum adsorption capacities for the Cu(II) ions were 0.169, 0.236, and 0.507 mmol/g, respectively, for H-NM, H-SA, and H-APTES indicating that the NH_2-functionalization rather than the surface area of the adsorbents was responsible for the enhanced adsorption. The adsorption capacities for RR-120 dye were found to be 9.64 μmol/g for H-NM, 75.76 μmol/g for H-SA, and 29.33 μmol/g for H-APTES. The results demonstrated that APTES-functionalization and sulfuric acid activation are promising modifications, and both modified halloysites have good application potential for heavy metals as well as for azo dye removal.

Keywords: adsorption; Reactive Red 120; copper ions; halloysite; chemical modification

1. Introduction

Intense economic growth, accompanied by the development of urban areas, industrial zones, communication and transport systems, and increased production due to higher demand and consumption, contributes to increased environmental pressures. Economic development itself brings many opportunities and benefits. However, it is also a threat to the environment. Most of these pollutants (inorganic and organic) are harmful to both living organisms and humans. Thus, for example, most heavy metals are elements that are a burden to the organism, due to their high toxicity and their presence in the organism negatively affects its functioning [1,2]. Among the organic pollutants that are quite common in the environment, dyes, which are used on a large scale in the textile industry, are an important group. The first observable effect caused by dye contamination is the coloration of the water, which affects its aesthetic qualities as well as its transparency and sunlight permeability by interfering with photosynthesis processes. The presence of dyes in water impedes the flow of water through fish gills, impairing gas exchange, which contributes to reduced growth rates and even death. In addition, dyes cause a decrease in oxygen diffusion in water bodies, leading to animal extinctions. In addition, some classes of dye, particularly azo dyes, were shown to have carcinogenic and mutagenic properties [3–5].

Such negative effects of pollution on the environment and living organisms make it necessary to take appropriate measures to eliminate both the causes of pollution and its effects. The main technological processes for restoring water to its optimal parameters are classified into biological (aerobic and anaerobic treatment processes), chemical (ozonation, electrocoagulation, electrochemical, and oxidation processes), and physical (coagulation, flocculation, filtration, and adsorption) [2,5]. Each of these methods has its advantages and disadvantages. The various methods differ in their effectiveness in removing contaminants and in the cost of the process. Among all the methods, adsorption is undoubtedly one of the most popular techniques used in water treatment [2,5]. It is recommended because of its low cost, high efficiency, simplicity, and versatility, removing both organic and inorganic materials. Of course, the efficiency of adsorption is closely related to the physical and chemical properties of the adsorbent used. Hence, the selection of an appropriate adsorbent for the removal of specific contaminants is a priority challenge that will affect the overall efficiency of the process. The effectiveness of the water treatment process can be maximized by selecting an optimum adsorbent, i.e., one with the highest adsorption capacity at the lowest price. Finding new adsorbents is, therefore, very important. Many research groups are working on this.

Activated carbons are the most widely used adsorbents because their specific physicochemical properties (high porosity and specific surface area) result in their high adsorption capacity and high efficiency in removing aqueous pollutants. However, several processes and their complexity required to produce activated carbon, as well as the need for its periodic regeneration, affect its high cost, which significantly limits its use and application. This has stimulated many researchers to obtain and develop alternative low-cost adsorbents. In addition to low cost, such materials should of course have high availability and be either renewable or waste. Based on the classification proposed by Crini et al. [6], such non-conventional sorbents can be categorized into the following groups: (1) natural materials including inorganic and siliceous materials as well as clays; (2) agricultural wastes such as sawdust, bark, solid wastes (rice husk, nutshell, corn cob, wheat straw); (3) industrial by-products and wastes (fly ash, red mud, sludge); (4) biosorbents such as chitin/chitosan, peat, fungi or bacterial biomass; and (5) miscellaneous adsorbents, e.g., cotton waste, hydrogels.

A fairly common group of potential, low-cost sorbents are clays and other clay minerals [7–10], among which halloysite seems to be a very promising material. Halloysite belongs to the group of common and widely distributed minerals; its large deposits can be found in Poland, the USA, Brazil, China, Turkey, and elsewhere [11]. The morphology of the halloysite grain is related to its origin and especially to the crystallization conditions in the geological environment. Halloysite consists of a layer of silicon tetrahedra and aluminum octahedra that form a porous three-dimensional structure. Individual plates are separated from each other by a free space, called a gallery, in which there may be absorbed ions and molecules (including water) loosely bound to the surface of the plate using mainly hydrogen bonds. In the unit cell of the mineral, there are four hydroxyl groups associated with the octahedral layer, which have different orientations to the plane. Three of them are located on the outer surface of the bundle and one –OH group is located inside the bundle. Adsorption and grafting are two types of reactions that can occur on the surface of halloysite grains. Due to its physicochemical properties, halloysite can be used as a filler-modifier for a wide range of polymers [12], in the biological field [13], as an adsorbent to remove pollutants from water and wastewater [9,10]. In addition, it can be easily modified, e.g., by thermal treatment, acid activation, or functionalization, to tailor the adsorption properties of the mineral for a specific pollutant [9,10]. The effectiveness of both unmodified and modified halloysite as adsorbents for the removal of both heavy metals [10,14–17] and various dyes [10,18–20] from water was reported in numerous papers.

The purpose of this paper is to evaluate the suitability and effectiveness of unmodified halloysite (H-NM) as well as halloysites modified with sulfuric acid (H-SA) and (3-aminopropyl)triethoxysilane (H-APTES) as potential alternative adsorbents for the removal of water pollutants. The copper(II) ions and Reactive Red 120 azo dye (RR-120) were selected as models of water contaminants. The effect of the method used to modify the adsorbent on its ability to adsorb the metal as well as the dye was investigated. The adsorption kinetics and adsorption under equilibrium conditions, as well as the effects of adsorbent dosage and solution pH, were evaluated.

2. Results and Discussion

2.1. Characteristics of the Halloysite Samples

SEM images of the raw and modified halloysite samples are shown in Figure 1 while the surface composition obtained from EDS analysis is summarized in Table 1.

Figure 1. Scanning electron microscopy (SEM) images of the halloysite samples (magnification 15,000×).

Table 1. The EDS analysis results of the halloysite samples.

Adsorbent	C	O	Al	Si	Fe	Ti	N
				at.%			
H-NM	-	63.5	16.0	18.4	1.8	0.3	-
H-SA	-	63.2	15.7	19.2	1.7	0.2	-
H-APTES	28.9	36.9	12.9	16.7	0.9	0.4	3.3

The SEM images of the studied halloysite samples show its modification results. One can observe visible differences in the surface morphology between the original sample H-NM and modified samples H-SA and H-APTES. Treating H-NM with sulfuric acid increases its roughness. The deposition of APTES glued together halloysite particles.

The presence of carbon on the surface of the halloysite modified with APTES was observed. The carbon content in the H-APTES sample was 28.9 at.%, while no carbon was found in the H-NM and H-SA samples. This difference is due to the use of an organic compound in the modification process. The N observed in this sample is derived from APTES molecules. Modification of raw halloysite with sulfuric acid does not cause significant changes in the O, Al, Si, Fe, and Ti contents.

The porous structure of the halloysite samples was characterized using low-temperature nitrogen adsorption–desorption isotherms (Figure 2).

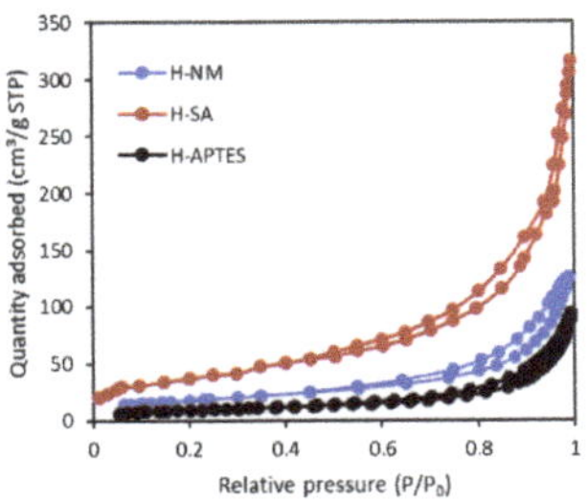

Figure 2. The nitrogen adsorption–desorption isotherms at 77 K for halloysite samples.

The specific surface areas (S_{BET}) and pore volumes calculated from the N_2 adsorption–desorption isotherms are listed in Table 2.

Table 2. Specific surface area and pore volumes of the halloysite samples.

Adsorbent	S_{BET} (m^2/g)	V_t (cm^3/g)	V_{mi} (cm^3/g)	V_{me} (cm^3/g)
H-NM	53	0.217	0.019	0.198
H-SA	129	0.399	0.057	0.342
H-APTES	34	0.108	0.0147	0.093

The porous structure of the halloysite samples was closely correlated with their modification methods. The use of sulfuric acid gave a significant increase in all porosity parameters. Their values were about 2 times greater for H-SA in comparison with the H-NM sample. For the halloysite modified with APTES observed effect was the opposite and values of porosity parameters were two times smaller for H-APTES in comparison with the H-NM sample.

The obtained from thermogravimetric analysis weight losses in characteristic temperatures (end of drying, maximum of DTG curves, end of weight losses) are listed in Table 3.

Table 3. Thermal properties of the row and modified halloysite samples.

	H-NM		H-SA		H-APTES	
	Temp. (°C)	Δm (%)	Temp. (°C)	Δm (%)	Temp. (°C)	Δm (%)
End of drying	180	2.1	200	2.6	185	9.5
Temp. of max of DTG	448	7.2	459	9.5	472	19.5
End of weight loss	680	9.4	700	11.6	715	25.8

The first step of the mass loss associated with the increase in temperature concerns the removal of water (the dehydration of physisorbed water and interlayer water) in the case of H-NM and H-SA samples and in the case of H-APTES-ethanol. APTES is not removed due to its boiling point of 223 °C. Further steps of the TG curves include dehydroxylation (hydroxyl groups in the [AlO_6] octahedral layer of halloysite are removed). All samples are at endothermic status during the whole heating process, attributed to continuous dehydration and dehydroxylation.

The infrared spectra for H-NM, H-SA, and H-APTES halloysite samples are presented in Figure 3 and the specific absorption bands revealed in the FTIR vibration spectra in the range of 4000–400 cm^{-1} are listed for comparison in Table 4.

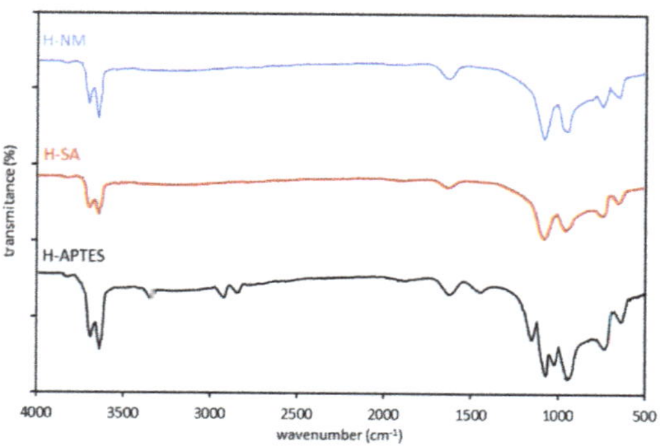

Figure 3. FTIR spectra of raw and modified halloysite samples.

Table 4. FTIR analysis of raw and modified halloysite samples.

H-NM	H-SA	H-APTES
	Wavenumber (cm^{-1})	
		3697
		3622
		3295
3697	3696	2930
3622	3621	2886
1634	1634	1631
1108	1107	1484
913	911	1140
791	791	1108
753	752	1089
		913
		791
		753

The infrared spectra of H-NM and H-SA halloysite samples show the following vibration bands: the bands at 3697 and 3622 cm^{-1}, assigned to two Al_2OH-stretching bands (vibrations of the inner OH and the OH located at the inner surface of halloysite tubes), the band at 1634 cm^{-1} which corresponds to strongly bending vibrations of the adsorbed water, the band at 1108 cm^{-1}, assigned to apical Si–O, the band observed at 914 cm^{-1} caused by the O–H deformation of inner-surface hydroxyl groups, and O–H deformation of inner hydroxyl groups, the bands at 791 and 753 cm^{-1} assigned to O–H translation vibrations of halloysite O–H units. Compared to the spectrum of raw halloysite (H-NM) with acid-treated (H-SA), one can observe that the spectra of the modified sample show a significant decrease in the intensity of all bands. In addition, for H-SA, the band at 1108 cm^{-1} is widened. FTIR measurements for the H-APTES halloysite sample exhibit several new bands in comparison with H-NM and H-SA samples. The FTIR spectra show several bands whose intensity is mainly due to APTES: peak located at 3295 cm^{-1} due to the N–H symmetric stretching of APTES, bands at 2928 and 2885 cm^{-1} corresponded to $\nu_A(C–H)_{CH2}$ and $\nu_S(C–H)_{CH3}$, respectively, the band at 1483 cm^{-1} corresponded to $\delta(NH_2)$, the siloxane stretching band, located at 1146 cm^{-1}, the band lying at 1089 cm^{-1} due to the C–O symmetric stretching of APTES.

2.2. *Adsorption Study*

2.2.1. Effect of Adsorbent Dose

To determine the optimal amounts of adsorbents for adsorption tests, the effect of halloysite dosage on adsorption efficiency was investigated. For this purpose, each amount of halloysite corresponding to a dose of 0.5 to 2.5 g/L for RR-120 or 0.5 to 5.0 g/L for Cu(II) was contacted with a dye solution (50 μmol/L), and a solution containing copper(II) ions (0.787 mmol/L). Figure 4 shows the effect of varying the dose of halloysite on the adsorption efficiency of the two adsorbates from water. A similar relationship of progressive increase in adsorption with increasing dose of halloysite is observed for both adsorbates. This is due to the increase in adsorption sites available for adsorbate, which is the result of an increase in the amount of adsorbent in the solution.

When the dose of halloysites was increased from 0.5 to 5 g/L, the removal efficiency of Cu(II) ions increased from 28.6% to 79.2% for H-NM, from 65.1% to 94.3% for H-APTES and from 31.7% to 77.7% for H-SA. For the next experiments, the adsorbent dose of 2.5 g/L was used. A similar increasing trend was observed for dye adsorption. When the dose of halloysite was 2.5 g/L, the efficiency of RR120 removal was found to be 40.2% for H-NM, 67.7% for H-APTES, and reached 99% for halloysite modified with sulfuric acid. Due to the large differences in the adsorption capacity of the different halloysites towards the

RR-120 dye, different doses of each adsorbent were used in further experiments—2.5 g/L for H-NM, 1.5 g/L for H-APTES and 1.0 g/L for H-SA.

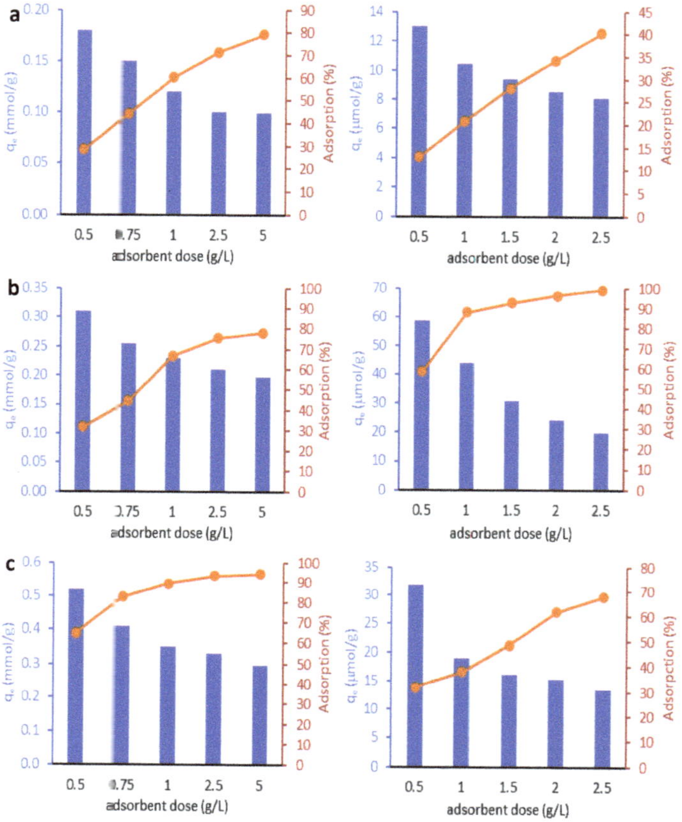

Figure 4. Effect of adsorbent dose on the adsorption of Cu(II) ions (left column) and RR-120 dye (right column) on the H-NM (**a**), H-SA (**b**), and H-APTES (**c**).

2.2.2. Adsorption Kinetics

The adsorption capacity as a function of contact time is shown in Figure 5. As can be seen, both the Cu(II) ions and the RR-120 dye were adsorbed very rapidly, with more than 80% of the initial metal load and more than 90% of the dye being adsorbed after 30 min, followed by a slow increase until equilibrium is reached after about 60 min.

The pseudo-first-order (PFO, Equation (1)) and pseudo-second-order (PSO, Equation (2)) kinetic equations were used to fit the adsorption data [21]:

$$\log(q_e - q_t) = \log q_e - \frac{k_1}{2.303}t \quad (1)$$

$$\frac{t}{q_t} = \frac{1}{k_2 q_e^2} + \frac{1}{q_e}t \quad (2)$$

where k_1—the pseudo-first-order adsorption rate constant (1/min); k_2—the pseudo-second-order adsorption rate constant (g/mmol·min for Cu(II) and g/μmol·min for RR-120); and t—time (min).

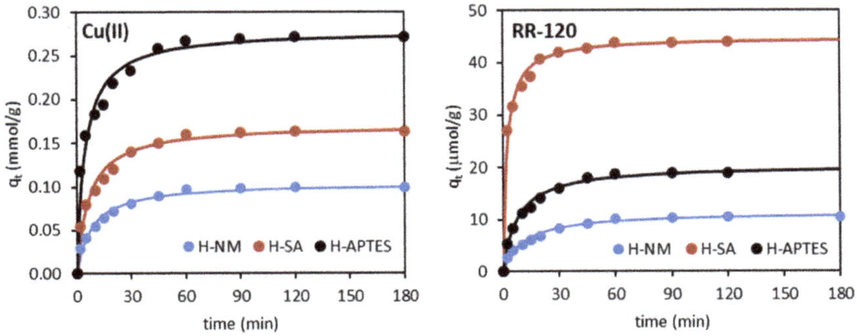

Figure 5. Adsorption kinetics of Cu(II) and RR-120 on the tested halloysites (line: fitting of a pseudo-second-order kinetic model).

The values of the adsorption rate constants k_1 and k_2, the sorption capacities $q_{e\,exp}$ (experimental) and $q_{e\,cal}$ (calculated), as well as the correlation coefficients R^2 describing the Cu(II) and RR-120 adsorption on all three halloysites were determined by the linear regression method. The obtained values are summarized and compared in Table 5.

Table 5. Kinetic modeling data for the adsorption of Cu(II) ions and RR-120 dye on halloysite samples.

Kinetic Model	Adsorbent/Adsorbate		
	H-NM	**H-SA**	**H-APTES**
		Cu(II)	
$q_{e\,exp}$ (mmol/g)	0.101	0.164	0.271
Pseudo-first-order			
k_1 (1/min)	0.0420	0.0428	0.0490
$q_{e1\,cal}$ (mmol/g)	0.051	0.104	0.151
R^2	0.948	0.966	0.986
Pseudo-second-order			
k_2 (g/mmol·min)	1.211	0.913	0.790
$q_{e2\,cal}$ (mmol/g)	0.105	0.171	0.279
R^2	0.999	0.999	0.998
		RR-120	
$q_{e\,exp}$ (μmol/g)	10.42	43.96	19.03
Pseudo-first-order			
k_1 (1/min)	0.0843	0.0615	0.0608
$q_{e1\,cal}$ (μmol/g)	15.73	25.01	17.87
R^2	0.905	0.911	0.921
Pseudo-second-order			
k_2 (g/μmol·min)	0.0083	0.0104	0.0064
$q_{e2\,cal}$ (μmol/g)	11.42	44.84	19.05
R^2	0.997	0.999	0.999

The choice of the equation best describing the adsorption of both adsorbates was made based on the higher values of the R^2 correlation coefficients. Thus, it can be concluded that the adsorption of both copper(II) ions and RR-120 dye followed the pseudo-second-order equation ($R^2 \geq 0.997$). This is also confirmed by the more closely related values of the adsorption capacities experimental ($q_{e\,exp}$) and calculated ($q_{e2\,cal}$) for the PSO model.

Copper(II) ions were adsorbed best on H-APTES and weakest on unmodified halloysite (H-NM < H-SA < H-APTES). The values of the adsorption rate constants k_2 increased in the order: H-APTES < H-SA < H-NM.

Adsorption of RR-120 dye was most efficient on halloysite H-SA (q_e = 44.84 µmol/g), less efficient on halloysite H-APTES (q_e = 19.05 µmol/g), and least efficient on halloysite H-NM (q_e = 11.42 µmol/g). The adsorption of the dye on halloysites was fastest on halloysite H-SA (k_2 = 0.0104 g/(µmol·min)), slower on H-NM (k_2 = 0.0083 g/(µmol·min)) and slowest on H-APTES (k_2 = 0.0064 g/(µmol·min)), so that, in contrast to Cu(II) ions, it correlated with the mesopore volume of each adsorbent. The specific adsorption takes place mainly in the micropores and partly in the mesopores. Thus, in general, the large volume of the micropores of an adsorbent is reflected in its large adsorption capacity. Mesopores also participate in adsorption but their primary function is transport—they are the "road/highway" by which the adsorbate is transported to the micropores. A well-developed mesoporous structure usually translates into a better kinetic capacity of the adsorbent (higher rate of adsorbate uptake in solution). The observed correlation of the adsorption rate of the dye with the mesopore volume of individual halloysites suggests that its adsorption depends more on the porous structure of the adsorbent than on the functionalization of its surface. In the case of Cu(II) ions, no correlation between adsorption and the porous structure of the adsorbents was observed, suggesting a completely different mechanism of adsorption of metal ions compared to dye molecules.

2.2.3. Adsorption Isotherms

The adsorption isotherms of the copper(II) ions and the RR-120 dye on H-NM, H-SA, and H-APTES from aqueous solutions are shown in Figure 6. Three isotherm models, namely the Freundlich, Langmuir, and Temkin, were used in this study to describe the adsorption data [22].

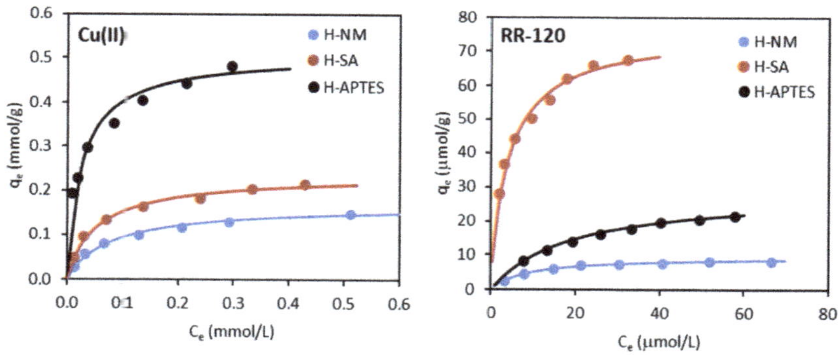

Figure 6. Adsorption isotherms of Cu(II) ions and RR-120 dye on the halloysite samples (line: fitting of Langmuir isotherm).

The Freundlich model assumes multilayer adsorption on heterogeneous and non-uniform surfaces and is given by Equation (3).

$$q_e = K_F C_e^{1/n} \quad (3)$$

where K_F ((mmol/g)(L/mmol)$^{1/n}$ for Cu(II) and (µmol/g)(L/µmol)$^{1/n}$ for RR-120) and n are the Freundlich constants.

The Langmuir model is based on the assumption of monolayer adsorption on homogeneous surfaces and is expressed by Equation (4)

$$q_e = \frac{q_m K_L C_e}{1 + K_L C_e} \quad (4)$$

where q_m is the maximum adsorption capacity (mmol/g form Cu(II) and µmol/g for RR-120) and K_L is the Langmuir constant (L/mmol or L/µmol for Cu(II) or RR-120, respectively).

The Temkin isotherm, which assumes that the free energy of adsorption is a function of the adsorbent surface coverage and also takes into account the effect of intermediate adsorbate–adsorbate interactions on the adsorption process, can be expressed by the following equation:

$$q_e = \frac{RT}{b_T}\ln(A_T C_e) \tag{5}$$

where b_T is the Temkin isotherm constant related to the heat of adsorption (J/mol); A_T is the equilibrium binding constant corresponding to the binding energy (L/g); R is the gas constant (8.314 J/mol·K); and T (K) is the absolute temperature.

The calculated parameters of the Freundlich, Langmuir, and Temkin isotherms describing the adsorption of Cu(II) ions and the dye RR-120 on halloysites, along with the correlation coefficients R^2, are summarized in Table 6.

Table 6. The Freundlich, Langmuir, and Temkin isotherm constants for the adsorption of Cu(II) ions and RR-120 dye on halloysite samples.

Isotherm Model	Adsorbent/Adsorbate		
	H-NM	**H-SA**	**H-APTES**
		Cu(II)	
Freundlich			
K_F ((mmol/g)(L/mmol)$^{1/n}$)	0.230	0.329	0.675
$1/n$	0.449	0.397	0.264
R^2	0.943	0.944	0.990
Langmuir			
q_m (mmol/g)	0.169	0.236	0.507
K_L (L/mmol)	12.50	18.16	39.01
R^2	0.995	0.996	0.995
Temkin			
b_T (kJ/mol)	72.87	53.39	29.78
A_T (L/g)	151.5	233.9	966.6
R^2	0.961	0.965	0.921
		RR-120	
Freundlich			
K_F ((μmol/g)(L/μmol)$^{1/n}$)	1.750	24.04	3.140
$1/n$	0.406	0.316	0.487
R^2	0.941	0.973	0.987
Langmuir			
q_m (μmol/g)	9.640	75.76	29.33
K_L (L/μmol)	0.111	0.241	0.051
R^2	0.999	0.997	0.997
Temkin			
b_T (kJ/mol)	1.198	0.169	0.361
A_T (L/g)	1.047	3.406	0.400
R^2	0.989	0.992	0.994

The data presented in Table 6 indicate that all three models describe the adsorption of copper(II) ions and the dye Reactive Red 120 from aqueous solutions on halloysites quite well. Comparing the correlation coefficient R^2, it can be seen that the Langmuir model best describes the adsorption of both adsorbates on halloysites, suggesting monolayer adsorption of both adsorbates on homogeneous halloysite surfaces. Comparing the calculated q_m and K_F parameters, it can be observed that both modifications of halloysite (with sulfuric acid and (3-aminopropyl)triethoxysilane) significantly improved their adsorption capacities. However, the effect of modification on adsorption was different for metal ions and dye.

The q_m values calculated from the Langmuir model for Cu(II) ions were 0.169 mmol/g for H-NM, 0.236 mmol/g for H-SA, and 0.507 mmol/g for APTES-modified halloysite, respectively. The adsorption capacity of H-APTES was ~3.0 times higher than that of H-NM and ~2.1 times higher than that of H-SA, although this adsorbent had the lowest BET-specific surface area (Table 2). The adsorption capacity of sulfuric acid-activated halloysite increased ~1.4 times compared to unmodified halloysite although its BET surface area increased significantly from 53 to 129 m^2/g. These results indicate that the NH_2 functionalization, rather than the specific surface area, was responsible for the enhanced adsorption of copper ions on the halloysite. These observations are generally in agreement with results reported by other authors who studied the adsorption of copper ions on various materials functionalized with amine groups. The presence of surface $-NH_2$ groups improved Cu(II) adsorption on, inter alia, modified MCM-41 [23], APTES-modified zeolite [24], amino-functionalized mesoporous silica [25], or magnetic nanoparticles [26].

The mechanism for the adsorption of Cu(II) ions on amino-functionalized materials includes surface complexation, electrostatic attraction, and ion exchange of amino hydrolysis by H-bonding between amino and hydroxyl groups, as proposed by Li et al. [26]. In solution, at the natural pH (~5–6), $-NH_2$ groups on the surface can undergo protonating and deprotonating reactions (Equation (6)).

$$-NH_2 + H^+ \leftrightarrow -NH_3^+ \tag{6}$$

At the same time, complexes of $Cu^{2+}(CuOH^+)$ with the amino groups on the adsorbent surface are formed via coordination interactions (Equations (7) and (8)).

$$-NH_2 + Cu^{2+} \leftrightarrow -NH_2Cu^{2+} \tag{7}$$

$$-NH_2 + Cu^{2+} + H_2O \leftrightarrow -NH_2Cu(OH)^+ + H^+ \tag{8}$$

Furthermore, an ion exchange between the amino groups and the hydroxyl groups from the solution takes place via hydrogen bonds (Equation (9)) and so formed $-NH_2OH^-$ groups can react with $Cu^{2+}(CuOH^+)$ via electrostatic attraction and/or surface complexation according to Equation (10).

$$-NH_2 + H_2O \leftrightarrow -NH_2OH^- + H^+ \tag{9}$$

$$-NH_2OH^- + Cu^{2+}(CuOH^+) \leftrightarrow -NH_2OH^- \cdots Cu^{2+}(-NH_2OH^- \cdots CuOH^+) \tag{10}$$

As one might guess, the adsorption of copper(II) ions on APTES-modified halloysite probably follows the same mechanism.

To fully assess the suitability of the materials used for the removal of copper ions, it is necessary to compare them with other adsorbents. Such a comparison is shown in Table 7, where the adsorption capacities of various other mineral adsorbents (raw and modified) are listed. For comparison, the q_m values obtained from the Langmuir model were used. These values were converted to the same units (mg/g) where necessary. The adsorption capacities of other non-mineral adsorbents can be found in the review paper [27]. The data in Table 7 show that the adsorption capacities of the halloysites used in this work are very good.

The RR-120 dye was also better adsorbed on modified halloysites than on H-NM, although in contrast to Cu(II), its adsorption was most effective on sulfuric acid-modified halloysites. This suggests that, unlike copper ions, the adsorption of RR-120 dye is determined by the textural properties (specific surface area) of the adsorbent rather than by chemical modification of its surface. This is supported by the results presented in Section 3.1. The FTIR spectra (Figure 3 and Table 4) for H-SA are almost identical to those for H-NM, indicating that the modification of halloysite with sulphuric acid has not changed the chemical character of its surface. The TG results (Table 3) also confirm this. The only observable change is a significant increase in the porosity of H-SA, including an increase in its specific surface area. The enhanced adsorption of RR-120 on H-SA is, therefore, the result of its higher BET surface area and, therefore, more active sites to which the dye molecules can attach.

Table 7. Comparison of the adsorption capacities of halloysites with some other mineral adsorbents for Cu(II) adsorption.

Adsorbent	q_m (mg/g)	Ref.
H-NM	10.73 *	This paper
H-SA	14.98 *	This paper
H-APTES	32.19 *	This paper
alginate/kaolin	0.339	[28]
bentonite clay	0.406	[29]
alginate/montmorillonite	0.680	[28]
clinoptilolite	5.269	[30]
zeolite modified with APTES	6.176	[24]
natural montmorillonite	12.22	[31]
palygorskite	12.53	[32]
urea modified halloysite (H-Mo)	13.87	[17]
melamine modified halloysite (H-Me)	15.24	[17]
H_2SO_4-activated montmorillonite	15.40	[31]
kaolinite	16.79	[33]
maghnite	21.78	[34]
natural smectite	28.82	[35]
natural Ca-bentonite	32.26	[36]
vermiculite	32.68	[32]
CH_3COONa-intercalated halloysite	52.30	[16]
$CaCl_2$ pretreated Algerian bentonite	55.47	[37]
acid-activated palygorskite	93.02	[38]

* converted from mmol/g.

The adsorption capacity determined from the Langmuir model was 9.64 μmol/g for the original halloysite, 29.33 μmol/g for H-APTES, and 75.76 μmol/g for H-SA. Thus, the activation of halloysite with sulfuric acid was the most favorable, significantly increasing its specific surface area from 53 m^2/g (H-NM) to 129 m^2/g (H-SA), as shown in Table 2. The increase in adsorption capacity of H-SA is, therefore, most likely due to its increased BET surface area and thus more active sites/centers available for the adsorbate molecules. Modification of halloysite with (3-aminopropyl)triethoxysilane also significantly increased the adsorption capacity of the resulting material compared to the original mineral. The adsorption capacity of H-APTES increased even though its specific surface area decreased as a result of the modification (from 53 m^2/g for the original halloysite to 34 m^2/g), indicating a different mechanism of dye adsorption on this material compared to H-SA.

The removal of dyes from aqueous solutions using various adsorbents is a very complex and demanding process. Specific adsorption mechanisms include van der Waals interactions, ion exchange, electrostatic interactions, surface complexation, hydrophobic and π-π interactions, or hydrogen bond formation [39]. In the case of H-SA, the improvement in its adsorption capacity is accompanied by an increase in its BET surface area. This suggests that the adsorption of RR-120 on its surface is mainly due to pore filling and hydrophobic, π-π, and hydrogen bonding interactions. The enhanced adsorption properties of halloysite H-APTES are mainly due to the presence of basic groups ($-NH_2$) derived from (3-aminopropyl)triethoxysilane on its surface. The surface amine groups are critical for anionic binding (in this case the anionic dye RR-120) and H-bond formation. Adsorption studies were carried out in solutions of the dye at a natural pH (~6.1). The mechanism of adsorption of the anionic dye in an acidic environment (pH 6.1) is related to the protonation of the amine groups of the adsorbent, as shown by the following reaction:

$$\text{Halloysite-NH}_2 + \text{H}^+ \leftrightarrow \text{Halloysite-NH}_3^+ \quad (11)$$

At the same time, the dye molecule dissociates:

$$RR\text{-}120\text{-}SO_3Na \rightarrow RR\text{-}120\text{-}SO_3^- + Na^+ \quad (12)$$

As a result of the electrostatic interaction between the adsorbent and the adsorbate molecule, adsorption takes place according to the equation:

$$Halloysite\text{-}NH_3^+ + RR\text{-}120\text{-}SO_3^- \leftrightarrow Halloysite\text{-}NH_3^+\text{-}SO_3^-\text{-}RR\text{-}120 \quad (13)$$

Table 8 compares the adsorption capacity of the halloysites with other raw and modified minerals reported in the literature. The adsorption capacities of the halloysites described in this paper are satisfactory. A comparison with other non-mineral adsorbents can be found in a recently published review paper [39].

Table 8. Comparison of the adsorption capacities of halloysites with some other mineral materials for the adsorption of Reactive Red 120.

Adsorbent	q_m (mg/g)	Ref.
H-NM	12.90 *	This paper
H-SA	101.4 *	This paper
H-APTES	39.24 *	This paper
pumice	0.320	[40]
kaolin	1.090	[41]
chitosan-montmorillonite composite	5.608	[42]
natural untreated clay (Tunisia)	29.94	[43]
CTAB-modified bentonite	51.28	[44]
raw Fouchana clay	54.64	[45]
CTAB-modified bentonite	81.97	[46]
zeolite	145.9	[47]
Fe_3O_4@zeolite composite	154.3	[47]
HDTMA-bromide modified clay	163.9	[45]
functionalized magnesium phyllosilicate	229.9	[48]

* converted from µmol/g.

2.2.4. Effect of Solution pH

The physicochemical properties of the solution, and in particular its pH, play an important role in the adsorption process as they affect the degree of ionization of the adsorbate (metal speciation) and also the charge that can appear on the surface of the adsorbent. An important parameter for determining the surface charge on the adsorbent is the pH point of zero charge (pH_{pzc}). Thus, in a solution with a $pH < pH_{pzc}$, a positive charge accumulates on the surface, whereas in a solution with a $pH > pH_{pzc}$, the surface of the adsorbent is negatively charged. The pH_{pzc} values for all three halloysite samples were determined using the pH drift method as described elsewhere [49] and were found to be 6.5 for H-NM, 6.9 for H-APTES, and 4.3 for H-SA.

The effect of the initial pH on the adsorption capacity of halloysites for Cu(II) was studied by changing the pH of the solution from 2 to 6, while for the RR-120 dye for solutions with pH in the range of 2–10 (pH of the solutions was adjusted with 0.01 mol/L NaOH and/or 0.01 mol/L HCl). Such a limited range of measurement was applied to copper ions since Cu^{2+} precipitation occurs at a pH higher than 6.0 [26,34,50].

The adsorption capacity of Cu(II) and RR-120 dye on halloysite samples at different pH is shown in Figure 7.

It can be seen that the adsorption of Cu(II) is strongly pH-dependent and increases with an increase in pH. As the pH increased from 2.0 to 6.0, the adsorption capacity of Cu(II) on H-NM, H-SA, and H-APTES increased from 0.051 to 0.101 mmol/g, from 0.089 to 0.164 mmol/g and from 0.120 to 0.271 mmol/g, respectively. The adsorption capacity was very low in a strongly acidic environment and increased with increasing pH values until it reached optimal efficiency in the pH range of 5.0–6.0. The relatively low adsorption in acidic media is probably due to the competition of positively charged H^+ ions with Cu^{2+} ions for the same adsorption sites and functional groups present on the adsorbent surface [50]. For example, in the case of H-APTES, hydrogen ions present in solution in large quantities competed with Cu(II) cations for ion exchange with NH_4^+ ions derived from APTES. As reported by Zhang et al. [51], H^+ ions may also inhibit the dissociation of protons from other surface functional groups (e.g., carboxyl and hydroxyl groups) necessary for ion exchange with Cu(II) cations. The observed successive increase in adsorption efficiency with increasing pH can, therefore, be explained by a decrease in competition between H^+ and Cu^{2+} ions for the same surface groups. Furthermore, with increasing pH, the number of negatively charged functional groups on the adsorbent surface also increases, which, as suggested by Rahman and Islam [50], increases the adsorption of positively charged cupric ions. As suggested by other authors [26,50], copper ions in solution at pH 6.0 can exist in three species (Cu^{2+}, $Cu(OH)^{+}$, and $Cu(OH)_2$), which can be adsorbed by ion exchange mechanism or by hydrogen bonding. A similar trend in the adsorption of copper(II) ions—an increase in adsorption efficiency with increasing pH—was reported by other authors, among others, on the maple wood sawdust [50], amino-functionalized magnetic nanoparticles [26], maghnite [34], poly(acrylic acid/chestnut shell pigment) hydrogel [51], amine-functionalized MCM-41 [23] or natural zeolite modified with APTES [24].

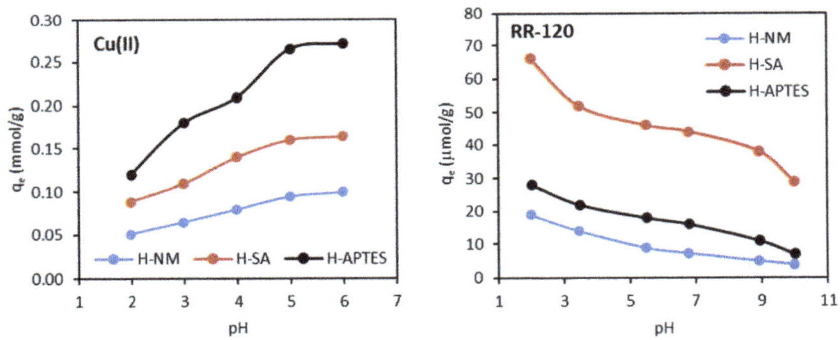

Figure 7. Effect of initial pH on the adsorption of Cu(II) ions and RR-120 dye on halloysite samples.

The effect of pH on the adsorption of RR-120 dye from aqueous solutions was studied in the pH range of 2 to 10 as is shown in Figure 7. It can be seen that the adsorption of RR-120 dye on halloysites is closely related to the initial pH of the solution. The adsorption capacity of RR-120 dye decreases as the pH of the solution increases from 2 to 10. For all three adsorbents, the maximum adsorption was found at pH 2, when 43, 56, and 68% of the initial amount of dye was adsorbed on H-NM, H-APTES, and H-SA, respectively. The RR-120 is an anionic dye and contains six sulfonate groups in its molecule, the pKa of which is 2.1 [52]. This means that it exists in the anionic form in practically the entire pH range over which the experiments were carried out (2–10), due to the transformation of the $-SO_3H$ group into a negative $-SO_3^-$ group in an aqueous medium. The pH effect on RR-120 adsorption can be explained by the pH_{pzc} concept. The H-NM, H-APTES, and H-SA had different values of pH_{pzc} as 6.5, 6.9, and 4.3, respectively. Consequently, the halloysites obtain a positively charged surface at the solution pH below pH_{pzc}, which favors the adsorption of an anionic dye such as RR-120 due to electrostatic attraction. The decrease in adsorption at high solution pH ($pH > pH_{pzc}$) results from a stronger electrostatic repulsion between the negatively charged surface of the halloysite and the anionic dye molecule. Similar results showing a decrease in adsorption with increasing pH were reported by other researchers [20,52–59].

3. Materials and Methods

3.1. Reagents and Materials

The Reactive Red 120 azo dye (RR, $C_{44}H_{24}Cl_2N_{14}Na_6O_{20}S_6$, CAS number: 61951-82-4) was purchased from Boruta-Zachem SA (Bydgoszcz, Poland), the (3-aminopropyl) triethoxysilane (APTES) was obtained from Sigma-Aldrich (St. Louis, MO, USA) while cuprizone (bis-(cyclohexanone) oxaldehydrozone)–synthetic chelating compound used for the determination of copper(II) ions was purchased from Acros Organics (Geel, Belgium). Other high-purity reagents including copper(II) sulfate pentahydrate ($CuSO_4{\cdot}5H_2O$) were received from Chempur (Piekary Śląskie, Poland). The halloysite samples were received from the "Dunino" mine (Intermark Company, Legnica, Poland). The morphology and structure of this raw mineral were described elsewhere [60,61].

3.2. Preparation and Characterization of Adsorbents

First, the native halloysite (H-NM) was mixed with ethanol in a weight ratio of 1:3, then it was subjected to ultrasound (35 kHz) for 3 h in an ultrasound bath IS-7S (Intersonic S.C., Olsztyn Poland), and at the same time, it was stirred with a mechanical stirrer. After this time, APTES was added to the mixture in an amount corresponding to 15% of the mass of halloysite and the process was continued for another 3 h under the same conditions as before. After the modification, the solvent was evaporated and the obtained product (H-APTES) was ground to a powder. Drying took place at 90 °C.

The native halloysite was also treated with sulfuric acid (98%) in proportion 92:8. The first stage involved the preliminary treatment of native halloysite fraction with ultrasounds of frequency 30 kHz for 2–4 h in an ultrasounds bath. The obtained product was separated from the unreacted reagent to obtain the final modified halloysite (H-SA) and then dried at 90 °C.

For both modified samples (H-APTES and H-SA) and also for native halloysite (H-NM) the Brunauer–Emmett–Teller (BET) specific surface areas as well as micro- and meso-pore volumes were calculated based on measured low-temperature nitrogen adsorption isotherms using a TriStar II 3020 V1.03 (Micromeritics Company, Norcross, GA, USA).

Determination of their surface chemical compositions was carried out using JSM-6490LV (JEOL Company, Peabody, MA, USA) scanning electron microscope coupled with an energy dispersive X-ray spectrometer (EDS).

Thermogravimetric analysis was performed using a thermobalance type TGA Q50 (Thermo Scientific, Waltham, MA, USA). The measurement was carried out at the heating rate of 10 °C/min in the temperature regime from 25 up to 800 °C in an inert gas (nitrogen) atmosphere.

For halloysite samples, the surface chemistry was additionally characterized by recording FTIR spectra using a Nicolet iS10 FTIR spectrometer (Thermo Scientific, Waltham, MA, USA).

3.3. Bath Adsorption Experiments

Adsorption studies of Cu(II) ions and Reactive Red 120 dye on all three halloysites were performed in Erlenmeyer flasks containing adsorbate solutions of appropriate concentrations and corresponding adsorbent weights. The flasks were shaken at a constant speed of 100 rpm at room temperature. All experiments were repeated twice and the average of the two measurements was used for further calculations. Adsorption kinetics and equilibrium adsorption (adsorption isotherms) were studied as well as the effects of adsorbent dose and solution pH.

The experimental procedures varied due to the individual characteristics of the metal ions and the dye. For example, for the adsorption of copper(II) ions, the solution volume was always 20 mL, and the initial concentrations ranged from 0.157 to 1.259 mmol/L (10 to 80 mg/L). Kinetic studies and experiments on the effect of adsorbent dose and solution pH were performed for Cu(II) solutions with initial concentrations of 0.787 mmol/L (50 mg/L). For the RR-120 dye, solution volumes were 10 mL and initial concentrations ranged from 10 to 100 μmol/L. The effect of adsorbent dosage, solution pH, and contact time on adsorption was performed for an initial dye concentration of 50 μmol/L.

The natural (original) pH of the adsorbate solutions (~5.7 for Cu(II) and ~6.1 for RR-120) was chosen for the adsorption studies.

The concentrations of both adsorbates in the solutions after adsorption were measured using a UV–Vis spectrophotometer (Varian Carry 3E series, Palo Alto, CA, USA). Copper ions were determined spectrophotometrically in the form of complexes with cuprizone, which forms a blue chelate with copper in a slightly alkaline environment. The Cu(II) ion concentration was measured at a wavelength of 600 nm corresponding to the absorption maximum of the complex formed. The calibration curve for the determination of copper(II) ions was linear (R^2 = 0.999) over the tested concentration range of 0.0016 to 0.0472 mmol/L (0.1–3.0 mg/L) and was described by the equation y = 15.563x + 0.0089. The concentration of RR-120 dye in the solutions was determined spectrophotometrically by measuring the absorbance at 512 nm. The resulting calibration curve (y = 0.0331x + 0.011) was linear over the entire range studied, from 1 to 100 μmol/L (R^2 = 0.999).

The adsorption efficiency (%), as well as the amount of adsorbate adsorbed at equilibrium (q_e) and at time t (q_t), were calculated by the following equations

$$Adsorption(\%) = \frac{(C_0 - C_e)}{C_0} \times 100 \quad (14)$$

$$q_t = \frac{(C_0 - C_t)V}{m} \quad (15)$$

$$q_e = \frac{(C_0 - C_e)V}{m} \quad (16)$$

where V is the volume of the solution (L), m is the mass of the halloysite sample (g), while C_0, C_t, and C_e are the initial adsorbate concentration, concentration after time t, and concentration at equilibrium, respectively. The concentrations of Cu(II) ions are expressed in mmol/L, while the concentrations of the RR-120 dye are expressed in μmol/L.

4. Conclusions

In this paper, the adsorption of copper(II) ions and Reactive Red 120 azo dye (RR-120) as models of water contaminants on unmodified (H-NM) and modified halloysites was studied. Both (3-aminopropyl)triethoxysilane (H-APTES) and sulfuric acid (H-SA) were used to modify the halloysite. Adsorption kinetics, adsorption isotherms as well as the effects of adsorbent dosage and solution pH were investigated. The results demonstrated that the adsorption of both the adsorbates increased with the increase in halloysite dosage. Adsorption was strongly pH-dependent and increased (Cu(II)) or decreased (RR-120) with an increase in solution pH. The adsorption kinetics followed the pseudo-second-order model. Adsorption isotherms were well described by the Langmuir model. A maximum capacity of Cu(II) adsorbed on natural halloysite as well as on H-SA and H-APTES at equilibrium was 0.169, 0.236, and 0.507 mmol/g, respectively. The maximum adsorption capacities for the RR-120 dye were 9.64, 75.76, and 29.33 μmol/g for H-NM, H-SA, and H-APTES, respectively. Both modification methods increased the adsorption of the copper(II) ions as well as the RR-120 dye. The results suggest that Cu(II) adsorption is rather affected by the surface chemistry of the halloysite, while the porous structure (specific surface area) of the adsorbent plays a major role in the adsorption of the dye. The results showed that APTES-functionalization and activation by sulfuric acid are promising modifications, and both modified halloysites have good application potential for heavy metals as well as azo dye removal.

Author Contributions: Conceptualization, K.K. and E.W.; methodology, K.K. and E.W.; software, K.K. and A.Ś.; validation, E.W., A.Ś. and K.K.; formal analysis, E.W. and K.K.; investigation, E.W. and K.K.; resources, A.Ś. and I.L.; data curation, E.W., K.K. and A.Ś.; writing—original draft preparation, K.K., E.W. and A.Ś.; writing—review and editing, A.Ś. and I.L.; visualization, K.K. and A.Ś.; supervision,

K.K. and A.Ś.; project administration, K.K.; funding acquisition, K.K. and E.W. All authors have read and agreed to the published version of the manuscript.

Funding: This research received no external funding.

Data Availability Statement: The data presented in this study are available on request from the corresponding author.

Conflicts of Interest: The authors declare no conflict of interest.

References

1. Rafique, M.; Hajra, S.; Tahir, M.B.; Gillani, S.S.A.; Irshad, M. A review on sources of heavy metals, their toxicity and removal technique using physico-chemical processes from wastewater. *Environ. Sci. Pollut. Res.* **2022**, *29*, 16772–16781. [CrossRef] [PubMed]
2. Vardhan, K.H.; Kumar, P.S.; Panda, R.C. A review on heavy metal pollution, toxicity and remedial measures: Current trends and future perspectives. *J. Mol. Liq.* **2019**, *290*, 111197. [CrossRef]
3. Mrabet, B.S.; Harfi, S. A review on classifications, recent synthesis and applications of textile dyes. *Inorg. Chem. Commun.* **2020**, *115*, 107891. [CrossRef]
4. Khattab, T.A.; Abdelrahman, M.S.; Rehan, M. Textile dyeing industry environmental impacts and remediation. *Environ. Sci. Pollut. Res.* **2020**, *27*, 3803–3818. [CrossRef]
5. Lellis, B.; Fávaro-Polonio, C.Z.; Pamphil, J.A.; Polonio, J.C. Effects of textile dyes on health and the environment and bioremediation potential of living organisms. *Biotechnol. Res. Innov.* **2019**, *3*, 275–290. [CrossRef]
6. Crini, G.; Lichtfouse, E.; Wilson, L.D.; Crini, N.M. Conventional and non conventional adsorbents for wastewater treatment. *Environ. Chem. Lett.* **2018**, *17*, 195–213. [CrossRef]
7. Iravani, R.; An, C.; Adamian, Y.; Mohammadi, M. A review on the use of nanoclay adsorbents in environmental pollution control. *Water Air Soil Pollut.* **2022**, *233*, 109. [CrossRef]
8. Kausar, A.; Iqbal, M.; Javed, A.; Aftab, K.; Nazli, Z.H.; Bhatti, H.M.; Nouren, S. Dyes adsorption using clay and modified clay: A review. *J. Mol. Liq.* **2018**, *256*, 395–407. [CrossRef]
9. Yu, L.; Wang, H.; Zhang, Y.; Zhang, B.; Liu, J. Recent advances in halloysite nanotube derived composites for water treatment. *Environ. Sci. Nano* **2015**, *3*, 28–44. [CrossRef]
10. Anastopoulos, I.; Mittal, A.; Usman, M.; Mittal, J.; Yu, G.; Núñez-Delgado, A.; Kornaros, M. A review on halloysite-based adsorbents to remove pollutants in water and wastewater. *J. Mol. Liq.* **2018**, *269*, 855–868. [CrossRef]
11. Wilson, I.; Keeling, J. Global occurrence, geology and characteristics of tubular halloysite deposits. *Clay Miner.* **2016**, *51*, 309–324. [CrossRef]
12. Cheng, C.; Song, W.; Zhao, Q.; Zhang, H. Halloysite nanotubes in polymer science: Purification, characterization, modification and applications. *Nanotechnol. Rev.* **2020**, *9*, 323–344. [CrossRef]
13. Biddeci, G.; Spinelli, G.; Colomba, P.; Di Blasi, F. Review nanomaterials: A review about halloysite nanotubes, properties, and application in the biological field. *Int. J. Mol. Sci.* **2022**, *23*, 11518. [CrossRef] [PubMed]
14. Matusik, J.; Wścisło, A. Enhanced heavy metal adsorption on functionalized nanotubular halloysite interlayer grafted with aminoalcohols. *Appl. Clay Sci.* **2014**, *100*, 50–59. [CrossRef]
15. Mellouk, S.; Cherifi, S.; Sassi, M.; Marouf-Khelifa, K.; Bengueddach, A.; Schott, J.; Khelifa, A. Intercalation of halloysite from Djebel Debagh (Algeria) and adsorption of copper ions. *Appl. Clay Sci.* **2009**, *44*, 230–236. [CrossRef]
16. Mellouk, S.; Belhakem, A.; Marouf-Khelifa, K.; Schott, J.; Khelifa, A. Cu(II) adsorption by halloysites intercalated with sodium acetate. *J. Colloid Interf. Sci.* **2011**, *360*, 716–724. [CrossRef]
17. Wierzbicka, E.; Legocka, I.; Kuśmierek, K.; Świątkowski, A. Modified polymers filler as an effective adsorbent of Cu(II) ions from an aqueous solution. *Polimery* **2021**, *66*, 239–244. [CrossRef]
18. Bessaha, F.; Marouf-Khelifa, K.; Batonneau-Gener, I.; Khelifa, A. Characterization and application of heat-treated and acid-leached halloysites in the removal of malachite green: Adsorption, desorption, and regeneration studies. *Desalin. Water Treat.* **2015**, *57*, 14609–14621. [CrossRef]
19. Kuśmierek, K.; Świątkowski, A.; Wierzbicka, E.; Legocka, I. Enhanced adsorption of Direct Orange 26 dye in aqueous solutions by modified halloysite. *Physicochem. Probl. Miner. Process.* **2020**, *56*, 693–701. [CrossRef]
20. Wierzbicka, E.; Kuśmierek, K.; Świątkowski, A.; Legocka, I. Efficient Rhodamine B dye removal from water by acid- and organo-modified halloysites. *Minerals* **2022**, *12*, 350. [CrossRef]
21. Tan, K.L.; Hameed, B.H. Insight into the adsorption kinetics models for the removal of contaminants from aqueous solutions. *J. Taiwan Inst. Chem. Eng.* **2017**, *74*, 25–48. [CrossRef]
22. Al-Ghouti, M.A.; Da'ana, D.A. Guidelines for the use and interpretation of adsorption isotherm models: A review. *J. Hazard. Mater.* **2020**, *393*, 122383. [CrossRef] [PubMed]
23. Parida, K.; Mishra, K.G.; Dash, S.K. Adsorption of copper(II) on NH_2-MCM-41 and its application for epoxidation of styrene. *Ind. Eng. Chem. Res.* **2012**, *51*, 2235–2246. [CrossRef]

24. Shirendev, N.; Bat-Amgalan, M.; Kano, N.; Kim, H.-J.; Gunchin, B.; Ganbat, B.; Yunden, G. A natural zeolite developed with 3-aminopropyltriethoxysilane and adsorption of Cu(II) from aqueous media. *Appl. Sci.* **2022**, *12*, 11344. [CrossRef]
25. Wei, J.; Chen, S.; Li, Y.; He, Z.; Geng, L.; Liao, L. Aqueous Cu(II) ion adsorption by amino functionalized mesoporous silica KIT-6. *RSC Adv.* **2020**, *10*, 20504–20514. [CrossRef]
26. Li, H.; Xiao, D.; He, H.; Lin, R.; Zuo, P. Adsorption behavior and adsorption mechanism of Cu(II) ions on amino-functionalized magnetic nanoparticles. *Trans. Nonferrous Met. Soc. China* **2013**, *23*, 2657–2665. [CrossRef]
27. Krstić, V.; Urošević, T.; Pešovski, B. A review on adsorbents for treatment of water and wastewaters containing copper ions. *Chem. Eng. Sci.* **2018**, *192*, 273–287. [CrossRef]
28. Râpa, M.; Turcanu, A.A.; Matei, E.; Predescu, A.M.; Pantilimon, M.C.; Coman, G.; Predescu, C. Adsorption of copper (II) from aqueous solutions with alginate/clay hybrid materials. *Materials* **2021**, *14*, 7187. [CrossRef]
29. Nassef, E.; Mahmoud, A.; Salah, H.; El-taweel, Y. Removal of copper ions from liquid wastes using adsorption technique. *Int. J. Res. Ind. Eng.* **2017**, *6*, 522–562. [CrossRef]
30. Zendelska, A.; Golomeova, M.; Blazev, K.; Krstev, B.; Golomeov, B.; Krstev, A. Adsorption of copper ions from aqueous solutions on natural zeolite. *Environ. Prot. Eng.* **2015**, *41*, 17–36. [CrossRef]
31. Djebbar, M.; Djafri, F. Adsorption of Cu(II) on natural and treated clays. *Water Qual. Res. J.* **2016**, *51*, 26–32. [CrossRef]
32. Bourliva, A.; Sikalidis, A.K.; Papadopoulou, L.; Betsiou, M.; Michailidis, K.; Sikalidis, C.; Filippidis, A. Removal of Cu^{2+} and Ni^{2+} ions from aqueous solutions by adsorption onto natural palygorskite and vermiculite. *Clay Miner.* **2018**, *53*, 1–15. [CrossRef]
33. Wang, X.-S.; Wang, J.; Sun, C. Removal of copper(II) ions from aqueous solutions using natural kaolinite. *Adsorpt. Sci. Technol.* **2006**, *24*, 517–530. [CrossRef]
34. Zenasni, M.A.; Benfarhi, S.; Merlin, A.; Molina, S.; George, B.; Meroufel, B. Adsorption of Cu(II) on maghnite from aqueous solution: Effects of pH, initial concentration, interaction time and temperature. *Nat. Sci.* **2012**, *4*, 856–868. [CrossRef]
35. Langama, P.L.M.; Anguile, J.J.; Ndong, A.N.M.M.; Bouraima, A.; Bissielou, C. Adsorption of copper (ii) ions in aqueous solution by a natural mouka smectite and an activated carbon prepared from kola nut shells by chemical activation with zinc chloride ($ZnCl_2$). *Open J. Inorg. Chem.* **2021**, *11*, 131–144. [CrossRef]
36. Anna, B.; Kleopas, M.; Constantine, S.; Anestis, F.; Maria, B. Adsorption of Cd(II), Cu(II), Ni(II) and Pb(II) onto natural bentonite: Study in mono- and multi-metal systems. *Environ. Earth Sci.* **2014**, *73*, 5435–5444. [CrossRef]
37. Ouakouak, A.; Rihani, K.; Youcef, L.; Hamdi, N.; Guergazi, S. Adsorption characteristics of Cu (II) onto $CaCl_2$ pretreated algerian bentonite. *Mater. Res. Express* **2020**, *7*, 025045. [CrossRef]
38. Chen, H.; Zhao, Y.; Wang, A. Removal of Cu(II) from aqueous solution by adsorption onto acid-activated palygorskite. *J. Hazard. Mater.* **2007**, *149*, 346–354. [CrossRef]
39. Anastopoulos, I.; Ahmed, M.J.; Ojukwu, V.E.; Danish, M.; Stylianou, M.; Ighalo, J.O. A comprehensive review on adsorption of Reactive Red 120 dye using various adsorbents. *J. Mol. Liq.* **2024**, *394*, 123719. [CrossRef]
40. Hossein, M.A.; Behzad, H. Removal of Reactive Red 120 and Direct Red 81 dyes from aqueous solutions by pumice. *Res. J. Chem. Environ.* **2012**, *16*, 62–68.
41. Aragaw, T.A.; Alene, A.N. A comparative study of acidic, basic, and reactive dyes adsorption from aqueous solution onto kaolin adsorbent: Effect of operating parameters, isotherms, kinetics, and thermodynamics. *Emerg. Contam.* **2022**, *8*, 59–74. [CrossRef]
42. Kittinaovarat, S.; Kansomwan, P.; Jiratumnukul, N. Chitosan/modified montmorillonite beads and adsorption Reactive Red 120. *Appl. Clay Sci.* **2010**, *48*, 87–91. [CrossRef]
43. Errais, E.; Duplay, J.; Darragi, F.; M'Rabet, I.; Aubert, A.; Huber, F.; Morvan, G. Efficient anionic dye adsorption on natural untreated clay: Kinetic study and thermodynamic parameters. *Desalination* **2011**, *275*, 74–81. [CrossRef]
44. Al Rubai, H.F.; Hassan, A.K.; Sultan, M.S.; Abood, W.M. Kinetics of adsorption of Reactive Red 120 using bentonite modified by CTAB and study the effect of salts. *Nat. Environ. Pollut. Technol.* **2021**, *20*, 281–289. [CrossRef]
45. Nejib, A.; Joelle, D.; Fadhila, A.; Sophie, G.; Malika, T.-A. Adsorption of anionic dye on natural and organophilic clays: Effect of textile dyeing additives. *Desalin. Water Treat.* **2015**, *54*, 1754–1769. [CrossRef]
46. Tabak, A.; Baltas, N.; Afsin, B.; Emirik, M.; Caglar, B.; Eren, E. Adsorption of Reactive Red 120 from aqueous solutions by cetylpyridinium-bentonite. *J. Chem. Technol. Biotechnol.* **2010**, *85*, 1199–1207. [CrossRef]
47. Ay, C.; Sarpaşar, Z. Using zeolite and Fe_3O_4@zeolite composites in removal of Reactive Red 120 from wastewater: Isotherm, kinetic, thermodynamic and adsorption behaviors. *J. Dispers. Sci. Technol.* **2023**, *44*, 370–381. [CrossRef]
48. Kim, Y.W.; Kim, J.H.; Moon, D.H.; Shin, H.J. Adsorption and precipitation of anionic dye Reactive Red 120 from aqueous solution by aminopropyl functionalized magnesium phyllosilicate. *Korean J. Chem. Eng.* **2019**, *36*, 101–108. [CrossRef]
49. Kuśmierek, K.; Sprynskyy, M.; Świątkowski, A. Raw lignite as an effective low-cost adsorbent to remove phenol and chlorophenols from aqueous solutions. *Sep. Sci. Technol.* **2020**, *55*, 1741–1751. [CrossRef]
50. Rahman, M.S.; Islam, M.R. Effects of pH on isotherms modeling for Cu(II) ions adsorption using maple wood sawdust. *Chem. Eng. J.* **2009**, *149*, 273–280. [CrossRef]
51. Zhang, H.; Li, G.-W.; Feng, W.; Yao, Z.-Y. Cu(II) Adsorption from aqueous solution onto poly(acrylic acid/chestnut shell pigment) hydrogel. *Water* **2022**, *14*, 3500. [CrossRef]
52. Navaei, A.; Yazdani, M.; Alidadi, H.; Dankoob, M.; Bonyadi, Z.; Dehghan, A.; Hosseini, A. Biosorption of Reactive Red 120 dye from aqueous solution using *Saccharomyces cerevisiae*: RSM analysis, isotherms and kinetic studies. *Desalin. Water Treat.* **2019**, *171*, 418–427. [CrossRef]

53. Çelekli, A.; Yavuzatmaca, M.; Bozkurt, H. Modeling the removal of Reactive Red 120 on pistachio husk. *Clean* **2010**, *38*, 173–180. [CrossRef]
54. Çelekli, A.; Ilgün, G.; Bozkurt, H. Sorption equilibrium, kinetic, thermodynamic, and desorption studies of Reactive Red 120 on *Chara contraria*. *Chem. Eng. J.* **2012**, *191*, 228–235. [CrossRef]
55. Hosseini, A.R.; Khorshid, A.R.; Mahvi, A.H. Decolorisation of Reactive Red 120 dye by using Single-Walled Carbon Nanotubes in aqueous solutions. *J. Chem.* **2013**, *2013*, 938374. [CrossRef]
56. Çelekli, A.; Al-Nuaimi, A.I.; Bozkurt, H. Adsorption kinetic and isotherms of Reactive Red 120 on *Moringa oleifera* seed as an eco-friendly process. *J. Mol. Struct.* **2019**, *1195*, 168–178. [CrossRef]
57. Jawad, A.H.; Mubarak, N.S.A.; Sabar, S. Adsorption and mechanism study for reactive red 120 dye removal by cross-linked chitosan-epichlorohydrin biobeads. *Desalin. Water Treat.* **2019**, *164*, 378–387. [CrossRef]
58. Munagapati, V.S.; Wen, H.-Y.; Wen, J.-C.; Gollakota, A.R.K.; Shu, H.-M.; Reddy, G.M. Characterization of protonated amine modified lotus (*Nelumbo nucifera*) stem powder and its application in the removal of textile (Reactive Red 120) dye from liquid phase. *J. Mol. Liq.* **2021**, *338*, 116486. [CrossRef]
59. Chakraborty, T.K.; Ghosh, G.C.; Akter, M.N.; Audhikary, K.; Islam, M.S.; Ghosh, P.; Zaman, S.; Habib, A.; Kabir, A.H.M.E. Biosorption of Reactive Red 120 dye from aqueous solutions by using Mahagoni (*Swietenia mahagoni*) wood and bark charcoal: Equilibrium, and kinetic studies. *Pollution* **2021**, *7*, 905–921. [CrossRef]
60. Wieczorek, M.; Tomaszewska, J.; Bajda, T.; Długosz, J. Effect of calcinated halloysite on structure and properties of rigid poly(vinyl chloride) composites. *Chem. Process Eng.* **2022**, *43*, 383–404. [CrossRef]
61. Filice, S.; Bongiorno, C.; Libertino, S.; Compagnini, G.; Gradon, L.; Iannazzo, D.; La Magna, A.; Scalese, S. Structural Characterization and Adsorption Properties of Dunino Raw Halloysite Mineral for Dye Removal from Water. *Materials* **2021**, *14*, 3676. [CrossRef]

molecules

MDPI

Article

Removal of Cr(VI) from Wastewater Using Acrylonitrile Grafted Cellulose Extracted from Sugarcane Bagasse

Idrees Khan [1], Ashraf Ali [2,3,*], Alia Naz [1,*], Zenab Tariq Baig [1], Wisal Shah [1], Zia Ur Rahman [1], Tawaf Ali Shah [4], Kotb A. Attia [5], Arif Ahmed Mohammed [5] and Yaser M. Hafez [6]

[1] Department of Environmental Science, Faculty of Physical & Applied Sciences, The University of Haripur, Haripur 22620, Pakistan; khanamazawal@gmail.com (I.K.); zenab.tariq@uoh.edu.pk (Z.T.B.); wisal@uoh.edu.pk (W.S.); zia@uoh.edu.pk (Z.U.R.)
[2] School of Chemistry & Chemical Engineering, Henan University of Technology, Zhengzhou 450001, China
[3] Department of Chemistry, Faculty of Physical & Applied Sciences, The University of Haripur, Haripur 22620, Pakistan
[4] College of Agriculture Engineering and Food Sciences, Shandong University of Technology, Zibo 255000, China; tawafbiotech@yahoo.com
[5] Department of Biochemistry, College of Science, King Saud University, P.O. Box 2455, Riyadh 11451, Saudi Arabia; kattia1.c@ksu.edu.sa (K.A.A.); moaahmed@ksu.edu.sa (A.A.M.)
[6] EPCRS Excellence Center, Plant Pathology and Biotechnology Laboratory, Agricultural Botany Department, Faculty of Agriculture, Kafrelsheikh University, Kafr El-Sheikh 33516, Egypt; hafezyasser@gmail.com
* Correspondence: ashrafaliswati@gmail.com (A.A.); alianaz@uoh.edu.pk (A.N.)

Abstract: A highly efficient low-cost adsorbent was prepared using raw and chemically modified cellulose isolated from sugarcane bagasse for decontamination of Cr(VI) from wastewater. First, cellulose pulp was isolated from sugarcane bagasse by subjecting it to acid hydrolysis, alkaline hydrolysis and bleaching with sodium chlorate ($NaClO_3$). Then, the bleached cellulose pulp was chemically modified with acrylonitrile monomer in the presence Fenton's reagent (Fe^{+2}/H_2O_2) to carry out grafting of acrylonitrile onto cellulose by atom transfer radical polymerization. The developed adsorbent (acrylonitrile grafted cellulose) was analyzed by X-ray diffraction analysis (XRD), scanning electron microscopy (SEM) and Fourier transform infrared spectroscopy (FT-IR). Both raw cellulose and acrylonitrile grafted cellulose were used for chromium removal from wastewater. The effects of metal ion concentration, pH, adsorbent dose and time were studied, and their values were optimized. The optimum conditions for the adsorption of Cr(VI) onto raw and chemically modified cellulose were: metal ion concentration: 50 ppm, adsorbent dose: 1 g, pH: 6, and time: 60 min. The maximum efficiencies of 73% and 94% and adsorption capacities of 125.95 mg/g and 267.93 mg/g were achieved for raw and acrylonitrile grafted cellulose, respectively. High removal efficiency was achieved, owing to high surface area of 79.92 m^2/g and functional active binding cites on grafted cellulose. Isotherm and kinetics studies show that the experimental data were fully fitted by the Freundlich isotherm model and pseudo first-order model. The adsorbent (acrylonitrile grafted cellulose) was regenerated using three different types of regenerating reagents and reused thirty times, and there was negligible decrease (19%) in removal efficiency after using it for 30 times. Hence, it is anticipated that acrylonitrile could be utilized as potential candidate material for commercial scale Cr(VI) removal from wastewater.

Keywords: cellulose; bagasse; grafting co-polymerization; water treatment; chromium

1. Introduction

Increase in human population and rapid industrialization have led to the addition of various pollutants into the environment which affect living organisms [1]. Heavy metals (HMs) are added to the environment by different industries such as fertilizers production, pesticides production, untreated sewage, and petrochemical industry [2,3]. Even small amounts of heavy metals are harmful and cause anemia, damage to kidneys, intestines, liver, and can even lead to cancer [4–6]. The most toxic HMs are chromium, lead, mercury, cadmium, and arsenic, which cause several disorders in humans and animals [7–10]. Various industries, such as electroplating and the leather, textile, and tanning industries discharge chromium into the environment [11–14]. Chromium exists in nature in different oxidation states, among which the oxidation states of three Cr(III) and six Cr(VI) are more common. Cr(VI) is more toxic than Cr(III) and produces numerous infections such as malignant lung disease, chronic respiratory disease, irascibility and liver damage [15,16].

Various methods could be used for chromium removal, such as coagulation, membrane separation, filtration, electrocoagulation, reverse osmosis, electrodialysis, ultrafiltration, chemical precipitation and adsorption [17–25]. Most of these approaches are not useful, owing to lower removal efficiency, high energy consumption, high operational cost, the use of large quantities of chemicals and the production of secondary waste [20,26,27]. Adsorption is a more useful method because of high removal efficiency, low operational cost and simple operational design [28–31]. Different adsorbents are used for the removal of HMs, such as activated carbon, porous materials, ion exchangers, etc., but the high cost of commercial adsorbents is an obstacle in its use for wastewater treatment. Agriculturally based materials like fruit peels [32], rice husks [33], sawdust [34], bagasse [35], sugar beets, and soybean hull [36–38] have also been utilized for decontamination of HMs. However, agricultural by-product-based adsorbents have lignin and pectin, which cover the porous structure of cellulose and decrease its adsorption capacity. To tackle this challenge, the lignin and pectin can be removed by treating these materials with acid/alkali, followed by bleaching [39,40]. The cellulose fibers thus obtained are free from viscous and peptic compounds and have hydroxy, carbonyl, amine, etc. groups which can bind metal ions by electrostatic force of attraction or intermolecular force of attraction [41–43].

Chemical modification of cellulose fibers with a suitable ligand can enhance their interaction with HM ions, thus increasing their removal efficiency [44–47]. Therefore, it is necessary to modify natural cellulosic materials to improve their adsorption efficiency and their propensity to bind heavy metals [48]. Natural cellulosic materials have a high content of hydrogen bonding within the cellulosic chains, which prevents their interaction with metal ions [48,49]. Grafted cellulose has additional side chains which can bind the adsorbate molecules and thus the removal efficiency of grafted cellulose is higher than that of non-grafted or raw cellulose [50,51]. Several researchers have used sugarcane bagasse as an adsorbent, either in raw form or chemically modified form, for the removal of heavy metals, dyes and other pollutants. Ezeonuegbu et al. [52] utilized sugarcane bagasse for the removal of Pb(II) and Ni(II) from wastewater and obtained removal efficiencies of 89.31% and 96.33% for lead and nickel, respectively. Gusmano et al. [53] utilized modified sugarcane bagasse for the removal of etherdiamine from aqueous solution, and adsorption capacities of 869.6 and 1203.5 mg/g were obtained for succinic anhydride modified sugarcane bagasse and EDTA dianhydride modified sugarcane bagasse, respectively. In another study, Al-Mokhalelati et al. [54] used chemically modified sugarcane bagasse for the removal of methylene blue from aqueous solutions and obtained 98% removal efficiency for MB. Similarly, Jiang et al. [55], utilized sugarcane bagasse modified with polyethylenimine for the removal of Cu(II) from aqueous solution and adsorption capacities up to 107.5 mg/g were reported. The removal efficiency of raw sugarcane bagasse is not high enough to remove heavy metal ions, dyes or other organic pollutants at low concentration, owing to the presence of several other compounds such as pectin, lignin, etc., in addition to cellulose. To remove these peptic compounds and enhance the removal efficiency of sugarcane bagasse, it was subjected to acid hydrolysis, alkaline hydrolysis, and bleaching. After hydrolysis and bleaching, clean white cellulose fibers were obtained, which were functionalized with acrylonitrile to incorporate a nitrile group. The insertion of the nitrile group further enhances the interaction of cellulose with heavy metals, dyes and other pollutants; therefore, chemical modification of cellulose was carried out.

2. Results and Discussion

2.1. Characterization

2.1.1. Surface Morphology

Figure 1 shows the SEM images of raw sugarcane bagasse and acrylonitrile grafted cellulose (GC) isolated from sugarcane bagasse at different magnifications. The SEM images of raw sugarcane bagasse (Figure 1(A1,B1,C1)) show a cluster of fibers that are stuck together, which could be owing to the presence of viscous substances (lignin & pectin). The SEM images in Figure 1(A2,B2,C2) show that after chemical treatment and grafting, the surface morphology of sugarcane bagasse has been changed extensively. The change in surface morphology of sugarcane bagasse after hydrolysis and bleaching is owing to the removal of peptic compounds and lignin, and the breaking down of long cellulose fibers into shorter ones. The SEM images in Figure 1(A2,B2,C2) show that the morphology of acrylonitrile grafted sugarcane bagasse is smoother than the raw bagasse, owing to chemical treatment and the attachment of nitrile monomer onto the cellulose surface. Similar changes in the surface morphology of cellulosic adsorbents were reported by Ali et al. [51].

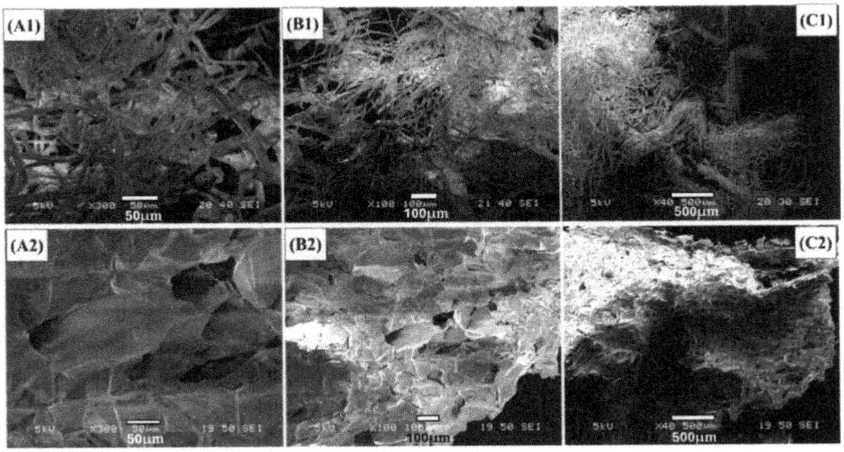

Figure 1. SEM images of raw sugarcane bagasse (**A1**,**B1**,**C1**), and grafted sugarcane bagasse (**A2**,**B2**,**C2**) in different resolutions.

2.1.2. XRD Analysis

Figure 2 shows the X-rays diffraction results of unmodified cellulose (A) and acrylonitrile grafted cellulose before adsorption (B) and after adsorption (C). In Figure 2, the XRD spectra of unmodified cellulose (A) and grafted cellulose before adsorption (B) show only small peaks at 13° and 13.2°, respectively, while the rest of the spectrum has no prominent peak in the range of 20–25°, as can be seen in Figure 2A,B. The results verify that following surface functionalization, the cellulose's crystallinity index (CrI) remains unchanged. The XRD spectra of grafted cellulose (Figure 2B) and unmodified cellulose (Figure 2A) showed no discernible variation in the XRD spectrum width. The primary crystalline cellulose I peak remained unaltered during grafting reactions, according to the XRD data. However, upon Cr(VI) adsorption, the XRD spectrum of the grafted cellulose showed noticeably wider widths than those of the ungrafted samples and had a sharp peak at 22.72° at 2θ value, indicating that Cr(VI) ions adsorbed onto GSB (Figure 2C). The XRD result of GSB confirms Cr(VI) adsorption on GSB, and the peaks at 22.72° correspond to Cr_2O_3 having a tetragonal crystalline phase, according to the inorganic crystalline library [56,57].

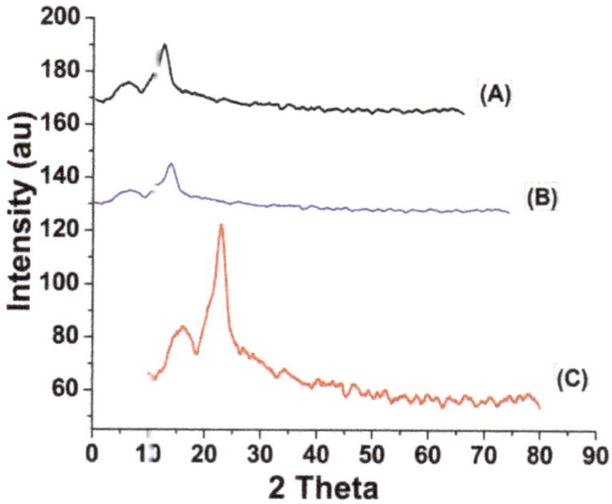

Figure 2. XRD analysis of unmodified cellulose isolated from SB (**A**), grafted cellulose isolated from SB before adsorption (**B**) and grafted cellulose isolated from SB after Cr(VI) adsorption (**C**).

2.1.3. FTIR Analysis

The FTIR analysis results of unmodified cellulose (UGC) and grafted cellulose isolated from sugarcane bagasse (GC) are presented in Figure 3A and B, respectively. The FTIR spectrum in Figure 3A shows that the peaks at 2870 cm^{-1} and 3391cm^{-1} are representing the C–H and O–H bond stretching vibration frequencies, respectively. In Figure 3B, which represents acrylonitrile grafted cellulose, the intensities of these peaks are very low, which shows that most of these groups have been modified with a CN group after grafting co-polymerization. Similarly, a prominent peak at 1656 cm^{-1} in Figure 3B accounts for the nitrile (–CN) group and confirms the insertion of a nitrile group onto the surface of the cellulose. There are some other minor peaks at 1690 cm^{-1} which represent the vibration frequency of water molecules (Figure 3A). In Figure 3B, another peak can be seen at 1510 cm^{-1}, which represents the carboxyl group stretching vibration [41], while this peak does not exist in raw sugarcane bagasse [42].

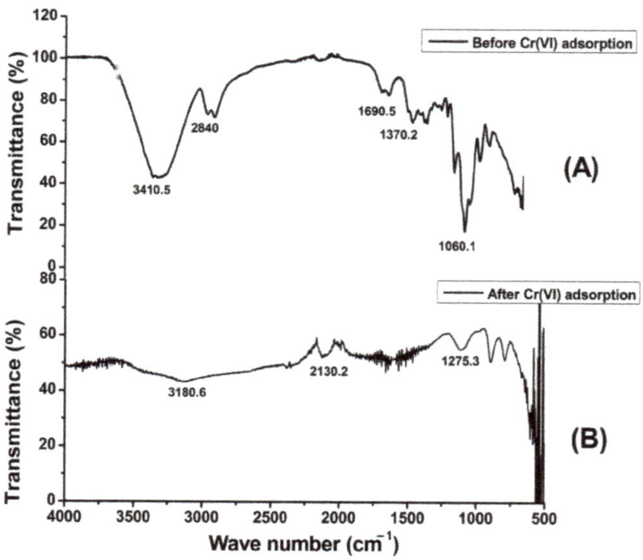

Figure 3. FT-IR analysis of unmodified cellulose isolated from sugarcane bagasse (**A**) and chemically modified cellulose isolated from sugarcane bagasse (**B**).

2.2. *Effect of Various Parameters on Cr(VI) Removal*

Effect of Metal Ion Concentration

The adsorption efficiency depends upon the concentration of adsorbate ions/molecules in a medium. The accumulation of analyte molecules/ions onto the surface of an adsorbent depends upon the available surface area and the number of active sites or functional groups. To explore the metal ion concentration effect on the removal efficiency of raw, unmodified and grafted sugarcane bagasse, Cr(VI) solutions of 50–300 ppm were checked during batch experiments while adsorbent dose, pH, time and temperature were held constant. In flasks, 100 mL chromium solutions of 50, 100, 150, 200, 250 and 300 ppm were taken, each with 1 g of adsorbent and pivoted at 300 rpm for 2h at 25 °C. After 2 h filtration was carried out to remove the adsorbent, the C(VI) ion concentration was determined by AAS. The same procedure was performed for raw sugarcane bagasse. Figure 4 shows that 70%, 82% and 94% removal efficiencies were obtained for raw sugarcane bagasse (RSB), un-grafted cellulose obtained from sugarcane bagasse (UGC) and grafted cellulose (GC), respectively, at lower concentrations, i.e., 50 ppm, and the removal efficiency decreased with increasing concentration. The removal efficiency of adsorbents declined with the increase in concentration and reached 35%, 48% and 65% at concentrations of 300 ppm for RSB, UGC and GC, respectively. Active sites on the adsorbent were sufficient to bind the metal ions when the concentration was low. By raising the concentration of HMs above the optimum values, then the adsorbent cannot adsorb more molecules, and thus, the removal efficiency decreases. The removal efficiency decreases at high concentrations owing to the saturation of active sites with metal ions, and no further ions can be picked up by adsorbents. The removal efficiency of GSB is higher than un-grafted cellulose (UGC) and raw sugarcane bagasse (RSB) at both lower and higher concentrations. The GC has 94% and 65% removal efficiency at 50 ppm and 300 ppm, respectively, while the removal efficiency of unmodified cellulose (UGC) is 82% and 48%, and for RSB, 70% and 35% at 50 ppm and 300 ppm concentration, respectively, which indicates the superior performance of GC over UGC and RSB. The finding of the current study matches well with other similar studies [46,47].

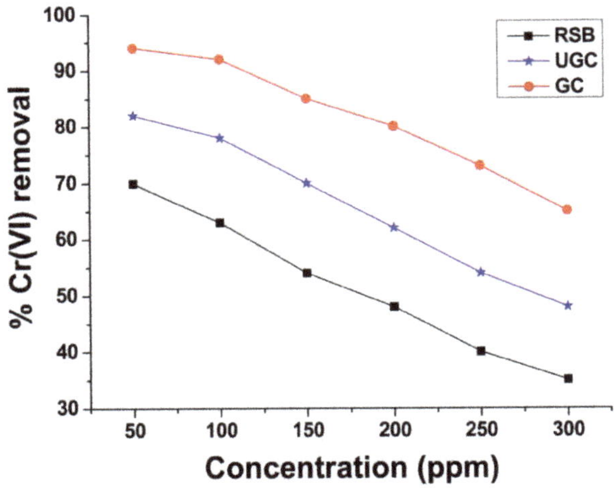

Figure 4. Effect of HM concentration on the removal efficiency of Cr(VI) by raw sugarcane bagasse (RSB), un-grafted cellulose (UGC) and grafted cellulose (GC) extracted from sugarcane bagasse.

2.3. *Effect of Adsorbent Mass*

In batch experiments, different quantities (0.1, 0.3, 0.5, 1, 2 and 3 g) of raw sugarcane bagasse (RSB), un-grafted cellulose (UGC) and grafted cellulose (GC) were taken in separate flasks, and 100 mL Cr(VI) solution (50 ppm) was taken in flasks. The pH was maintained at 4, and the flasks were agitated at 300 rpm for 60 min and filtered, and Cr(VI) was measured by AAS. Figure 5 shows the relationship of the adsorbent dose and removal efficiency of RSB, UGC and GC. The results show that at 0.1 g adsorbent dose, removal efficiencies of 38%, 50 and 67% efficiency were obtained for RSB, UGC and GC, respectively. Figure 5 shows that Cr(VI) removal was enhanced with the increase in adsorbent mass from 0.1 g and reached maximum values of 70%, 81% and 94% for RSB, UGC and GC, respectively. Figure 5 shows that there was no change in % removal when the adsorbent quantity was increased beyond 1.0 g and a straight line was obtained. The removal efficiency was lower when the adsorbent quantity was lower than the optimum dose because the mass of adsorbent was inadequate for capturing metal ions, and thus, low removal efficiency was recorded for RSB, UGC and GC. Initially, the percentage removal efficiency (%R) was increased because the number of active sites was high enough for the capturing of Cr(VI) ions. By enhancing the adsorbent mass from 1 to 3 g, no further increase was observed in removal efficiency because the number of active sites of adsorbent at this quantity (1 g) was high enough for picking Cr(VI) ions (50 ppm concentration). A similar trend was observed in other studies on heavy metal removal [50].

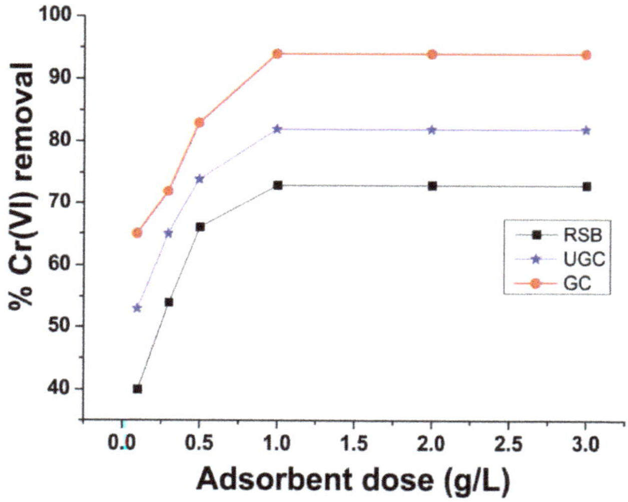

Figure 5. Effect of adsorbent mass (g/L) on Cr(VI) removal by raw sugarcane bagasse (RSB), un-grafted cellulose (UGC) and grafted cellulose (GC) extracted from sugarcane bagasse.

2.4. Effect of Time

The Cr(VI) removal by raw sugarcane bagasse (RSB), un-grafted cellulose (UGC) and grafted cellulose (GC) at different times is shown in Figure 6. Chromium solution (50 ppm) was taken in seven conical flasks (100 mL in each flask) and stirred for 10, 20, 30, 60, 120, 150 and 180 min for each adsorbent. Initially, when the adsorption time was increased, the Cr(VI) removal efficiency was increased, until 120 min. At 120 min, the removal efficiency reached maximum values of 67%, 81% and 94% for RSB, UGC and GC, respectively. Increasing adsorption time to 180 min does not change the removal efficiency, and the same efficiency was recorded because that adsorption equilibrium was achieved within 120 min. The removal efficiency was not enhanced beyond 120 min because the active sites of the adsorbent were occupied by metal ions within 120 min. The rate of metal ions removal was faster at the beginning owing to the large number of vacant active sites on the adsorbent [58].

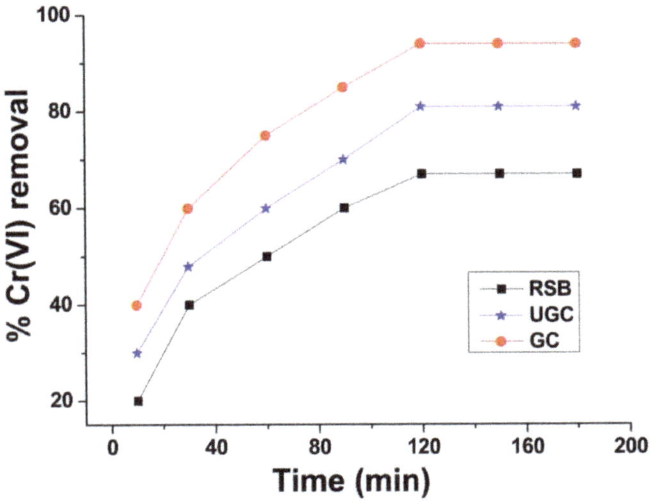

Figure 6. Effect of contact time on Cr(VI) adsorption onto raw sugarcane bagasse (RSB), un-grafted cellulose (UGC) and grafted cellulose extracted from sugarcane bagasse (GC).

2.5. *Effect of pH on the Removal of Cr(VI)*

Figure 7 shows the adsorption of Cr(VI) onto raw sugarcane bagasse (RSB), un-grafted cellulose (UGC) and grafted cellulose obtained from sugarcane bagasse (GC) at various pH values. The removal efficiency is low at pH 2–4, and increases sharply to a maximum value of 94% at pH 6. Further elevating the pH beyond 6 decreases the Cr(VI) removal, which reached a minimum value of 17 and 20% for raw and grafted sugarcane bagasse, respectively, at pH 12. At pH more than 6, the Cr(VI) removal decreased owing to OH^- ions present in the medium, which interact with adsorbent instead of $CrO_2{}^{-2}$ ions and occupy the positively charge sites on the adsorbents. Similarly, at pH below 6, abundant H^+ ions interrupt the ability of Cr(VI) cations to adsorb over raw and GC, and thus, the efficiency of metal ion removal decreases [59].

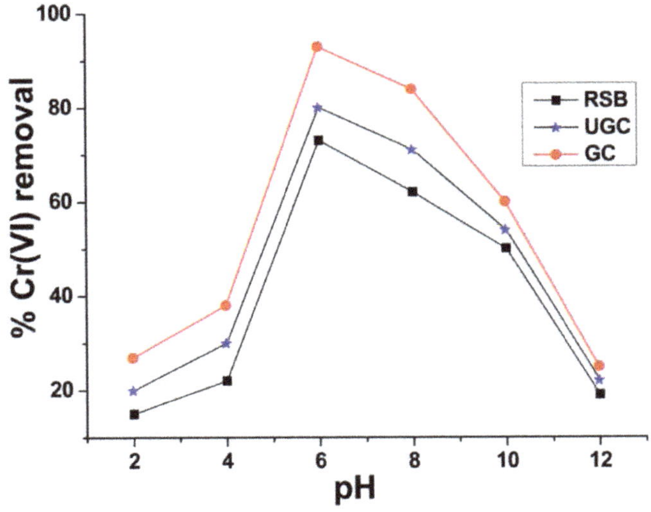

Figure 7. Effect of pH Cr(VI) removal by raw sugarcane bagasse (RSB), un-grafted cellulose (UGC) and grafted cellulose (GC) extracted from sugarcane bagasse.

2.6. Isotherm Study

Isotherm models show the process of adsorbate accumulation on the adsorbent's surface and provide the information about the type of adsorption: chemical or physical, monolayer or multilayer, etc. [60]. The experimental data of Cr(VI) accumulation on raw sugarcane bagasse (RSB), un-grafted cellulose (UGC) and grafted cellulose (GC) isolated from sugarcane bagasse was computed by the Freundlich model and Langmuir model [61].

2.6.1. Freundlich Model

The Freundlich isotherm model in linear form can be represented as given in Equation (1).

$$\ln q = \ln K \frac{1}{n} C_e \quad (1)$$

According to this model, the adsorption takes place over a heterogeneous surface, forming multiple layers of adsorbate molecules, and increases with the increase in concentration of an adsorbate. The adsorption data of Cr(VI) adsorption onto RSB & GSB was computed by the Freundlich isotherm model. The numerical values of "q" and "C_e" were calculated and $\ln q_e$ was plotted against $\ln C_e$. From the linear plot, the values of 1/n (slop), K (L mg^{-1}) and "n" were calculated as given in Equation (1). The results in Figure 8A show that the adsorbents (RSB, UGC and GC) are effective for Cr(VI) removal [62]. The numerical values of "1/n", KF and R^2 obtained from Cr(VI) removal on RSB and GSB are presented in Table 1. The "1/n" values for RSB, UGC and GC are 42.45, 64.30 and 124.36, respectively. The values of the Freundlich constant (KF) of 2.821, 4.72 and 9.786 and R^2 of 0.97, 0.97 and 0.98 were recorded as shown in Table 1.

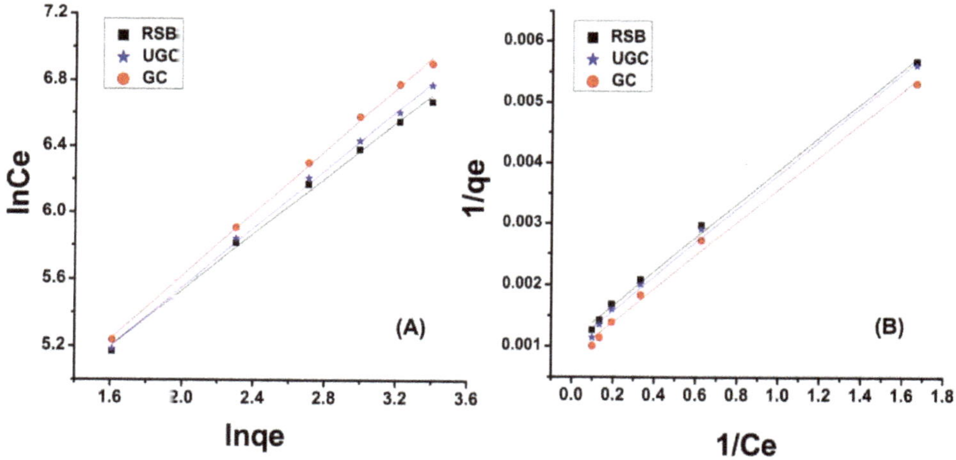

Figure 8. Freundlich (**A**) & Langmuir (**B**) isotherm model for Cr(VI) adsorption onto raw sugarcane bagasse (RSB), un-grafted cellulose (UGC) and grafted cellulose (GC) isolated from sugarcane bagasse.

Table 1. Numerical values of Freundlich & Langmuir isotherm model for raw sugarcane bagasse (RSB), un-grafted cellulose (UGC) and grafted cellulose (GC) isolated from sugarcane bagasse.

Adsorbent	Freundlich Model			Langmuir Model		
	1/n	K_F (mg/g)	R^2	q_{max} (mg/g)	K (L/mg)	R^2
RSB	42.45	2.8210	0.97	125.95	4.5808	0.91
UGC	63.30	4.7237	0.97	202.37	4.1733	0.89
GC	124.36	9.7863	0.98	267.93	3.9563	0.84

2.6.2. Langmuir Isotherm Model

This model predicts the homogeneous surface of adsorbent with a limited number of active sites, and the adsorption of adsorbate molecules/ions is monolayer. Equation (2) shows the mathematical form of the Langmuir isotherm model.

$$\frac{C_e}{q} = \frac{1}{q_{max}}K + \frac{C_e}{q_{max}} \tag{2}$$

where "C_e" is the equilibrium concentration and "q_e" is the quantity of adsorbate adsorbed per unit mass of adsorbent (g). The adsorption capacity is represented by "q_{max} (mg g^{-1})" and KL (L mg g^{-1}) is the constant of adsorption energy [24]. The q_{max}, K and KL (L mg g^{-1}) values were calculated from C_e/q versus C_e plot and linear regression. The values of "q_{max}", KL, R^2 and (RL) were calculated using Equation (3) and a plot of $1/C_e$ vs. $1/q_e$ as shown in Figure 8B.

$$RL = \frac{1}{(1 + KLC_0)} \tag{3}$$

The values of "q_{max}", KL and R^2 were found by plotting $1/C_e$ vs. $1/q_e$ and the data are given in Table 1. The numerical values of q_{max} were 125.95 mg/g, 202.37 mg/g and 267.93 mg/g; of K (L/mg), 4.5808, 4.1733 and 3.9563; and of R^2, 0.91, 0.89 and 0.84 for GSB, UGC and GC, respectively (Table 1). The results suggest that the adsorption of Cr(VI) onto RSB, UGC and GC fits more closely with the Freundlich model. The R^2 of the Freundlich and Langmuir models for RSB, UGC and GC are given in Table 1, which indicates that the adsorption data of Cr(VI) is fitted will with the Freundlich isotherm model.

2.7. Adsorption Kinetics

The mechanism and rate-limiting step of reaction are determined from the kinetics study. In the current study, batch adsorption experiments were conducted with different times to find the optimum time where the equilibrium established and maximum adsorption of Cr(VI) takes place over RSB and GSB. The kinetics results show that Cr(VI) adsorption was slow at initial stages and reached equilibrium in 120 min, which indicates that the adsorption was chemisorption, and the adsorbate interacts with the active sites of the adsorbent by chemical interactions. Moreover, to find the rate-determining step, the data were computed by various kinetics models [63].

2.7.1. Pseudo First-Order Model

The pseudo first-order kinetic model can be represented as

$$\ln(q_e - q_t) = lnq_e - K_1 t \tag{4}$$

In Equation (4), "q_t" and "q_e" are the quantity of Cr(VI) ions adsorbed at specific time and equilibrium, respectively, while "K_1" is the constant [60]. Cr(VI) solutions of 100, 200, 300 and 400 ppm were taken in different containers, each containing 0.4 g of adsorbent (GC), and the adsorption was checked from 10 min to 180 min. Adsorption capacity at a specific time (q_t) and equilibrium time (q_e), and rate constant K_1 (min^{-1}), were calculated from a (q_e/q_t) vs. (t) graph (Figure 9A).

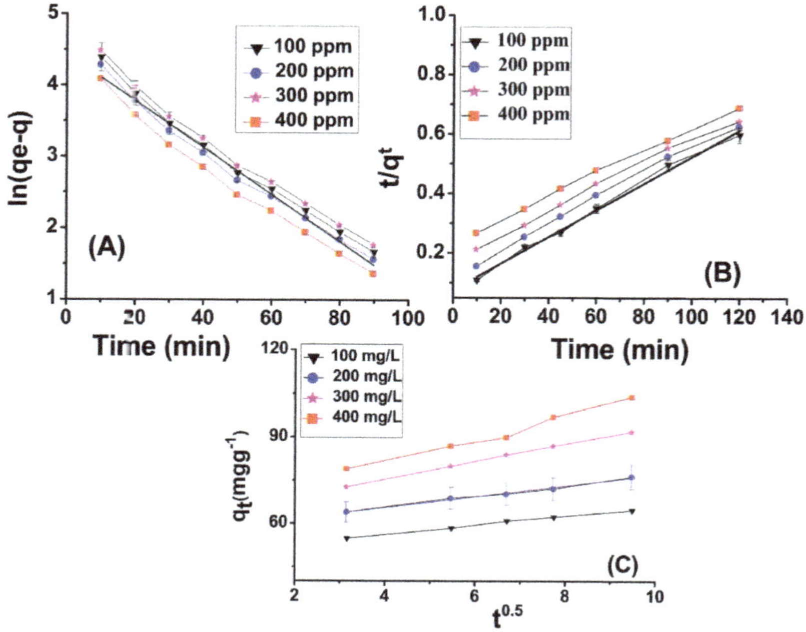

Figure 9. Pseudo first-order model (**A**), pseudo second-order model (**B**) and "intra-particle diffusion model" (**C**) for adsorption of Cr(VI) onto grafted cellulose (GC) isolated from sugarcane bagasse.

2.7.2. Pseudo Second-Order Model

Equation (5) shows the pseudo second-order kinetic model.

$$\frac{1}{q_t} = \frac{1}{k_2 q_e^2} + \frac{1}{q_e} \quad (5)$$

In Equation (5), "K_2" is the rate constant and "q_t" and "q_e" are the quantity of Cr(VI) adsorbed at specific time (t) and equilibrium time, respectively. Cr(VI) solutions of 100, 200, 300 and 400 ppm were taken in different flasks containing GSB and the adsorption was checked from 10 min to 180 min. The value of q_{max} at specific time (qt) and equilibrium time (q_e), and K_2 (gmg^{-1} min^{-1}) were calculated (t/q_t) vs. t plot [64].

2.7.3. Intra-Particle Diffusion

HM ions from the mixture/medium transfer to adsorbent by intraparticle diffusion [65]. The mathematical form of the intra-particle diffusion model can be represented as Equation (6).

$$q_t = K_p t0.5 + C \quad (6)$$

where q_t (mgg^{-1}) is Cr(VI) adsorbed at specific time (t), K_p (mgg^{-1} $min^{0.5}$) is a constant (diffusion within the particle) and C is the point of intersection. During batch experiments, four standard solutions of Cr(VI) of specific concentration (100, 200, 300, 400 ppm) were prepared, and adsorption over GSB was carried out for different time intervals (10 min to 180 min). The adsorption capacity at a specific time (q_t) was plotted against the square root of time (t0.5). A plot of (q_t) vs. (t0.5) is given in Figure 9C [27]. The numerical values of different constants of all three models, as well as their regression square (R^2), are shown in Table 2. The R^2 value of the pseudo second-order model is 0.99, which the highest among all three models.

Table 2. Pseudo first-order, pseudo second-order model and "intra-particle diffusion model" values for Cr(VI) adsorption onto GSB.

Concentration (ppm)	First-Order			Second-Order			IPD Model	
	q_e (mgg^{-1})	K_1	R^2	q_e (mgg^{-1})	K_2	R^2	R^2	K_i ($mgg^{-1}min^{-1}$)
100	188	0.644	0.938	176	0.0137	0.94	0.9713	2.0828
200	368	0.66	0.978	336	0.0242	0.95	0.9005	2.4641
300	546	0.637	0.936	480	0.0451	0.97	0.9281	3.2259
400	720	0.681	0.989	592	0.0550	0.99	0.9172	4.0492

2.8. Regeneration and Reuse of Spent Adsorbent (GSB)

The spent adsorbent (GC) collected from batch Cr(VI) adsorption experiments was divided into three equal parts, each weighted 100 g, and each portion was treated with 0.1 M NaOH, 0.1 M H_2SO_4 and 1 M ethylenediaminetetraacetic acid (EDTA) solutions. In a typical experiment, 100 g of spent GSB was added into a container containing 700 mL of 1 M NaOH and stirred in a reciprocal shaker at 45 °C for 3 h, followed by filtration. The percent recovery of chromium (VI) ions was determined from Equation (7). A similar method was used for the desorption of chromium from spent adsorbent (GSB) using other desorbing reagents, such as sulfuric acid and ethylenediaminetetraacetic acid (EDTA) solutions.

$$\% \text{ Cr(VI) desorption} = (\text{Mass of Cr(VI) desorbed (mg/g)})/(\text{Mass of Cr(VI) adsorbed (mg/g)}) \times 100 \quad (7)$$

The metal ion removal efficiency of regenerated GSB is presented in Figure 10. The maximum Cr(VI) removal efficiency (average of 85.5%) was obtained for NaOH-treated spent adsorbent (GC) after 30 cycles of regeneration and reuse, followed by H_2SO_4 (81.2%) and EDTA (74.4%) regenerated adsorbent. The results suggest that grafted sugarcane bagasse could be used many times after regeneration.

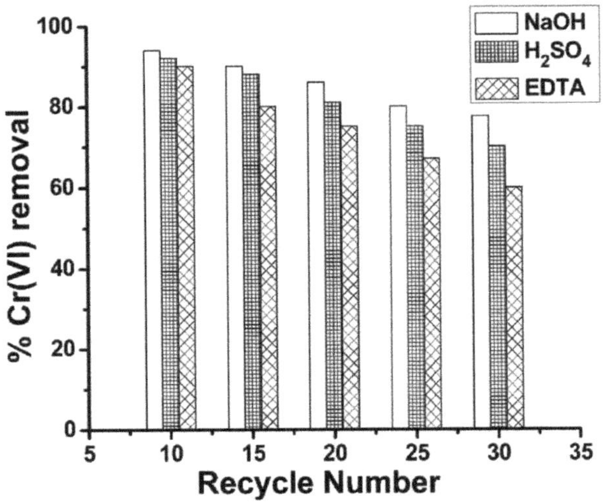

Figure 10. Average removal efficiency of spent adsorbent (GSB) after regeneration several times with different reagents.

3. Material and Methods

3.1. Adsorbent Preparation

Sugarcane bagasse was taken from Haripur, Pakistan, washed and dried. Then the dried sugarcane bagasse was subjected to acid hydrolysis, alkali hydrolysis, bleaching and grafting with acrylonitrile [51]. The preparation of acrylonitrile grafted cellulose is discussed below in detail and the illustration of cellulose extraction from sugarcane bagasse is shown in Figure S1.

3.1.1. Acid Hydrolysis

Crude sugarcane bagasse (200 g) was taken in a flask and 2% 1 M HCl (500 mL) was added into it. The contents were boiled for 3 h on a hotplate. After acid hydrolysis, the material was washed several times.

3.1.2. Alkaline Hydrolysis

After acid hydrolysis, the sample was further hydrolyzed with sodium hydroxide solution. Acid-hydrolyzed sugarcane bagasse was heated with 2% NaOH solution for 3 h at 100 °C, followed by washing with distilled water until neutralized.

3.1.3. Bleaching

The cellulose pulp obtained after alkaline hydrolysis was treated with sodium chlorate (0.1 mg) and glacial acetic acid (0.1 mL) at 40 °C. The reddish-brown color of cellulose pulp disappeared after bleaching, and the white pulp obtained after bleaching was dried and stored.

3.1.4. Grafting Co-Polymerization

Bleached sugarcane bagasse (100 g) was placed in a conical flask; 0.5 g $FeSO_4$, deionized water (200 mL), hydrogen peroxide (1 mL) and acrylonitrile (0.3 mL) were added into it and stirred at 40 °C for 1h, followed by washing with DW and drying. Soxhlet extraction of the acrylonitrile grafted cellulose pulp in anhydrous toluene was performed at 100 °C to eliminate the unreacted monomers.

3.2. Characterization

SEM provides information about the sorbent's morphology and the surface area that can be used for adsorption. An FT-IR spectrometer (TENSOR II, Bruker, Bremen, Germany) was used to measure the functional groups. The crystallinity was confirmed by X-ray diffraction examination utilizing an DW-XRD-Y300 X-ray diffractometer (Chongqing, China).

3.3. Adsorption of Cr(VI) on RSB and GSB

A standard solution of Cr(VI) (1000 ppm) was diluted to 50, 100, 150, 200, 250 and 300 ppm for use in batch adsorption experiments. Cr(VI) solution (100 mL) from every dilution was taken in flasks, followed by the addition of 1 g adsorbent (RSB and GSB) and pivoted at 300 rpm for 60 min. In the filtrate, the Cr(VI) ion concentration was measured by ASS. Equation (8) was used to find out the adsorption capacity (q_e).

$$q_e = \frac{V(C_0 - C_e)}{W} \tag{8}$$

where "q_e" is the mass of heavy metals, "V" is the volume of the solution in liters, "W" is the mass of adsorbent in grams, C_0 is the initial concentration of HMs and C_e is the equilibrium concentration. The % removal efficiency was found using Equation (9).

$$\text{Removal efficiency } (\%) = \frac{(C_0 - C_e)}{C_0} \times 100 \tag{9}$$

3.4. Isotherms Study

To find the mechanism of interaction of adsorbate with adsorbent, and the surface heterogeneity of the adsorbent, the experimental data were computed by different isotherm models, such as Freundlich model and Langmuir model. The linear regression was performed using MS Excel 2020 and Origin pro 8.5 software.

3.5. Adsorption Kinetics

A kinetics study was performed in order to check the time at which equilibrium is established for Cr(VI) adsorption at different concentrations and to find the rate determining step [48]. The experimental data (q_e, q_t, etc.) obtained at different concentrations and different time were computed by first-order, second-order and intraparticle diffusion models [49].

4. Conclusions

In conclusion, the developed adsorbents raw sugarcane bagasse (RSB) and nitrile grafted sugarcane bagasse (GSB) can effectively remove Cr(VI) from wastewater. Cellulose fibers were isolated from sugarcane bagasse, which is a cheap, abundant agricultural byproduct. The removal efficiency of raw sugarcane bagasse, un-grafted cellulose and grafted cellulose isolated from sugarcane bagasse were 73%, 82% and 94%, respectively. The effects of concentration, pH, adsorbent amount, and time were checked, and the optimum values were determined. The adsorption data of Cr(VI) were fitted with the Freundlich isotherm model and follow the pseudo second-order kinetics model. The spent adsorbent was regenerated using different regenerating reagents such as sulfuric acid, sodium hydroxide and ethylenediamine tetra acetic acid, and reused for Cr(VI) ion removal from wastewater. The removal efficiency of the regenerated absorbent was 85.5% for NaOH-regenerated adsorbent, 81.2% for H_2SO_4-regenerated adsorbent and 74.4% for EDTA-regenerated adsorbent after thirty cycles. The high removal efficiency of both newly prepared and regenerated adsorbent (GC), as well as the low cost, suggest that it could be utilized as an efficient alternative adsorbent for the water treatment. In -future, GC will be utilized for the removal of other HMs and dyes to check its performance for the removal of other analytes.

Supplementary Materials: The following supporting information can be downloaded at: https://www.mdpi.com/article/10.3390/molecules29102207/s1, Figure S1: Isolation of cellulose from sugarcane bagasse by acid hydrolysis, alkaline hydrolysis, bleaching and grafting.

Author Contributions: Conceptualization, I.K. and A.A.; methodology, A.N.; software, T.A.S. and W.S.; validation, Z.T.B., Z.U.R. and T.A.S.; formal analysis, A.A.M. and Z.U.R.; investigation, A.N.; resources, K.A.A., A.A.M. and Y.M.H.; data curation, A.A.M.; writing—original draft preparation, I.K. and A.A.; writing—review and editing, A.A. and A.N.; visualization, Y.M.H.; supervision, A.A. and A.N.; project administration, Z.U.R. and A.N.; funding acquisition, A.A.M., K.A.A. and Y.M.H. All authors have read and agreed to the published version of the manuscript.

Funding: This research was funded by King Saud University, Riyadh, Saudi Arabia, grant number RSP-2024 R369. The authors are also thankful to the Higher Education Commission (HEC), Pakistan for supporting this work through project No. 202015733/NRPU/R&D/HEC/2021.

Institutional Review Board Statement: Not applicable.

Informed Consent Statement: Not applicable.

Data Availability Statement: The data are available from the corresponding author and may be provided upon written request.

Acknowledgments: The authors extend their appreciation to Researchers Supporting Project number (RSP-2024 R369), King Saud University, Riyadh, Saudi Arabia. The authors are also thankful to the University of Haripur, Pakistan for providing laboratory facilities to conduct this work.

Conflicts of Interest: The authors declare no conflicts of interest.

References

1. Singh, A.; Sharma, A.; Verma, R.K.; Chopade, R.L.; Pandit, P.P.; Nagar, V.; Aseri, V.; Choudhary, S.K.; Awasthi, G.; Awasthi, K.K. *Heavy Metal Contamination of Water and Their Toxic Effect on Living Organisms, The Toxicity of Environmental Pollutants*; IntechOpen: Rijeka, Croatia, 2022.
2. Sharma, A.; Grewal, A.S.; Sharma, D.; Srivastav, A.L. *Heavy Metal Contamination in Water: Consequences on Human Health and Environment, Metals in Water*; Elsevier: Amsterdam, The Netherlands, 2023; pp. 39–52.
3. Thanigaivel, S.; Vickram, S.; Dey, N.; Jeyanthi, P.; Subbaiya, R.; Kim, W.; Govarthanan, M.; Karmegam, N. Ecological disturbances and abundance of anthropogenic pollutants in the aquatic ecosystem: Critical review of impact assessment on the aquatic animals. *Chemosphere* **2023**, *313*, 137475. [CrossRef]
4. Elnabi, M.K.A.; Elkaliny, N.E.; Elyazied, M.M.; Azab, S.H.; Elkhalifa, S.A.; Elmasry, S.; Mouhamed, M.S.; Shalamesh, E.M.; Alhorieny, N.A.; Elaty, A.E.A.; et al. Toxicity of heavy metals and recent advances in their removal: A review. *Toxics* **2023**, *11*, 580. [CrossRef]
5. Nivetha, N.; Srivarshine, B.; Sowmya, B.; Rajendiran, M.; Saravanan, P.; Rajeshkannan, R.; Rajasimman, M.; Pham, T.H.T.; Shanmugam, V.; Dragoi, E.-N. A comprehensive review on bio-stimulation and bio-enhancement towards remediation of heavy metals degeneration. *Chemosphere* **2023**, *312*, 137099. [CrossRef] [PubMed]
6. Mukherjee, A.G.; Renu, K.; Gopalakrishnan, A.V.; Veeraraghavan, V.P.; Vinayagam, S.; Paz-Montelongo, S.; Dey, A.; Vellingiri, B.; George, A.; Madhyastha, H.; et al. Heavy Metal and Metalloid Contamination in Food and Emerging Technologies for Its Detection. *Sustainability* **2023**, *15*, 1195. [CrossRef]
7. Balali-Mood, M.; Naseri, K.; Tahergorabi, Z.; Khazdair, M.R.; Sadeghi, M. Toxic mechanisms of five heavy metals: Mercury, lead, chromium, cadmium, and arsenic. *Front. Pharmacol.* **2021**, *12*, 643972. [CrossRef] [PubMed]
8. Mitra, S.; Chakraborty, A.J.; Tareq, A.M.; Bin Emran, T.; Nainu, F.; Khusro, A.; Idris, A.M.; Khandaker, M.U.; Osman, H.; Alhumaydhi, F.A.; et al. Impact of heavy metals on the environment and human health: Novel therapeutic insights to counter the toxicity. *J. King Saud Univ. Sci.* **2022**, *34*, 101865. [CrossRef]
9. Fu, Z.; Xi, S. The effects of heavy metals on human metabolism. *Toxicol. Mech. Methods* **2020**, *30*, 167–176. [CrossRef]
10. Zaynab, M.; Al-Yahyai, R.; Ameen, A.; Sharif, Y.; Ali, L.; Fatima, M.; Khan, K.A.; Li, S. Health and environmental effects of heavy metals. *J. King Saud Univ. Sci.* **2022**, *34*, 101653. [CrossRef]
11. Hoang, H.-G.; Lin, C.; Tran, H.-T.; Chiang, C.-F.; Bui, X.-T.; Cheruiyot, N.K.; Shern, C.-C.; Lee, C.-W. Heavy metal contamination trends in surface water and sediments of a river in a highly-industrialized region. *Environ. Technol. Innov.* **2020**, *20*, 101043. [CrossRef]
12. Saha, B.; Orvig, C. Biosorbents for hexavalent chromium elimination from industrial and municipal effluents. *Coord. Chem. Rev.* **2010**, *254*, 2959–2972. [CrossRef]
13. Saravanan, A.; Kumar, P.S.; Yashwanthraj, M. Sequestration of toxic Cr(VI) ions from industrial wastewater using waste biomass: A review. *Desalination Water Treat.* **2017**, *68*, 245–266. [CrossRef]
14. Samuel, M.S.; Selvarajan, E.; Chidambaram, R.; Patel, H.; Brindhadevi, K. Clean approach for chromium removal in aqueous environments and role of nanomaterials in bioremediation: Present research and future perspective. *Chemosphere* **2021**, *284*, 131368. [CrossRef]
15. Sharma, P.; Singh, S.P.; Parakh, S.K.; Tong, Y.W. Health hazards of hexavalent chromium (Cr (VI)) and its microbial reduction. *Bioengineered* **2022**, *13*, 4923–4938. [CrossRef]
16. Yang, W.; Lei, G.; Quan, S.; Zhang, L.; Wang, B.; Hu, H.; Li, L.; Ma, H.; Yin, C.; Feng, F. The Removal of Cr (VI) from Aqueous Solutions with Corn Stalk Biochar. *Int. J. Environ. Res. Public Health* **2022**, *19*, 14188. [CrossRef]
17. Thaçi, B.S.; Gashi, S.T. Reverse osmosis removal of heavy metals from wastewater effluents using biowaste materials pretreatment. *Pol. J. Environ. Stud.* **2018**, *28*, 337–341. [CrossRef]
18. Ates, N.; Uzal, N. Removal of heavy metals from aluminum anodic oxidation wastewaters by membrane filtration. *Environ. Sci. Pollut. Res.* **2018**, *25*, 22259–22272. [CrossRef]
19. Bashir, A.; Malik, L.A.; Ahad, S.; Manzoor, T.; Bhat, M.A.; Dar, G.N.; Pandith, A.H. Removal of heavy metal ions from aqueous system by ion-exchange and biosorption methods. *Environ. Chem. Lett.* **2019**, *17*, 729–754. [CrossRef]
20. Zhang, Y.; Duan, X. Chemical precipitation of heavy metals from wastewater by using the synthetical magnesium hydroxy carbonate. *Water Sci. Technol.* **2020**, *81*, 1130–1136. [CrossRef]
21. Smara, A.; Delimi, R.; Chainet, E.; Sandeaux, J. Removal of heavy metals from diluted mixtures by a hybrid ion-exchange/electrodialysis process. *Sep. Purif. Technol.* **2007**, *57*, 103–110. [CrossRef]
22. Mungray, A.A.; Kulkarni, S.V.; Mungray, A.K. Removal of heavy metals from wastewater using micellar enhanced ultrafiltration technique: A review. *Cent. Eur. J. Chem.* **2012**, *10*, 27–46. [CrossRef]
23. Tang, X.; Zheng, H.; Teng, H.; Sun, Y.; Guo, J.; Xie, W.; Yang, Q.; Chen, W. Chemical coagulation process for the removal of heavy metals from water: A review. *Desalination Water Treat.* **2016**, *57*, 1733–1748. [CrossRef]
24. Choumane, R.; Peulon, S. Development of an efficient electrochemical process for removing and separating soluble Pb(II) in aqueous solutions in presence of other heavy metals: Studies of key parameters. *Chem. Eng. J.* **2021**, *423*, 130161. [CrossRef]
25. Zhang, Y.; Zhao, M.; Cheng, Q.; Wang, C.; Li, H.; Han, X.; Fan, Z.; Su, G.; Pan, D.; Li, Z. Research progress of adsorption and removal of heavy metals by chitosan and its derivatives: A review. *Chemosphere* **2021**, *279*, 130927. [CrossRef]

26. Gautam, R.K.; Mudhoo, A.; Lofrano, G.; Chattopadhyaya, M.C. Biomass-derived biosorbents for metal ions sequestration: Adsorbent modification and activation methods and adsorbent regeneration. *J. Environ. Chem. Eng.* **2014**, *2*, 239–259. [CrossRef]
27. Georgaki, M.-N.; Charalambous, M.; Kazakis, N.; Talias, M.A.; Georgakis, C.; Papamitsou, T.; Mytiglaki, C. Chromium in water and carcinogenic human health risk. *Environments* **2023**, *10*, 33. [CrossRef]
28. Bilal, M.; Ihsanullah, I.; Younas, M.; Shah, M.U.H. Recent advances in applications of low-cost adsorbents for the removal of heavy metals from water: A critical review. *Sep. Purif. Technol.* **2021**, *278*, 119510. [CrossRef]
29. Bhatnagar, A.; Sillanpää, M.; Witek-Krowiak, A. Agricultural waste peels as versatile biomass for water purification—A review. *Chem. Eng. J.* **2015**, *270*, 244–271. [CrossRef]
30. Chakraborty, R.; Asthana, A.; Singh, A.K.; Jain, B.; Susan, A.B.H. Adsorption of heavy metal ions by various low-cost adsorbents: A review. *Int. J. Environ. Anal. Chem.* **2022**, *102*, 342–379. [CrossRef]
31. Ali, I.; Asim, M.; Khan, T.A. Low cost adsorbents for the removal of organic pollutants from wastewater. *J. Environ. Manag.* **2012**, *113*, 170–183. [CrossRef] [PubMed]
32. Phuengphai, P.; Singjanusong, T.; Kheangkhun, N.; Wattanakornsiri, A. Removal of copper(II) from aqueous solution using chemically modified fruit peels as efficient low-cost biosorbents. *Water Sci. Eng.* **2021**, *14*, 286–294. [CrossRef]
33. Batagarawa, S.M.; Ajibola, A.K. Comparative evaluation for the adsorption of toxic heavy metals on to millet, corn and rice husks as adsorbents. *J. Anal. Pharm. Res.* **2019**, *8*, 119–125. [CrossRef]
34. Meez, E.; Rahdar, A.; Kyzas, G.Z. Sawdust for the Removal of Heavy Metals from Water: A Review. *Molecules* **2021**, *26*, 4318. [CrossRef]
35. Iwuozor, K.O.; Oyekunle, I.P.; Oladunjoye, I.O.; Ibitogbe, E.M.; Olorunfemi, T.S. A review on the mitigation of heavy metals from aqueous solution using sugarcane bagasse. *Sugar Tech* **2021**, *24*, 1167–1185. [CrossRef]
36. Shahrokhi-Shahraki, R.; Benally, C.; El-Din, M.G.; Park, J. High efficiency removal of heavy metals using tire-derived activated carbon vs commercial activated carbon: Insights into the adsorption mechanisms. *Chemosphere* **2021**, *264*, 128455. [CrossRef] [PubMed]
37. Wu, J.; Wang, T.; Wang, J.; Zhang, Y.; Pan, W.-P. A novel modified method for the efficient removal of Pb and Cd from wastewater by biochar: Enhanced the ion exchange and precipitation capacity. *Sci. Total. Environ.* **2021**, *754*, 142150. [CrossRef] [PubMed]
38. Maharana, M.; Manna, M.; Sardar, M.; Sen, S. Heavy Metal Removal by Low-Cost Adsorbents. In *Green Adsorbents to Remove Metals, Dyes and Boron from Polluted Water*; Springer: Berlin/Heidelberg, Germany, 2021; pp. 245–272.
39. Shi, S.-C.; Liu, G.-T. Cellulose nanocrystal extraction from rice straw using a chlorine-free bleaching process. *Cellulose* **2021**, *28*, 1–12. [CrossRef] [PubMed]
40. Ali, A.; Saeed, K. Phenol removal from aqueous medium using chemically modified banana peels as low-cost adsorbent. *Desalination Water Treat.* **2016**, *57*, 11242–11254. [CrossRef]
41. Sudhaik, A.; Raizada, P.; Ahamad, T.; Alshehri, S.M.; Nguyen, V.-H.; Van Le, Q.; Thakur, S.; Thakur, V.K.; Selvasembian, R.; Singh, P. Recent advances in cellulose supported photocatalysis for pollutant mitigation: A review. *Int. J. Biol. Macromol.* **2022**, *226*, 1284–1308. [CrossRef] [PubMed]
42. Ortega, F.; Versino, F.; López, O.V.; García, M.A. Biobased composites from agro-industrial wastes and by-products. *Emergent Mater.* **2022**, *5*, 873–921. [CrossRef]
43. Etale, A.; Onyianta, A.J.; Turner, S.R.; Eichhorn, S.J. Cellulose: A review of water interactions, applications in composites, and water treatment. *Chem. Rev.* **2023**, *123*, 2016–2048. [CrossRef]
44. Kaur, J.; Sengupta, P.; Mukhopadhyay, S. Critical review of bioadsorption on modified cellulose and removal of divalent heavy metals (Cd, Pb, and Cu). *Ind. Eng. Chem. Res.* **2022**, *61*, 1921–1954. [CrossRef]
45. Rahman, L.; Fui, C.J.; Ting, T.X.; Sarjadi, M.S.; Arshad, S.E.; Musta, B. Polymer ligands derived from jute fiber for heavy metal removal from electroplating wastewater. *Polymers* **2020**, *12*, 2521. [CrossRef]
46. Daochalermwong, A.; Chanka, N.; Songsrirote, K.; Dittanet, P.; Niamnuy, C.; Seubsai, A. Removal of heavy metal ions using modified celluloses prepared from pineapple leaf fiber. *ACS Omega* **2020**, *5*, 5285–5296. [CrossRef] [PubMed]
47. Ali, A. Removal of Mn(II) from water using chemically modified banana peels as efficient adsorbent. Environ. *Nanotechnol. Monit. Manag.* **2017**, *7*, 57–63. [CrossRef]
48. Hasani, N.; Selimi, T.; Mele, A.; Thaçi, V.; Halili, J.; Berisha, A.; Sadiku, M. Theoretical, equilibrium, kinetics and thermodynamic investigations of methylene blue adsorption onto lignite coal. *Molecules* **2022**, *27*, 1856. [CrossRef]
49. Edebali, S. *Advanced Sorption Process Applications*; Intech Open: London, UK, 2019.
50. Pérez, B.F.; Espina, J.A.; González, M.d.L.F. Adsorption of Heavy Metals Ions from Mining Metallurgical Tailings Leachate Using a Shell-Based Adsorbent: Characterization, Kinetics and Isotherm Studies. *Materials* **2022**, *15*, 5315. [CrossRef]
51. Ali, A.; Saeed, K.; Mabood, F. Removal of chromium (VI) from aqueous medium using chemically modified banana peels as efficient low-cost adsorbent. *Alex. Eng. J.* **2016**, *55*, 2933–2942. [CrossRef]
52. Ezeonuegbu, B.A.; Machido, D.A.; Whong, C.M.; Japhet, W.S.; Alexiou, A.; Elazab, S.T.; Qusty, N.; Yaro, C.A.; Batiha, G.E.-S. Agricultural waste of sugarcane bagasse as efficient adsorbent for lead and nickel removal from untreated wastewater: Biosorption, equilibrium isotherms, kinetics and desorption studies. *Biotechnol. Rep.* **2021**, *30*, e00614. [CrossRef] [PubMed]
53. Gusmão, K.A.G.; Gurgel, L.V.A.; Melo, T.M.S.; Carvalho, C.d.F.; Gil, L.F. Adsorption studies of etherdiamine onto modified sugarcane bagasses in aqueous solution. *J. Environ. Manag.* **2014**, *133*, 332–342. [CrossRef]

54. Al-Mokhalelati, K.; Al-Bakri, I.; Wattar, N.A.S.A. Adsorption of methylene blue onto sugarcane bagasse-based adsorbent materials. *J. Phys. Org. Chem.* **2021**, *34*, e4193. [CrossRef]
55. Jiang, W.; Xing, Y.; Zhang, L.; Guo, X.; Lu, Y.; Yang, M.; Wang, J.; Wei, G. Polyethylenimine-modified sugarcane bagasse cellulose as an effective adsorbent for removing Cu(II) from aqueous solution. *J. Appl. Polym. Sci.* **2021**, *138*, 49830. [CrossRef]
56. Yan, L.; Dong, F.-X.; Lin, X.; Zhou, X.-H.; Kong, L.-J.; Chu, W.; Diao, Z.-H. Insights into the removal of Cr (VI) by a biochar–iron composite from aqueous solution: Reactivity, kinetics and mechanism. *Environ. Technol. Innov.* **2021**, *24*, 102057. [CrossRef]
57. Tsegay, M.; Gebretinsae, H.; Nuru, Z. Structural and optical properties of green synthesized Cr2O3 nanoparticles. *Mater. Today Proc.* **2021**, *36*, 587–590. [CrossRef]
58. Xia, Y.; Li, Y.; Xu, Y. Adsorption of Pb (II) and Cr (VI) from aqueous solution by synthetic allophane suspension: Isotherm, kinetics, and mechanisms. *Toxics* **2022**, *10*, 291. [CrossRef] [PubMed]
59. Silva, B.A.N.A.; Figueiredo, H.; Neves, I.C.; Tavares, T. The role of pH on Cr (VI) reduction and removal by Arthrobacter viscosus. *Int. J. Chem. Biol. Eng.* **2009**, *2*, 100–103.
60. Abin-Bazaine, A.; Trujillo, A.C.; Olmos-Marquez, M. Adsorption isotherms: Enlightenment of the phenomenon of adsorption. *Wastewater Treat.* **2022**, 1–15. [CrossRef]
61. Bazaine, A.; Trujillo, A.; Marquez, M. *Adsorption Isotherms: Enlightenment of the Phenomenon of Adsorption, Wastewater Treatment*; IntechOpen: Rijeka, Croatia, 2022; Volume 2.
62. Khamwichit, A.; Dechapanya, W.; Dechapanya, W. Adsorption kinetics and isotherms of binary metal ion aqueous solution using untreated venus shell. *Heliyon* **2022**, *8*, e09610. [CrossRef]
63. Ali, A.; Alharthi, S.; Al-Shaalan, N.H.; Naz, A.; Fan, H.-J.S. Efficient Removal of Hexavalent Chromium (Cr(VI)) from Wastewater Using Amide-Modified Biochar. *Molecules* **2023**, *28*, 5146. [CrossRef]
64. Long, W.; Chen, Z.; Chen, X.; Zhong, Z. Investigation of the Adsorption Process of Chromium (VI) Ions from Petrochemical Wastewater Using Nanomagnetic Carbon Materials. *Nanomaterials* **2022**, *12*, 3815. [CrossRef] [PubMed]
65. Pholosi, A.; Naidoo, E.B.; Ofomaja, A.E. Intraparticle diffusion of Cr(VI) through biomass and magnetite coated biomass: A comparative kinetic and diffusion study. *S. Afr. J. Chem. Eng.* **2020**, *32*, 39–55. [CrossRef]

molecules

MDPI

Article

Catalytic Ozonation of Reactive Black 5 in Aqueous Solution Using Iron-Loaded Dead Leaf Ash for Wastewater Remediation

Latif Hussain [1], Farhan Javed [1], Muhammad Wasim Tahir [1], Hafiz Muhammad Shahzad Munir [2,*], Amir Ikhlaq [3] and Anna Wołowicz [4,*]

1 Department of Chemical Engineering, University of Engineering and Technology, Lahore 54890, Pakistan; latifshifai@gmail.com (L.H.); farhan.javed@uet.edu.pk (F.J.); wasim.tahir@uet.edu.pk (M.W.T.)

2 Department of Chemical Engineering, Khawaja Fareed University of Engineering and Information Technology, Rahim Yar Khan 64200, Pakistan

3 Institute of Environmental Engineering and Research, University of Engineering and Technology, Lahore 54890, Pakistan; aamirikhlaq@uet.edu.pk

4 Department of Inorganic Chemistry, Faculty of Chemistry, Institute of Chemical Sciences, Maria Curie Sklodowska University, Maria Curie-Sklodowska Square 2, 20-031 Lublin, Poland

* Correspondence: shahzad.munir@kfueit.edu.pk (H.M.S.M.); anna.wolowicz@mail.umcs.pl (A.W.); Tel.: +48-81-537-57-27 (A.W.)

Abstract: In the current study, iron-loaded dead leaf ash (Fe-DLA) was used as a novel catalyst in the heterogeneous catalytic ozonation process (HCOP) for textile wastewater containing Reactive Black 5 (RB-5). The research demonstrates a significant boost in removal efficiency, reaching 98.76% with 1.0 g/min O_3 and 0.5 g/L catalyst dose, by investigating key variables such as pH, ozone and catalyst doses, initial concentration, and the presence of scavengers in 1 L wastewater. The addition of tert-butyl alcohol (TBA) reduced RB-5 elimination, indicating the involvement of OH radicals. Catalyst reusability decreased slightly (2.05% in the second run; 4.35% in the third), which was attributed to iron leaching. A comparison of single ozonation (Fe-DLA) adsorption and catalytic ozonation processes (Fe-DLA/O_3) revealed that the combined process improved dye degradation by 25%, with removal rates ranking as Fe-DLA adsorption O_3 Fe-DLA/O_3, with an impressive 76.44% COD removal. These results strongly support RB-5 removal using Fe-DLA and HCOP at a basic pH, highlighting the catalyst's utility in practical wastewater treatment.

Keywords: catalytic ozonation; dead leaves ash; Reactive Black 5; wastewater

1. Introduction

The rapid and expansive surge in industrialization over the past few decades has triggered a substantial increase in the demand for freshwater resources. The discharge of a considerable volume of non-biodegradable and highly toxic organic pollutants into various aquatic ecosystems causes a significant deterioration in water quality and reduces the amount of water available for drinking. These pollutants encompass a broad spectrum of harmful substances, including pharmaceutical compounds [1,2], different types of dyes used in industrial processes [3,4], and potent pesticides employed in agriculture and pest control activities [5]. This multifaceted issue underscores the critical need to address the intricate interplay between industrial expansion, water consumption, and the pervasive release of hazardous substances that pose significant ecological and public health concerns [6–8]. The textile industry is a major global industrial sector, which uses a huge amount of water during dyeing and finishing. Wastewater from the textile industry contributes almost 80% of the overall industrial wastewater discharge and includes different types of dyes that are highly hazardous to the environment (annual dyes production: >100,000 types; weight > 700,000 tons) [7–9]. Two-hundred-and-eighty-thousand tons of the global production of dyes directly enters wastewater [8]. Reactive dyes are the most applied due to their brightness, ease of application, fastness, and the fact that this type of dye is recommended for cellulosic materials, which still form the largest percentage of materials used in the textile industry [9]. Reactive dyes can efficiently covalently bond with the hydroxyl groups in the cellulose, resulting in the best and fastest color [10]. Reactive Black 5 is one such dye, which is extensively applied in the textile industries [11]. These organic contaminants contribute to odor and color and enhance the chemical oxygen demand (COD) in the water effluent [7].

During the dyeing process, some of the dyes are hydrolyzed before reaction with the cellulose and are physically attached to the fiber, either by adsorption or by absorption. Such dyes need extensive cleaning during the wash-off step, or else they may leach even during the product's usage, putting the end-user's health at risk. After the dyeing process, the fabrics are extensively rinsed in the wash-off process, which is considered to be the most water-consuming step in the textile manufacturing industry [12]. The wash-off process is completed in many cyclic steps to ensure that the fabric is thoroughly cleaned [10]. The resultant wastewater produced in this step contains residual dyes, additives, and other chemicals, which are hazardous to aquatic life and other living beings. Such contaminants, if discharged without proper treatment, may lead to the severe deterioration of our ecosystem, affecting human lives [13]. Stringent regulations have been established, along with restrictive environmental policies, to cope with the scarcity of fresh water, leading to the introduction of sustainable manufacturing methods in several industries, including the textile sector. Many conventional methods have been adopted to conserve water and reduce wastewater generation, including electrocoagulation, adsorption, and ozonation [14,15] (Figure S1). There are various advantages of using ozone as an oxidant, including its high oxidation potential and ease of dosing as a gas [16]. The treatment capacity is enhanced when the ozonation process is aided by a heterogeneous catalyst. In recent years, some advanced oxidation processes (AOPs) were developed that appear promising for the removal of recalcitrant contaminants [7,17–19].

The heterogeneous catalytic ozonation process is one of the treatment processes that has been effectively applied for the removal of different contaminants from industrial wastewater, among other dyes, and this process has shown great potential for further development. Ozone is a strong oxidant, but it is preferred not to use it alone due to its selectivity and the formation of intermediates. Adding a catalyst to the ozone treatment leads to the formation of hydroxyl radicals, which are more efficient [20–23]. The selection of a suitable catalyst plays a major role in the optimization of the process because it affects the ozone's decomposition and the formation of hydroxyl radicals [24–27]. Pakistan, being an agriculture-rich country, produces a huge amount of waste biomass annually, which is mostly burnt and discarded. Alternatively, it can be converted from waste to another resource, providing a sustainable solution. The abundant availability of *Sapindus mukorossi* trees in the local area and the regular collection of dead waste leaves make them a key potential resource [28].

To the best of the authors' knowledge and as the literature data indicate (Table S1) [29–32], the present study is the first application of the iron-loaded *Sapindus mukorossi* dead leaf ash catalyst in Reactive Black 5 treatment in the heterogeneous catalytic ozonation process. In previous studies involving the catalytic ozonation process, extensive research was conducted utilizing various inorganic materials as catalysts, such as activated carbon [33], clays, zeolites [34], metal oxides [35], and graphene oxides [36]. However, these materials are expensive and have some limitations in their application. Therefore, it is indeed important to investigate the application of novel materials as catalysts in such processes. Waste-biomass-based materials offer an economic and environmentally friendly solution with abundant availability. There has been limited research reporting on the application of a biomass-ash-based catalytic ozonation process for the elimination of dyes. The potential of the *Sapindus mukorossi* biosorbent was previously shown for wastewater containing Reactive Red 214 (C_0 = 20–80 mg/L) decolorization via the electro-coagulation/biosorption process (the total organic carbon removal efficiency was $\%R_{TOC}$ = 99.56% (C_0 = 50 mg/L), $\%R_{TOC}$ = 99.9% (C_0 = 20 mg/L), 91.75%

(C_0 = 80 mg/L)) with the reaction time = 20 min, current density = 21.7 mA/cm^2, pH 7, and biosorbent dosage = 8 g/L [37]. In the presented investigations, the primary emphasis was on the application capabilities and efficiency of the iron-loaded dead leaf ash (Fe-DLA) as a catalyst with renewable attributes and economic viability. This assessment pertains specifically to its role within the catalytic ozonation process, and a method was employed for the eradication of persistent dye pollutants such as Reactive Black 5.

The present research focused on the application of a novel catalyst of iron-loaded dead leaves of *Sapindus mukorosi* trees in the heterogeneous catalytic ozonation process for Reactive Black 5 treatment of an aqueous solution. Various parameters, like pH, ozone dose, Fe-DLA dose, initial concentration, TBA effect, and reusability, were taken into account.

The experimental endeavors undertaken in this study yielded noteworthy outcomes in terms of decolorization efficiency. This accomplishment holds significant promise, acting as a foundational stepping-stone toward the potential integration and utilization of Fe-DLA in authentic wastewater treatment scenarios. The combination of Fe-DLA and the catalytic ozonation process presents a potentially groundbreaking approach to address the challenge of removing Reactive Black 5 from contaminated water sources, underscoring the practical application of this catalyst and process combination in the domain of environmental remediation.

2. Results and Discussion

2.1. Catalyst Characterization

An Attenuated Total Reflectance Fourier Transform Infrared Spectroscopy (ATR-FTIR) analysis of the dead leaf ash (DLA) and the iron-loaded dead leaf ash (Fe-DLA) was conducted to determine the functional groups present on the catalyst's surface. The FTIR spectra recorded in the wavelength region from 400 to 4000 cm^{-1} are depicted in Figure 1a, whereas their zoom at wavelengths 400–1600 cm^{-1} are presented in Figure 1c,d.

The DLA and Fe-DLA spectra reveal multiple peaks that can be associated with both stretching (symmetric, asymmetric) and/or bending (scissoring, rocking, wagging, twisting) vibrations. The peaks at 447 cm^{-1} and 484 cm^{-1} represent the O-Si-O bending vibrations of silica. The peaks at 518 cm^{-1} and 558 cm^{-1} represent the Al-Si-O bending vibrations related to the presence of silica. The peaks at 719 cm^{-1} and 926 cm^{-1} represent the Si-O-Si symmetric stretching vibrations. Peaks at 1032 cm^{-1} and 1076 cm^{-1} indicate the Si-O asymmetric stretching vibrations [38,39]. In the previous study [40,41] and the study of various biomass ashes [42], some similar peaks were found.

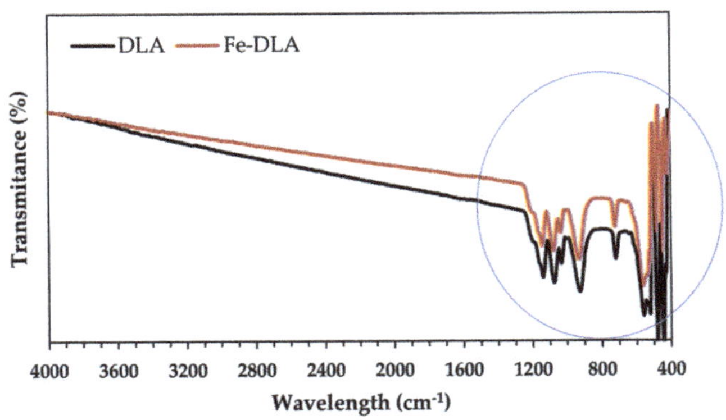

(a)

Figure 1. *Cont.*

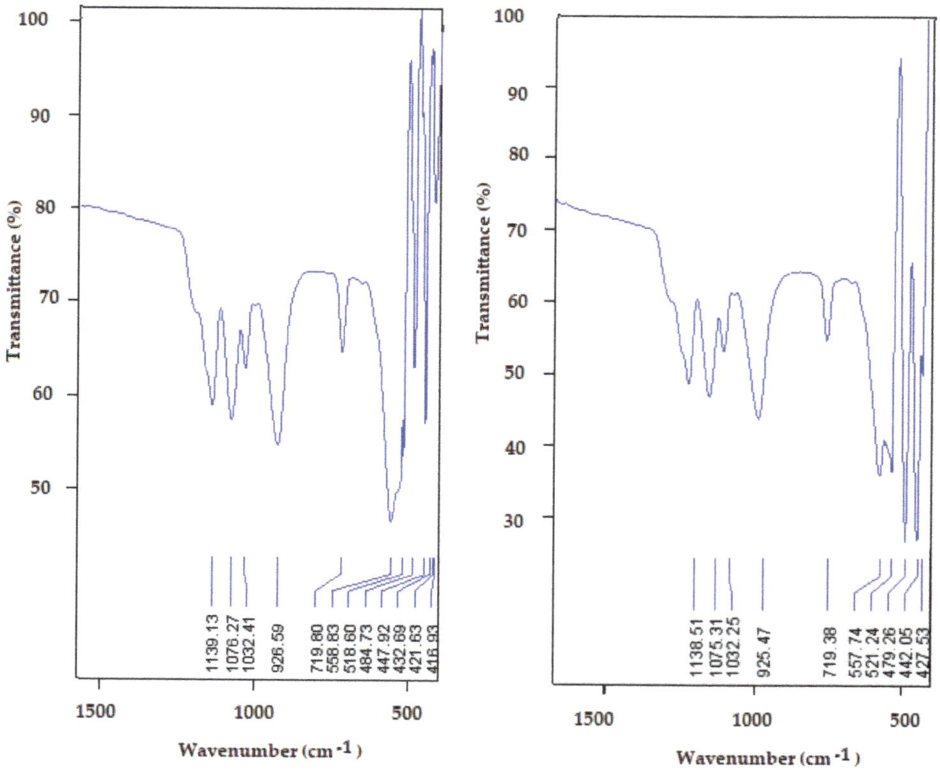

(b) (c)

Figure 1. The FTIR spectra of DLA (**a**,**c**) and Fe-DLA catalysts (**a**,**b**).

A surface structural analysis of the DLA and Fe-DLA catalysts was conducted to elucidate the morphological changes that occurred before and after Fe loading. Scanning Electron Microscopic (SEM) images at a magnification of 50,000 were taken, and are shown in Figure 2. The surface morphologies of both the DLA and the iron-doped catalyst look similar. The images clearly depict a porous spongy surface in both cases. The surface contains many cavities, and the structural surface shows a potential dye attachment ability, and the possibility of proceeding with the catalysis. No major visible difference was found in the SEM images of DLA and Fe-DLA.

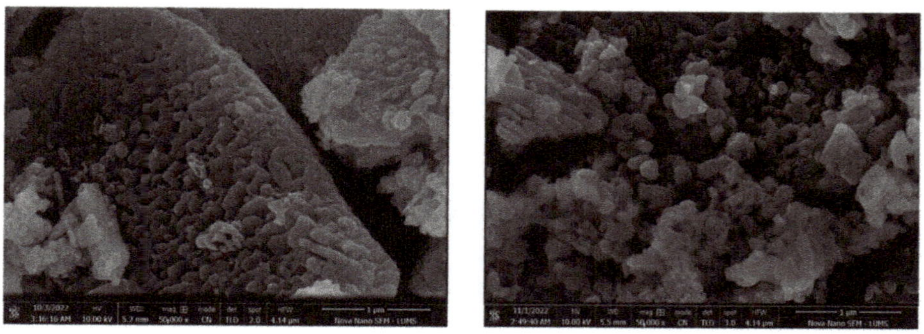

(a) (b)

Figure 2. SEM image of (**a**) DLA and (**b**) Fe-DLA.

The elemental analysis of the DLA and Fe-DLA was conducted using an energy-dispersive X-ray spectroscopy (EDX) analysis, which displayed some clear peaks in Fe in the case of Fe-DLA. The EDX spectra of DLA and Fe-DLA are presented in Figure S2, whereas the elemental analysis is presented in Table 1.

Table 1. EDX composition of DLA and Fe-DLA catalysts applied in the RB-5 wastewater treatment.

Element	Weight (%)	
	DLA	Fe-DLA
C	8.39	5.79
O	30.68	29.83
Si	1.10	1.00
P	35.90	27.7
Ca	21.71	18.37
Fe	1.10	17.01
Cu	1.12	0.30

The EDX results revealed the presence of 1.1% iron in the DLA catalyst. The EDX analysis of DLA showed its composition, and C (8.39%), O (30.68%), Si (1.10%), P (35.9%), Ca (21.71%), and Fe (1.1%) were found. The EDX spectrum of Fe-DLA showed the presence of 17.01% iron, confirming the loading of iron on the DLA. The amount of loaded Fe is satisfactory when using the standard method if compared with the previous works in the literature and can provide an efficient base for the conversion of O_3 to OH radicals [43].

The surface area and pore size of the prepared catalyst was found using the Bruner–Emmet–Teller (BET) method (Table 2).

Table 2. Catalyst surface characteristics.

Material	Surface Area	Pore Size
DLA	14.20 m^2/g	1.75 nm
Fe-DLA	11.30 m^2/g	1.81 nm

The specific surface area was not very large, because it assumes a small value of 14.20 m^2/g, whereas the mean pore size was equal to 1.75 nm for DLA. The nitrogen adsorption–desorption hysteresis loop shapes were employed to determine the specific pore structures and explore the textural properties of the DLA and Fe-DLA catalysts. The low-temperature nitrogen adsorption–desorption isotherms for the DLA and Fe-DLA catalysts are depicted in Figure 3a,b. The DLA catalyst has a higher adsorption capacity than Fe-DLA. The adsorption amount at a p/p_0 of around 0.995 is equal to 27.89 cm^3/g for the DLA catalyst and 22.63 cm^3/g for the Fe-DLA catalyst. The N_2 adsorption content in the DLA catalyst in the low-relative-pressure ranges (<0.02) is around two times higher compared to the Fe-DLA catalyst. According to the IUPAC classification, which distinguished the six types of isotherms (I–VI), the shape of low-temperature nitrogen adsorption/desorption isotherms is characteristic of type II (nonporous or microporous), which behaves like a porous media containing macropores and mesopores. The obtained average pore size values confirm the presence of micropores (<2 nm) in the catalysts' structure. For these samples, N_2 molecules first diffuse into the micropore under low pressure, then, with the increase in relative pressure, are adsorbed in a monolayer state and in subsequent layers as a condensation state [44]. The low-temperature N_2 adsorption isotherms show weaker hysteresis loop effects, and the hysteresis phenomenon is usually associated with capillary condensation in mesopore structures. The obtained hysteresis loops belong to the H3 type, indicating that slit-shaped pores are expected in these samples [45].

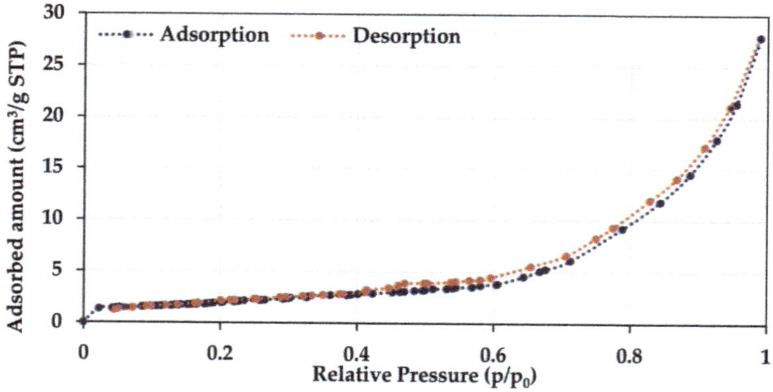

(a)

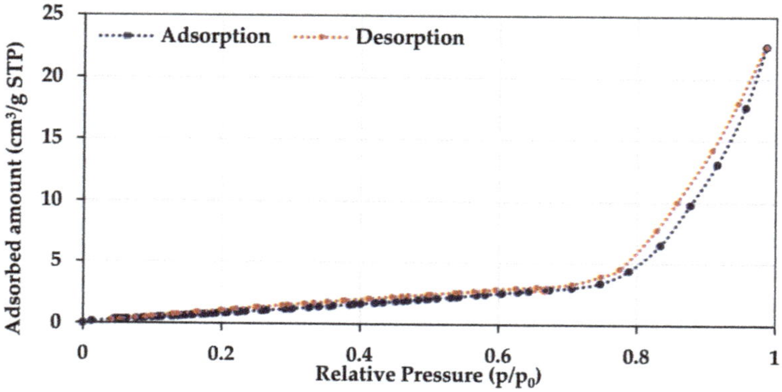

(b)

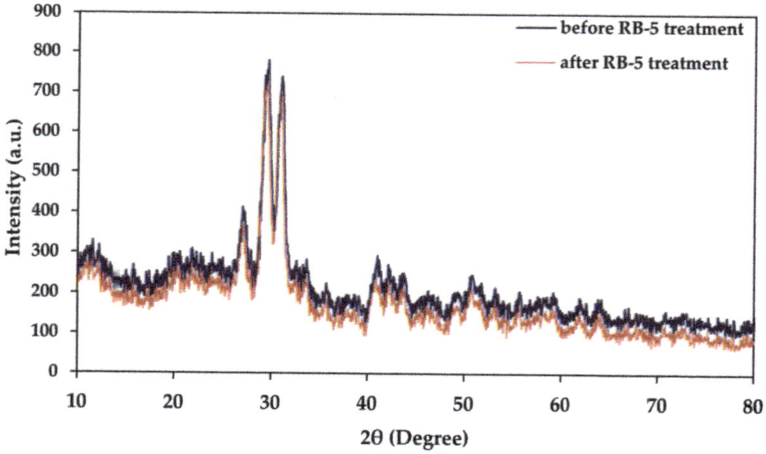

(c)

Figure 3. (**a**,**b**) The low-temperature nitrogen adsorption–desorption isotherms and (**c**) the XRD spectra for (**a**) DLA, (**b**) Fe-DLA, and (**c**) Fe-DLA catalysts before and after the treatment of wastewater containing RB-5 dye.

The results showed that after Fe was loaded on DLA, its surface area decreased from 14.2 m^2/g to 11.3 m^2/g and a slight increase could be seen in the pore size, from 1.75 nm to 1.81 nm. The BET surface area of the DLA catalyst decreased after the Fe, perhaps due to surface capping by Fe(II) metal ions, as revealed by previous studies [42,43]. The Fe-loaded catalyst plays an important role in the catalytic oxidation process. Previous findings [46,47] indicate that Fe may interact with aqueous ozone, leading to the production of OH radicals. OH radical formation is important in such an oxidation process to achieve significant dye removal. Therefore, the Fe loading enhances the catalytic activity in this type of heterogeneous catalytic ozonation process.

The XRD spectra of the Fe-DLA catalyst before and after its addition to wastewater containing the RB-5 dye treatment are shown in Figure 3c. The spectra before treatment indicate the amorphous nature of the material. A major peak traced at 2 theta of 29.66 represents the silica phase, which is present in the ash as a major constituent. The Fe^{2+} peak is traced at 2 theta of 41.03, which indicates the presence of iron, whose valency did not change after its addition to wastewater containing the RB-5 treatment. The peaks obtained at 31.03 and 27.22 degrees correspond to other major constituents in ash, such as Ca and P, respectively [48,49].

The points of zero charge (pH_{pzc}) in DLA and Fe-DLA achieved using the mass titration method were found to be 8.0 $\pm$ 0.2 and 7.1 $\pm$ 0.1, respectively.

2.2. *Parameters Effect Study*

2.2.1. Effect of Ozone Dose in the Single Ozonation Process

To set a suitable ozone dose in the system, the ozone dose was studied using a single ozonation process for RB-5 removal. The oxidizing power of O_3 makes it an efficient oxidant; hence, it can be used as an effective oxidant for the mineralization of dyes. The ozone dose varied in the range from 0.6 mg/min to 1.0 mg/min and decolorization wastewater containing RB-5 was analyzed. The results (Figure 4) clearly show an effective increase in color removal with an increase in the ozone dose, and maximum removal was achieved at 1.0 mg/min ozone dose. After 40 min of treatment, 41.08% dye removal was achieved at 6 mg/min dose, while removal yield being 54.08% and 75.13% were achieved at 0.8 mg/min and 1.0 mg/min, respectively. At an alkaline pH, a high ozone dose causes the rapid generation of quickly oxidizing OH radicals, which enhances the mineralization rate of the organic dye molecules, thereby accelerating the rate of removal [43,50,51].

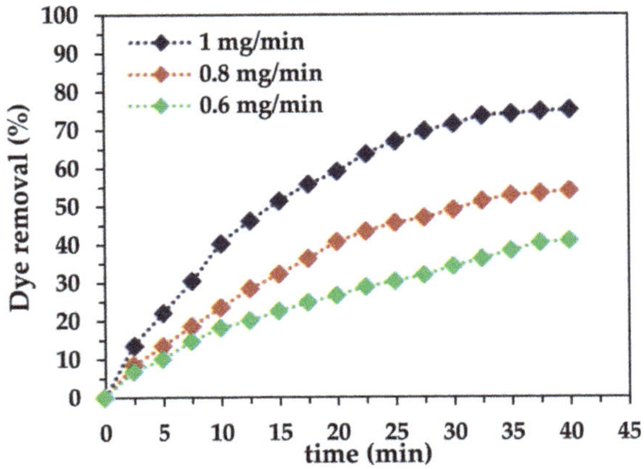

Figure 4. Effect of ozone dose in the simple ozonation process (pH 5.6, T = 25 °C, C_0 = 50 mg/L, V (volume) = 1.0 L, catalyst dose = 0 g/L).

2.2.2. Effect of pH

The pH of the dye solution was adjusted at 2, 7, and 9 using 0.1 mol/L HCl and 0.1 mol/L NaOH to study its effect on the ozonation process. The effect of pH on the percentage removal of RB-5 dye is depicted in Figure 5.

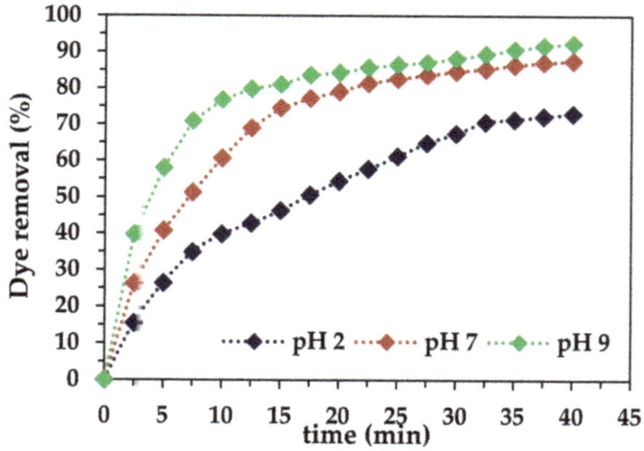

Figure 5. Effect of pH on dye removal by O_3 (t(time) = 40 min, pH 2, 7, 9, ozone dose = 1.0 mg/min; T = 25 °C; C_0 = 50 mg/L, V = 1.0 L, catalyst dose = 0 g/L).

In this simple ozonation, the presented results clearly show an increase in the removal of RB-5 dye as the pH shifts from the acidic to basic. At pH 2, the maximum removal was achieved at 40 min of treatment and was equal to 72.9%. The graph is not very steep at any stage of the treatment, and displays a slow and steady process of dye removal. This is because only the molecular O_3 reacts in the acidic medium, and RB-5 molecules, being complex molecules, do not degrade abruptly [11]. In a neutral medium, the rate of dye removal was higher because, in the presence of hydroxyl ions, some of the O_3 is converted into hydroxyl radicals. Such radicals have a high (2.8 V) oxidation potential to degrade dye molecules, enhancing the rate, and after 40 min of treatment, 87.5% of the dye was removed. After a further increase in the pH to 9, the concentration of hydroxyl ions in the basic pH increases, and the conversion of molecular O_3 into OH radicals with a much higher oxidation potential than O_3 was observed [52]. Consequently, the utilization of an alkaline medium results in significant enhanceent in the overall efficiency of wastewater treatment. Additionally, within a remarkably short treatment time, equal to 15 min, a substantial elimination of approximately 81.0% of the dye was achieved. Furthermore, extending the treatment duration to 40 min yielded even more satisfactory results, with a substantial RB-5 removal of 92.4%. These results indicate the significant role of the alkaline medium in wastewater treatment containing dye, and highlight its potential for rapid and efficient dye elimination [52]. The reason for the better removal efficiency in a basic medium is due to the presence of hydroxyl radicals in the solution, and the maximum removal was achieved at pH 9. Previous studies have shown that the HCOP has maximum activity at a pH close to that of wastewater [25].

2.2.3. Effect of Fe-DLA Dose

At the selected ozone dose and pH, the catalyst dose was varied, at 0.25 g/L 0.50 g/L, and 1.0 g/L, to elucidate its effect on the decolorization of wastewater containing RB-5 dye (Figure 6). When 0.25 g/L of the Fe-DLA catalyst was used, the removal rate was too slow in the first 10 min of treatment, i.e., 78.48%, and after 40 min, 94.31% of RB-5 dye was removed. When using a 0.5 g/L dose of the catalyst, the dye removal efficiency was 86.53% after 10 min and 98.75% after 40 min, whereas using of 1.0 g/L of the catalyst led to a dye removal yield of 90.99% (10 min) and 99.01% (40 min). For that specific volume, concentration, and ozone dose, the results were not as good with a 0.25 g/L catalyst dose;

upon increasing the concentration to 1.0 g/L, the removal rate increased. Hence, 0.5 g/L was selected as a comparatively better dose, considering both the cost and efficiency of the treatment process. The augmentation in the quantity of the catalyst introduced to the system corresponds to the creation of a more extensive platform, offering an increased surface area. This augmented surface area facilitates the accelerated attachment of dye molecules. As a result, the subsequent interaction with hydroxyl (OH) radicals is expedited. This heightened interaction, in turn, serves to bolster the pace at which dye molecules undergo mineralization. In essence, the escalation in catalyst dose plays a pivotal role in fostering a more conducive environment for the swift degradation of dye molecules, ultimately contributing to the accelerated and efficient mineralization of the dye [43,50].

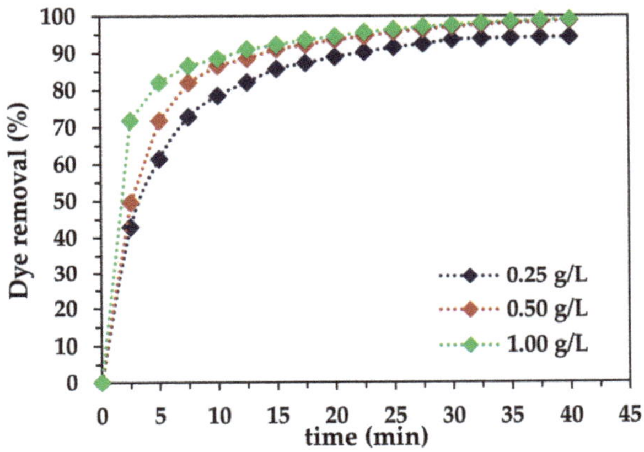

Figure 6. Effect of catalyst dose on HCOP (t = 40 min, pH 9, T = 25 °C, C_0 = 50 mg/L, ozone dose = 1.0 mg/min, V = 1.0 L, catalyst dose = 0.25–1.0 g/L).

2.2.4. Effect of RB-5 Initial Concentration

The initial concentration of RB-5 was altered in the range of 10–90 mg/L to check its effect on the efficiency of pollutant removal using the Fe-DLA/O_3 process. The results depicted in Figure 7 indicate that with the increase in the initial concentration of dye, the removal efficiency decreases. After 40 min of treatment, the highest RB-5 removal, equal to 99.95%, was achieved at an initial concentration of 10 mg/L and the lowest removal of 76.53% was observed at 90 mg/L. RB-5 removal efficiencies of 99.8%, 98.8%, and 95.3% were achieved at 30, 50, and 70 mg/L, respectively. Figure 7 shows that the curves of dye removal (%) vs. t are steeper in the first 10–15 min of treatment; after that, $\%R$ increases slightly over time [43]. The findings of the study unveil a discernible disparity in the pace of dye removal when using higher and lower concentrations of RB-5. Notably, the removal process appears to exhibit a relatively sluggish performance at elevated RB-5 concentrations in contrast to lower concentrations. This observed divergence in behavior could potentially stem from a limitation in the availability of active sites on the catalyst surface at higher dye concentrations. This scarcity of reactive sites may impede the swift interaction between the dye molecules and the catalyst, consequently leading to a decelerated removal rate [25].

2.2.5. Process Comparison Study

A comparative study of Fe-DLA adsorption, simple ozonation (O_3), and catalytic ozonation (Fe-DLA/O_3) was performed at the following optimum parameters: pH 9, ozone dose 1.0 mg/min, catalyst dose 0.5 g/L, and 50 mg/L RB-5. The results are shown in Figure 8. The process' efficiency could be arranged in the following order: Fe-DLA/O_3 > O_3 > Fe-DLA adsorption. Simple ozonation at optimum parameters removed 92.38% of the dye, and when the dye solution was kept for adsorption on the Fe-DLA catalyst with continuous stirring with a magnetic stirrer, 18.99% of the RB-5 dye was removed after

40 min of treatment [42]. For comparison, the Fe-DLA/O_3 process was the most efficient of the three processes, removing 99.89% of the dye in 40 min. Most importantly, in the case of HCOP, the curve is initially steeper, showing that the rate of conversion is much higher compared to the other processes [20]. In comparison, the percent of RB-5 removal in the first 10 min of treatment with the Fe-DLA/O_3 process was 86.53%; in treatment with simple O_3, it was 76.82%; and in the Fe-DLA adsorption treatment, it was just 7.03%. The rapid rate of RB-5 removal and overall higher efficiency clearly suggest that HCOP using the Fe-DLA catalyst may be an efficient treatment method for the continuous removal of effluents from textile wastewater containing dyes, and may be helpful in recycling and conserving water [19,24,25].

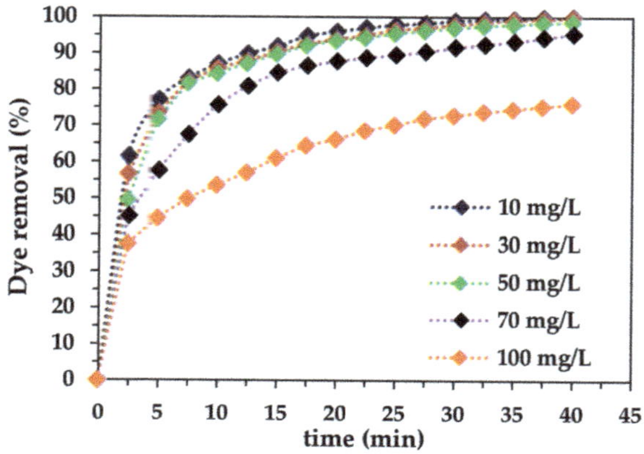

Figure 7. Effect of initial RB-5 concentration on HCOP (t = 40 min, pH 9, T = 25 °C, C_0 = 10–100 mg/L, ozone dose = 1.0 mg/min, V = 1.0 L, catalyst dose = 0.50 g/L).

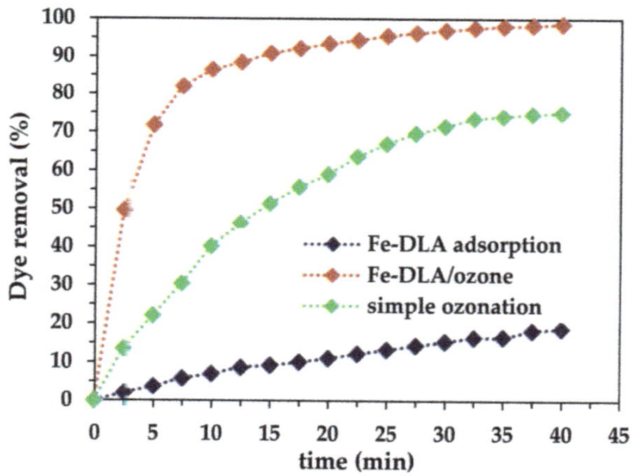

Figure 8. Comparative study of dye removal by various processes (t = 40 min, pH 9, T = 25 °C, C_0 = 50 mg/L, ozone dose = 1.0 mg/min, V = 1.0 L, catalyst dose = 0.50 g/L).

Previous research works [53,54] indicate the effective application of leaf-based ash as a low-cost biosorbent for the elimination of textile dyes. Most of the studies conducted on biomass ash involve the adsorption of dyes; for example, Duan et al. [53] investigated the performance of neem leaf ash as an economical green adsorbent for the treatment of various dyes. A stronger adsorption of cationic dyes was achieved, with an adsorption yield of up to 91.98%, and a highly consistent reusability performance for up to five cycles. Alam et al. [54] studied the removal of Methylene Blue on indigenous biosorbents such as banana leaf ash and rice husk ash. A high removal of up to 93.75% of the dye was achieved by banana leaf ash under optimum conditions. These studies revealed that leaf ash has potential in Methylene Blue removal.

However, limited research has been reported on the application of a biomass-ash-based catalytic ozonation process for the elimination of dyes. The present study is the first investigation on the use of dead leaf ash as a renewable, low-cost catalyst in the catalytic ozonation process for the removal of RB-5. Babar et al. [55] investigated the removal of Methylene Blue by iron-loaded, biochar-based heterogeneous catalytic ozonation. The research showed a high catalytic performance of biochar ash, achieving 95.82% dye elimination at a higher catalyst dose of 5 g/L, as compared to the present study, which showed a high ash performance, with 98.76% dye removal, at even a low dose of 0.5 g/L. Asgari et al. [56] reported the elimination of RB-5 using the catalytic ozonation process on modified bone char ash decorated with MgO–$FeNO_3$. Up to 90% of dye was degraded at a low catalyst dose of 0.1 g/L. As compared to the present study, the optimum results were achieved at pH 10, and the dead leaves' abundant availability makes the catalyst readily available compared to bones. Therefore, the dead-leaf-based catalyst offers an eco-friendly low-cost solution to dye-contaminated wastewater as compared to the materials studied in the literature.

2.2.6. COD Removal Study

COD concentration before and after treatment was studied using the standard vial reagent methods and simple ozonation and catalytic ozonation. The results (Figure 9) show a 12.0% increase in the removal of COD in the case of Fe-DLA/O_3 catalytic ozonation in comparison to the simple ozonation process. The addition of the catalyst facilitates the generation of hydroxyl radicals, which mineralize the contaminant more efficiently compared to the simple ozone method, resulting in more COD removal in the defined time. The COD removal efficiency reached 76.44% in the case of the Fe-DLA/O_3 process, whereas the removal efficiency for simple ozonation was equal to 64.04% after 40 min of treatment [49,57,58].

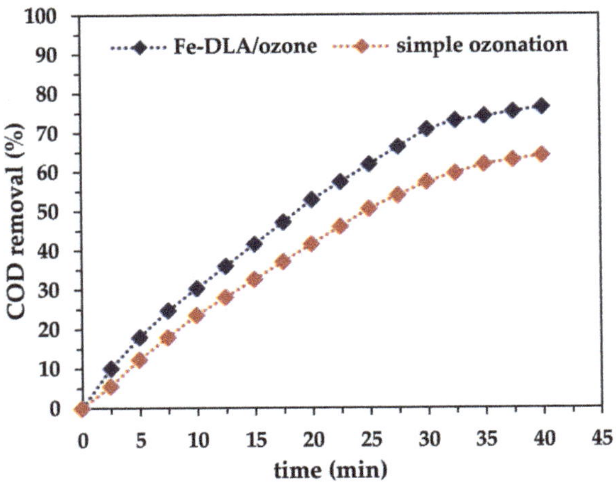

Figure 9. COD removal by HCOP at optimum parameters (t = 40 min, pH 9, T = 25 °C, C_0 = 50 mg/L, ozone dose = 1.0 mg/min and 0 mg/min, V = 1.0 L, catalyst dose = 0.50 g/L).

2.2.7. TBA Effect

Catalytic ozonation aims to generate OH radicals by breaking down the O_3 on the catalyst surface. Tert-butyl alcohol (TBA) was used as a hydroxyl radical scavenger to confirm the generation of OH radicals in this process. The TBA effect on the RB-5 removal efficiency was studied under optimum conditions. The plot in Figure 10 clearly depicts the greater decrease found in the removal efficiency when the TBA was used, and the reduction yield is higher in the case of the Fe-DLA/O_3 process compared to the simple ozonation process. Additionally, the dye removal efficiency decreased from 75.13% to 71.35% and from 98.74% to 90.22% for the simple O_3 and HCOP, respectively, when using the Fe-DLA processes. The considerable decrease in removal in the case of the catalytic process clearly suggests the generation of OH radicals in the Fe-DLA/O_3 process.

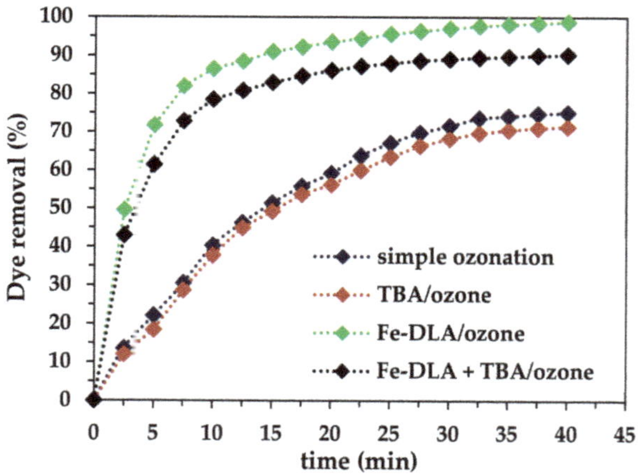

Figure 10. TBA effect on dye removal (t = 40 min, pH 9, T = 25 °C, C_0 = 50 mg/L, ozone dose = 1.0 mg/min, V = 1.0 L, catalyst dose = 0.50 g/L, TBA = 20 mg/L).

2.2.8. Catalyst Reusability Study

The catalyst's performance and cost-effectiveness are concluded in the reusability study. The Fe-DLA catalyst was reused for up to three runs consecutively in the Fe-DLA/O_3 process, and Figure 11 shows the results. For the second run, the catalyst was washed with distilled water and reused. The removal efficiency dropped in the following runs, from 98.72% (first run), through 90.42% (second run), to 80.80% (third run). Despite of the catalyst efficiency dropped the catalyst has still ability for the dye removal. This reduction in the catalyst's performance may be attributed to the dye molecules becoming attached to the pores and surface of the catalyst, and the iron leaching out during the treatment. The reusability study shows that the catalyst is stable enough to use for a few cycles at an industrial scale.

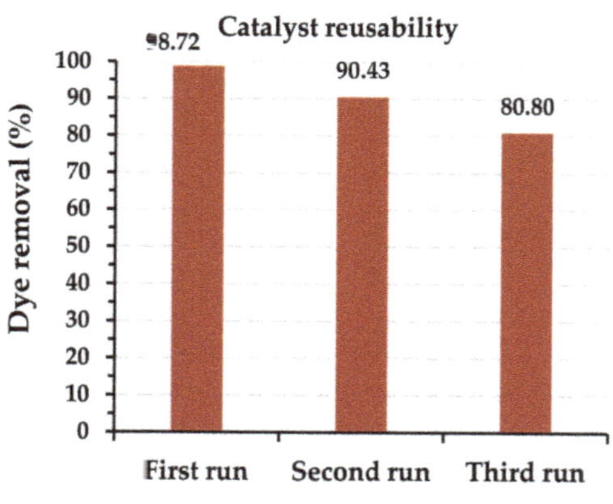

Figure 11. Catalyst reusability study (t = 40 min, pH 9, T = 25 °C, C_0 = 50 mg/L, ozone dose = 1.0 mg/min, V = 1.0 L, catalyst dose = 0.50 g/L).

Each experimental run was conducted by loading Fe-DLA via the RB-5 dye, whose concentration in wastewaters was 5 mg/L. At the end of each run, the catalyst was recovered by filtration, with an average recovery of 98.0 ± 0.3%. Iron leaching tests were conducted, and the average iron leaching rate was determined to be 0.091 ± 0.01 mg/L, which was not considered significant.

2.2.9. Proposed Mechanism

The proposed mechanism of RB-5 removal via Fe-DLA catalytic ozonation is shown in Figure 12. At pH 9, the OH radicals are readily generated from ozone and adopt their positions on the surface platform of the Fe-DLA catalyst. It is hypothesized that aqueous ozone interactions with surface hydroxyl groups on Fe-DLA lead to the generation of superoxide ion radicals [46,47]. These radicals further interact with the aqueous ozone to generate faster oxidant hydroxyl radicals, as shown in Figure 12. The hydroxyl radicals rapidly attack the dye molecules and mineralize them. The hypothesis concerning the radical-based mechanism in this study seems to be possible due to the TBA effect shown in Figure 10, but the OH radicals' dominant role could not be definitely confirmed. The process comparison study (Figure 8) indicates both a slight adsorption effect of the catalyst and the involvement of hydroxyl radicals in the dye's degradation [46,47].

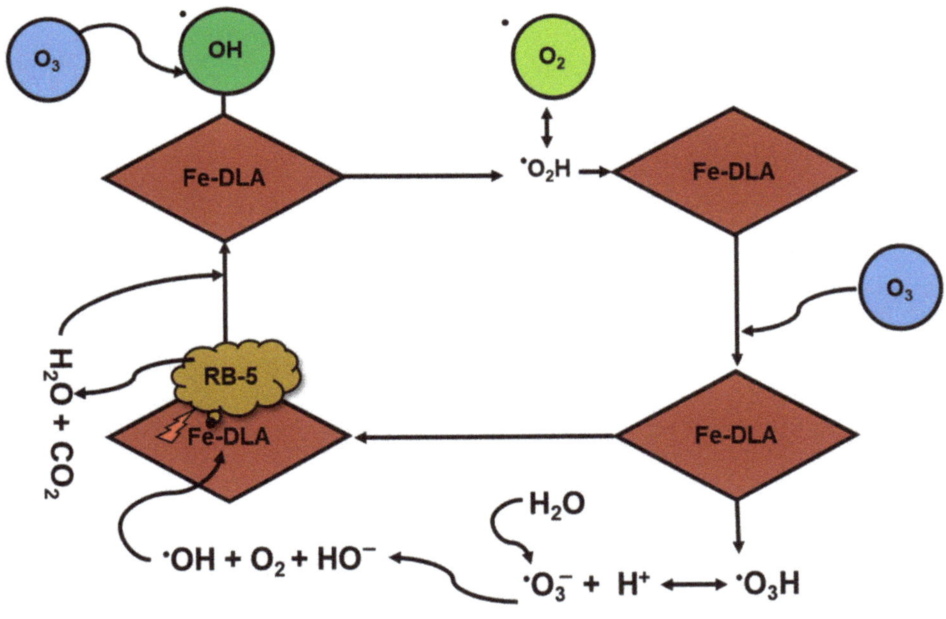

Figure 12. Proposed mechanism of Fe-DLA catalytic ozonation for RB-5 removal.

3. Materials and Methods

3.1. Material and Reagents

The Reactive Black 5 (RB-5) was supplied by Sigma Aldrich, Gillingham, UK (Table 3). Various chemicals used for the Fe-DLA preparation, including $FeSO_4$, HNO_3, HCl, NaOH, $Na_2S_2O_7$, KI, $NaHCO_3$, and Na_2CO_3, were provided by Merck, Darmstadt, Germany. All chemicals used were of analytical grade, and were used without further purification. The standard solutions were prepared through standard methods.

Table 3. General characteristics of Reactive Black 5.

IUPAC name	4-amino-5-hydroxy-3,6-bis((4-((2-(sulfooxy) ethyl)sulfonyl) phenyl)azo)-2,7-naphthalenedisulfonic acid tetrasodium salt
Synonyms	C.I. Reactive Black 5, Reactive Black 5, Reactive Black B, Remazol Black 5, Remazol Black B, Drimaren Black R/K-3B
Chemical structure 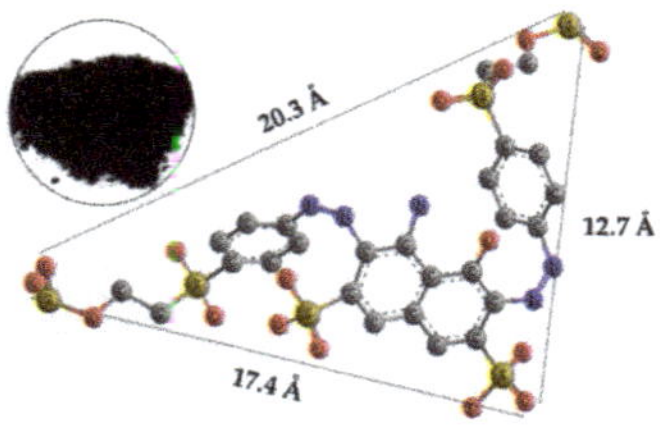	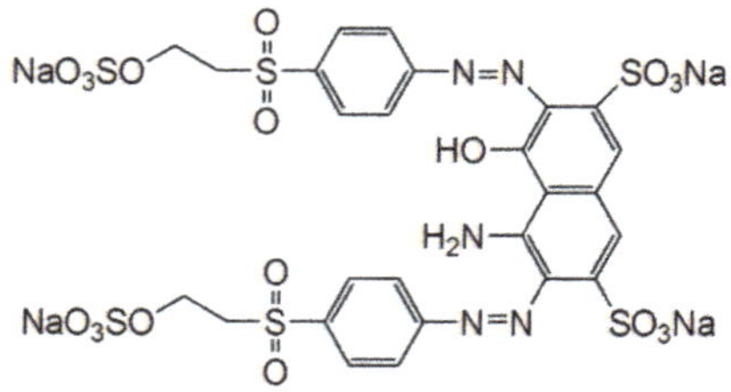
Molecular formula	$C_{26}H_{21}N_5Na_4O_{19}S_6$
Molecular weight	991.82 g/mol
Dye content	≥50%
Color index (C.I.) name	C.I. Reactive Black 5
Color index (C.I.) number	C.I. 20505
Chemical class	Azo
Application class	Cotton
CAS register number	17095-24-8
λ_{max}	596 nm

3.2. Preparation of Fe-DLA

The dead waste leaves of *Sapindus mukorossi* were collected from Begum pora, Lahore, and washed multiple times with distilled water to remove dirt and attached contaminants. The washing procedure was repeated until a constant pH was observed. The dead leaves were converted to powdered biomass using the method adopted from [37,59].

The prepared biomass powder was placed in a furnace at 650 °C for conversion into dead leaf ash (DLA). The formed DLA was then immersed in 0.1 mol/L HNO_3 for more than 24 h, and then filtered through suction filtration and washed with deionized water until the pH of the water was neutralized. Then, it was kept in the oven at 110 °C overnight [13]. Then, the 0.176 mol/L solution of $FeSO_4$ was prepared by dissolving 4.96 g of $FeSO_4$ in 100 mL of deionized water. The impregnation method was used to load iron on the DLA, for which 6.0 g of the DLA was taken and dissolved in 30 mL of the prepared solution. The solution was then kept on a hot plate at 60 °C and stirred for 6 h until the water evaporated. After that, it was kept in the oven at 100 °C for 6 h. The obtained ash was then washed with deionized water and dried.

3.3. Catalytic Ozonation Experiments

The stock solution of Reactive Black 5 dye of 1000 mg/L was meticulously concocted by solubilizing a proper amount of dye in the distilled water. After that, this solution was applied in the preparation of working solutions of various concentrations of RB-5 dye, using the standard dilution method. The setup for RB-5 removal using the HPOC method is depicted in Figure 13.

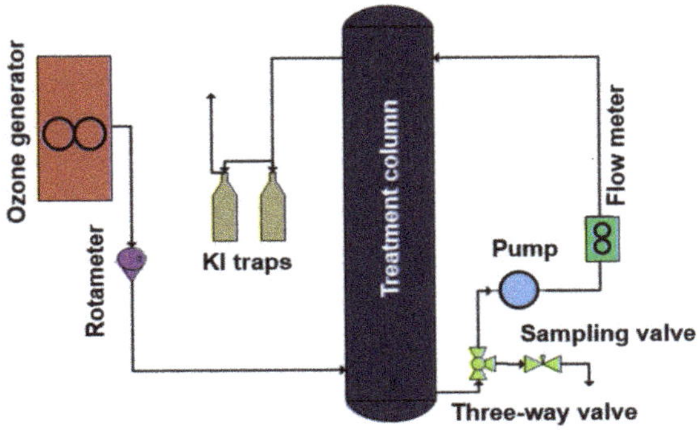

Figure 13. Experimental setup of the heterogeneous catalytic ozonation process (HCOP).

This experimental scheme provides an essential visual aid for comprehending the configuration and arrangement of the apparatus, enhancing the clarity and understanding of the experimental procedure employed in this study. A glass column with a height of 914.5 mm and an inner diameter of 63.4 mm was used as a reactor. Ozone was generated from atmospheric air using a locally manufactured ozone generator. Two gas wash bottles were used, containing 250 mL of 2.0% KI solutions dissolved in deionized water. A total of 1.0 L of the dye solution was put into the reaction column for each run of a different initial concentration, with proper parameter adjustments. Varying amounts of the catalyst were introduced in the column from the top by removing the lid. Ozone was introduced into the column through a porous diffuser from the bottom of the column. A recirculation pump (Iwaki, Japan, Model: MD-6K) with a capacity of up to 3 L/min was attached to the bottom of the column. The solution was recirculated in the column through the pump at a constant rate of 0.5 L/min, controlled by the flow control valve on the flowmeter. This recirculation not only enhances the contact between the ozone and dye solution via turbulence, but also reduces the amount of unreacted ozone that escapes to KI traps via direct interactions. The ozone produced by the ozone generator was sparged into the solution at the bottom of the ozonation column using a ceramic sparger. The ozone, in the form of tiny bubbles with an average size of 1–3 mm, was continuously injected into the solution during the reaction period. The bubbles caused a high level of turbulence and the fast degradation of the dye molecules, while the off gas was engaged in KI traps. A radical scavenger study was conducted using Tert-Butyl Alcohol (TBA) as a hydroxyl radical scavenger. For scavenger studies, at the start of the experimental run, after loading the catalyst dose and setting the ozone flow, 20 mg/L of TBA was added to the reaction column and samples were withdrawn and analyzed after fixed time intervals. Samples collected from the sample collection valve at different intervals were quenched with 0.025 mol/L Na_2CO_3 to remove the residual O_3, given 10 min of settling time, filtered, and finally analyzed to obtain the RB-5 concentration [13,57].

3.4. Analytical Methods

The concentration of dye in the solutions after treatment was obtained with the absorbance analysis using the Perkin Emler lambda 35 double-beam UV-Vis Spectrophotometer. The absorbance of all samples was measured at a maximum absorbance wavelength λ_{max} = 596 nm [52].

The ozone dose was measured using the standard iodometric method. The ozone produced from the ozone generator was captured in 2.0% of KI solution in distilled water, and then quenched by the addition of 5 mL of 1 mol/L H_2SO_4 solution. The solution was then titrated with 0.0025 mol/L $Na_2S_2O_3$, and starch was used as an indicator. Using the runtime of the ozone generator, the total volume of sodium thiosulphate consumed, and its concentration, the ozone concentration was found.

The Chemical Oxidation Demand (COD) of the samples extracted under optimum conditions was measured with the help of standard digestion methods using the COD vials, and the percent removal was measured by relating the decrease in COD with the initial COD of the 50 mg/L aqueous solution. The initial COD of the 50 mg/L RB-5 solution was determined to be 135 mg/L.

3.5. Catalyst Characterization Techniques

The surface morphology of the prepared novel catalyst was studied using Scanning Electron Microscopy (SEM), and its elemental composition was found via Energy-Dispersive X-ray Spectroscopy (EDX) using the Nova NanoSEM 450 analyzer. Images were captured with the electron beam at an acceleration voltage of 10 kV. The surface area and pore size were analyzed with the Brunauer–Emmet–Teller (BET) method using the Micrometrics-ASAP-2020 analyzer. Samples were degassed under vacuum at 350 °C for 8 h. The 99.99% pure nitrogen adsorption–desorption isotherms were measured at 77 K. Functional groups present in the Fe-DLA were analyzed by FTIR-ATR spectrum using the FTIR Spectrometer Alpha-E, Bruker, in the wavelength region from 400 to 4000 cm^{-1} at a resolution of 2 cm^{-1}, using diamond single-reflection attenuated total reflectance. The X-Ray diffraction analysis (XRD) was applied to evaluate the crystallinity and chemical composition of the solid catalyst. The XRD spectrum of the Fe-DLA catalyst was analyzed using the Bruker D2 Phraser X-ray diffractometer in the 2θ range of 10–80, with X-ray generation at 30 kV and 10 mA.

4. Conclusions

The DLA-based catalytic ozonation technique has demonstrated remarkable efficiency in degrading toxic Reactive Black 5 molecules. The HPOC results show substantial degradation rates of RB-5 (reaction time of 40 min) and impressive removal percentages (99.89%), highlighting the efficiency of this approach. The first application of low-cost dead-leaf-ash-based catalyst Fe-DLA in HCOP for RB-5 removal in an aqueous solution, in comparison with other methods, such as simple ozonation and Fe-DLA adsorption, shows the significant advantages of the HPOC method over other methods, as follows: Fe-DLA/O_3 > O_3 > Fe-DLA adsorption. At optimal conditions, such as 50 mg RB-5/L, pH 9, an ozone dose of 1.0 mg/min, and the Fe-DLA dose of 0.5 g/L, up to 99% of RB-5 was removed, with COD removal being 76.44% in the Fe-DLA/O_3 process. The addition of a TBA scavenger to the system caused a slight reduction in dye removal in the studied time, which proved that OH radicals participate in the degradation of RB-5. By effectively tackling the removal of toxic Reactive Black 5, this technique shows the potential to contribute significantly to the sustainable treatment of textile-dye-contaminated wastewater, ultimately safeguarding our environment and human well-being. As we continue to refine and expand this innovative approach, we remain committed to the pursuit of cleaner and healthier water resources for future generations.

Supplementary Materials: The following supporting information can be downloaded at: https://www.mdpi.com/article/10.3390/molecules29040836/s1, Figure S1. Comparison of advantageous and disadvantageous of dyes containing wastewater decolorization; Figure S2. EDX spectra of (a) DLA and (b) Fe-DLA; Table S1. The catalytic ozonation process for Reactive Black 5 dye containing wastewater decolorization.

Author Contributions: Conceptualization, F.J., H.M.S.M. and A.W.; methodology, A.I., L.H. and F.J.; validation, A.I., M.W.T. and F.J.; formal analysis, H.M.S.M. and A.W.; investigation, L.H.; resources, M.W.T.; data curation, L.H.; writing—original draft preparation, A.W., F.J., L.H. and H.M.S.M., writing—review and editing, A.W., A.I. and F.J.; visualization, A.W. and F.J.; supervision, M.W.T. All authors have read and agreed to the published version of the manuscript.

Funding: This research received no external funding.

Institutional Review Board Statement: Not applicable.

Informed Consent Statement: Not applicable.

Data Availability Statement: Data is contained within the article or Supplementary Materials.

Acknowledgments: The authors are thankful to the University of Engineering and Technology, Lahore, Pakistan for the technical support.

Conflicts of Interest: The authors declare no conflict of interest.

References

1. Frascaroli, G.; Reid, D.; Hunter, C.; Roberts, J.; Helwig, K.; Spencer, J.; Escudero, A. Pharmaceuticals in wastewater treatment plants: A systematic review on the substances of greatest concern responsible for the development of antimicrobial resistance. *Appl. Sci.* **2021**, *11*, 6670. [CrossRef]
2. Gadipelly, C.; Pérez-González, A.; Yadav, G.D.; Ortiz, I.; Ibáñez, R.; Rathod, V.K.; Marathe, K.V. Pharmaceutical industry wastewater: Review of the technologies for water treatment and reuse. *Ind. Eng. Chem. Res.* **2014**, *53*, 11571–11592. [CrossRef]
3. Al-Tohamy, R.; Ali, S.S.; Li, F.; Okasha, K.M.; Mahmoud, Y.A.-G.; Elsamahy, T.; Jiao, X.; Fu, Y.; Sun, J. A critical review on the treatment of dye-containing wastewater: Ecotoxicological and health concerns of textile dyes and possible remediation approaches for environmental safety. *Ecotox. Environ. Safe.* **2022**, *231*, 113160. [CrossRef]
4. Teo, S.H.; Ng, C.H.; Islam, A.; Abdulkareem-Alsultan, G.; Joseph, C.G.; Janaun, J.; Taufiq-Yap, Y.H.; Khandaker, S.; Islam, G.J.; Znad, H.; et al. Sustainable toxic dyes removal with advanced materials for clean water production: A comprehensive review. *J. Clean. Prod.* **2022**, *332*, 30039. [CrossRef]
5. Saleh, I.A.; Zouari, N.; Al-Ghouti, M.A. Removal of pesticides from water and wastewater: Chemical, physical and biological treatment approaches. *Environ. Technol. Inno.* **2020**, *19*, 101026. [CrossRef]
6. Li, X.; Chen, W.; Ma, L.; Wang, H.; Fan, J. Industrial wastewater advanced treatment via catalytic ozonation with an Fe-based catalyst. *Chemosphere* **2018**, *195*, 336–343. [CrossRef]
7. Rahmani, A.; Rahmani, H.; Rahmani, K. Degradation Reactive Black 5 dye from aqueous solution using ozonation with pumice and pumices modified by nanoscale zero valent iron(nZVI). *Glob. Nest J.* **2020**, *22*, 336–341. [CrossRef]
8. Hu, E.; Shang, S.; Chiu, K.-L. Removal of reactive dyes in textile effluents by catalytic czonation pursuing on-site effluent Rrecycling. *Molecules* **2019**, *24*, 2755. [CrossRef]
9. Pisitsak, P.; Tunsombatvisit, N.; Singanu, K. Utilization of waste protein from Antarctic krill oil production and natural dye to impart durable UV-properties to cotton textiles. *J. Clean. Prod.* **2018**, *174*, 1215–1223. [CrossRef]
10. Tara, N.; Arslan, M.; Hussain, Z.; Iqbal, M.; Khan, Q.M.; Afzal, M. One-site performance of floating treatment wetland macrocosms augmented with dye-degrading bacteria for the remediation of textile industry wastewater. *J. Clean. Prod.* **2019**, *217*, 541–548. [CrossRef]
11. Wantoputri, N.I.; Notodarmojo, S.; Helmy, Q. Reactive Black-5 removal by ozonation as post treatment. *IOP Conf. Ser. Mater. Sci. Eng.* **2019**, *536*, 012069. [CrossRef]
12. Yaseen, D.A.; Scholz, M. Textile dye wastewater characteristics and constituents of synthetic effluents: A critical review. *Int. J. Environ. Sci. Technol.* **2019**, *16*, 1193–1226. [CrossRef]
13. Lin, J.; Ye, W.; Xie, M.; Seo, D.H.; Luo, J.; Wan, Y.; der Bruggen, B.V. Environmental impacts and remediation of dye-containing wastewater. *Nat. Rev. Earth. Environ.* **2023**, *4*, 785–803. [CrossRef]
14. Peramune, D.; Manatunga, D.C.; Dassanayake, R.S.; Premalal, V.; Liyanage, R.N.; Gunathilake, C.; Abidi, N. Recent advances in biopolymer-based advanced oxidation processes for dye removal applications: A review. *Environ. Res.* **2022**, *215*, 114242. [CrossRef]
15. Ismail, G.A.; Sakai, H. Review on effect of different type of dyes on advanced oxidation processes (AOPs) for textile color removal. *Chemosphere* **2022**, *291*, 132906. [CrossRef]
16. Rekhate, C.V.; Srivastava, J.K. Recent advances in ozone-based advanced oxidation processes for treatment of wastewater- A review. *Chem. Eng. J. Adv.* **2020**, *3*, 100031. [CrossRef]
17. Bilińska, L.; Blus, K.; Foszpańczyk, M.; Gmurek, M.; Ledakowicz, S. Catalytic ozonation of textile wastewater as a polishing step after industrial scale electrocoagulation. *J. Environ. Manage.* **2020**, *265*, 11050. [CrossRef]
18. Psaltou, S.; Zouboulis, A. Catalytic ozonation and membrane contactors—A review concerning fouling occurrence and pollutant removal. *Water* **2020**, *12*, 2964. [CrossRef]
19. Hu, E.; Shang, S.; Tao, X.-M.; Jiang, S.; Chiu, K.-l. Regeneration and reuse of highly polluting textile dyeing effluents through catalytic ozonation with carbon aerogel catalyst. *J. Clean. Prod.* **2016**, *137*, 1055–1065. [CrossRef]
20. Asgari, G.; Faramal, J.; Nasab, H.Z.; Ehsani, H. Catalytic ozonation of indutrial textile wastewater using modified C-doped MgO eggshell membrane powder. *Adv. Powder Technol.* **2019**, *30*, 1297–1311. [CrossRef]
21. Bagheri, M.; Roshanaei, G.; Asghari, G.; Chavashi, S.; Ghasemi, M. Application of carbon-doped nano-magnesium oxide for catalytic ozonation of real textile wastewater: Fractional design and application. *Desalin. Water Treat.* **2020**, *175*, 79–89. [CrossRef]
22. Wang, J.; Chen, H. Catalytic ozonation for water and wastewater treatment: Recent advances and perspectives. *Sci. Total Environ.* **2019**, *704*, 135249. [CrossRef]

23. An, W.; Tian, L.; Hu, J.; Liu, L.; Cui, W.; Liang, Y. Efficient degradation of organic pollutants by catalytic ozonation and photocatalysis synergy system using double functional MgO/g-C_3N_4 catalyst. *Appl. Surf. Sci.* **2020**, *534*, 147518. [CrossRef]
24. Rame, R.; Purwanto, P.; Sudarno, S. Potential of catalytic ozonation in treatment of industrial textile wastewater in Indonesia. *Jurnel Ris. Teknol. Pencegah Pencemaran Industri.* **2020**, *11*, 1–11. [CrossRef]
25. Ikhlaq, A.; Shabbir, S.; Javed, F.; Kazmi, M.; Yasir, A.; Qazi, U.Y.; Zafar, M.; Qi, F. Catalytic ozonation combined electroflocculation for the removal of Reactive Black 5 in aqueous solution using $CuMn_2O_4$/RGO coated zeolite. *Desalin. Water Treat.* **2022**, *259*, 221–229. [CrossRef]
26. Morali, E.K.; Uzal, N.; Yetis, U. Ozonation pre and post-treatment of denim textile mill effluent: Effect of cleaner production measures. *J. Clean. Prod.* **2016**, *137*, 1–9. [CrossRef]
27. Wang, J.; Bai, Z. Fe based catalyst for heterogenous catalytic ozonation of emerging contaminants in water and wastewater. *Chem. Eng. J.* **2017**, *312*, 79–98. [CrossRef]
28. Khan, S.; Nisar, A.; Wu, B.; Zhu, Q.-L.; Wang, Y.-W.; Hu, G.-Q.; He, M. Bioenergy production in Pakistan: Potential, progress, and prospect. *Sci. Total Environ.* **2022**, *814*, 152872. [CrossRef]
29. Sharma, S.; Chokshi, N.P.; Ruparelia, J.P. Comparative studies for the degradation of Reactive Black 5 dye employing ozone-based AOPs. *Nanotechnol. Environ. Eng.* **2022**, *7*, 745–752. [CrossRef]
30. Chokshi, N.P.; Ruparelia, J.P. Catalytic ozonation of Reactive Black 5 over silver–cobalt composite oxide catalyst. *J. Inst. Eng. India Ser. A* **2020**, *101*, 433–443. [CrossRef]
31. Ikhlaq, A.; Zafar, M.; Javed Yasar, A.; Akram, A.; Shabbir, S.; Qi, F. Catalytic ozonation for the removal of Reactive Black 5 (RB-5) dye using zeolites modified with $CuMn_2O_4$/gC_3N_4 in a synergic electro flocculation-catalytic ozonation process. *Water Sci Technol.* **2021**, *84*, 1943–1953. [CrossRef]
32. Song, S.; He, Z.; Qiu, J.; Xu, L.; Chen, J. Ozone assisted electrocoagulation for decolorization of C.I. Reactive Black 5 in aqueous solution: An investigation of the effect of operational parameters. *Sep. Purif. Technol.* **2007**, *55*, 238–245. [CrossRef]
33. Yang, J.; Fu, L.; Wu, F.; Chen, X.; Wu, C.; Wang, Q. Recent Developments in Activated Carbon Catalysts Based on Pore Size Regulation in the Application of Catalytic Ozonation. *Catalysts* **2022**, *12*, 1085. [CrossRef]
34. Inchaurrondo, N.S.; Font, J. Clay, Zeolite and Oxide Minerals: Natural Catalytic Materials for the Ozonation of Organic Pollutants. *Molecules* **2022**, *27*, 2151. [CrossRef]
35. Lv, B.-H.; Liu, X.-Y.; Zhou, Z.-M.; Jing, G.-H. Catalytic Ozonation of Reactive Red 195 in Aqueous Solution over a Cobalt/Aluminum Oxide-Ceria Catalyst. *Processes* **2023**, *11*, 2141. [CrossRef]
36. Beltrán, F.J.; Álvarez, P.M.; Gimeno, O. Graphene-Based Catalysts for Ozone Processes to Decontaminate Water. *Molecules* **2019**, *24*, 3438. [CrossRef]
37. Javed, F.; Feroze, N.; Ramzan, N.; Ikhlaq, A.; Kazmi, M.; Munir, H.M.S. Treatment of Reactive Red 241 dye by electro coagulation/biosorption coupled process in a new hybrid reactor. *Desalin. Water Treat.* **2019**, *166*, 83–91. [CrossRef]
38. Kalembkiewicz, J.; Galas, D.; Sitarz-Palczak, E. The Physicochemical Properties and Composition of Biomass Ash and Evaluating Directions of its Applications. *Polish J. Environ. Studies* **2018**, *27*, 2593–2603. [CrossRef]
39. Nandiyanto, A.B.D.; Rahman, T.; Fadhlulloh, M.A.; Abdullah, A.G.; Hamidah, I.; Mulyanti, B. Synthesis of silica particles from rice straw waste using a simple extraction method. *IOP Conf. Ser. Mater. Sci. Eng.* **2016**, *128*, 012040. [CrossRef]
40. Yusuf, M.O. Bond Characterization in Cementitious Material Binders Using Fourier-Transform Infrared Spectroscopy. *Appl. Sci.* **2023**, *13*, 3353. [CrossRef]
41. Saikia, T.C.; Iraqui, S.; Khan, A.; Rashid, M.H. *Sapindus mukorossi* seed shell extract mediated green synthesis of CuO nanostructures: An efficient catalyst for C–N bond-forming reactions. *Mater. Adv.* **2022**, *3*, 1115. [CrossRef]
42. Trivedi, N.S.; Mandavgane, S.A.; Mehetre, S.; Kulkarni, B.D. Characterization and valorization of biomass ashes. *Environ. Sci. Pollut. Res.* **2016**, *23*, 20243–20256. [CrossRef]
43. Javed, F.; Tariq, A.; Ikhlaq, A.; Rizvi, O.S.; Ikhlaq, U.; Masood, Z.; Qazi, U.Y.; Qi, F. Application of Laboratory-Grade Recycled Borosilicate Glass Coated with Iron and Cobalt for the Removal of Methylene Blue by Catalytic Ozonation Process. *Arab. J. Sci. Eng.* **2023**, *48*, 8753–8768. [CrossRef]
44. Qi, L.; Tang, X.; Wang, Z.; Peng, X. Pore characterization of different types of coal from coal and gas outburst disaster sites using low temperature nitrogen adsorption approach. *Inter. J. Mining Sci. Technol.* **2017**, *27*, 371–377. [CrossRef]
45. ALOthman, Z.A. A Review: Fundamental Aspects of Silicate Mesoporous Materials. *Materials* **2012**, *5*, 2874–2902. [CrossRef]
46. Wang, J.; Yuan, S.; Dai, X.; Dong, B. Application, mechanism and prospects of Fe-based/Fe-biochar catalysts in heterogenous ozonation process: A review. *Chemosphere* **2023**, *319*, 138018. [CrossRef]
47. Li, L.; Fu, R.; Zou, J.; Wang, S.; Ding, J.; Han, J.; Zhao, M. Research Progress of Iron-Based Catalysts in Ozonation Wastewater Treatment. *ACS EST Water* **2023**, *3*, 908–922. [CrossRef]
48. Ma, X.; He, J.; Liu, Y.; Bai, X.; Leng, J.; Zhao, Y.; Chen, D.; Wang, J. Plant Photocatalysts: Photoinduced Oxidation and Reduction Abilities of Plant Leaf Ashes under Solar Light. *Nanomaterials* **2023**, *13*, 2260. [CrossRef]
49. Villar Cociña, E.; Savastano, H.; Rodier, L.; Lefran, M.; Frías, M. Pozzolanic Characterization of Cuban Bamboo Leaf Ash: Calcining Temperature and Kinetic Parameters. *Waste Biomass Valor.* **2018**, *9*, 691–699. [CrossRef]
50. Masood, Z.; Ikhlaq, A.; Farooq, U.; Qi, F.; Javed, F.; Aziz, H.A. Removal of antibiotics from veterinary pharmaceutical wastewater using combined electroflocculation and Fe-Zn zeolite 5A based catalytic ozonation process. *J. Water Process. Eng.* **2022**, *49*, 103039. [CrossRef]

51. Ikhlaq, A.; Aslam, T.; Zafar, A.M.; Javed, F.; Munir, H.M.S. Combined ozonation and adsorption system for the removal of heavy metals from municipal wastewater: Effect of COD removal. *Desalin. Water Treat.* **2019**, *159*, 304–309. [CrossRef]
52. Chokshi, N.P.; Bhutia, H.; Chotalia, A.; Kadiwala, A.; Ruperelia, J.P. Heterogeneous catalytic ozonation of Reactive Black 5 with cobalt oxide. *Int. J. Chemtech. Res.* **2017**, *10*, 402–409.
53. Duan, Y.-T.; Yao, Y.; Ameta, R.K. Removal and recovering of anionic and cationic dyes using Neem leaf ash prepared at 250; 500 and 750 °C: Analyzed by adsorption isotherm and physicochemical parameters. *J. Mol. Liq.* **2023**, *370*, 121012. [CrossRef]
54. Alam, M.Z.; Bari, M.N.; Kawsari, S. Statistical optimization of Methylene Blue dye removal from a synthetic textile wastewater using indigenous adsorbents. *Environ. Sustain. Indic.* **2022**, *14*, 100176. [CrossRef]
55. Babar, M.; Munir, H.M.S.; Nawaz, A.; Ramzan, N.; Azhar, U.; Sagir, M.; Tahir, M.S.; Ikhlaq, A.; Mubashir, M.; Khoo, K.S.; et al. Comparative study of ozonation and ozonation catalyzed by Fe-loaded biochar as catalyst to remove methylene blue from aqueous solution. *Chemosphere* **2022**, *307*, 135738. [CrossRef]
56. Asgari, G.; Akbari, S.; Mohammadi, A.M.S.; Poormohammadi, A.; Ramavandi, D. Preparation and catalytic activity of bone-char ash decorated with MgO-FeNO3 for ozonation of Reactive Black 5 dye from aqueous solution: Taguchi optimization data. *Data Brief* **2017**, *13*, 132–136. [CrossRef]
57. Munir, H.M.S.; Feroze, N.; Ikhlaq, A.; Kazmi, M.; Javed, F.; Mukhtar, H. Removal of colour and COD from paper and pulp industry wastewater by ozone and combined ozone/UV process. *Desalin. Water Treat.* **2019**, *137*, 154–161. [CrossRef]
58. Kruanak, K.; Jarusutthirak, C. Degradation of 2;4;6-trichlophenol in synthetic wastewater by catalytic ozonation using alumina supported nickel oxides. *J. Environ. Chem. Eng.* **2019**, *7*, 102825. [CrossRef]
59. Javed, F.; Feroze, N.; Ikhlaq, A.; Kazmi, M.; Ahmad, S.W.; Munir, H.M.S. Biosorption potential of *Sapindus mukorossi* dead leaves as a novel biosorbent for the treatment of Reactive Red 241 in aqueous solution. *Desalin. Water Treat.* **2019**, *137*, 345–357. [CrossRef]

molecules

MDPI

Article

Biosorption of Pb(II) Using Natural and Treated *Ardisia compressa* K. Leaves: Simulation Framework Extended through the Application of Artificial Neural Network and Genetic Algorithm

Alma Y. Vázquez-Sánchez [1], Eder C. Lima [2], Mohamed Abatal [3,*], Rasikh Tariq [4], Arlette A. Santiago [5], Ismeli Alfonso [6], Claudia Aguilar [7] and América R. Vazquez-Olmos [8]

1 Área Agroindustrial Alimentaria, Universidad Tecnológica de Xicotepec de Juárez, Av. Universidad Tecnológica No. 1000. Col. Tierra Negra Xicotepec de Juárez, Puebla 73080, Mexico; alma.vazquez@utxicotepec.edu.mx

2 Institute of Chemistry, Federal University of Rio Grande do Sul (UFRGS), Av. Bento Goncalves 9500, P.O. Box 15003, Porto Alegre 91501-970, RS, Brazil; eder.lima@ufrgs.br

3 Facultad de Ingeniería, Universidad Autónoma del Carmen, Campeche 24115, Mexico

4 Institute for the Future of Education, Tecnologico de Monterrey, Monterrey 64849, Mexico; rasikhtariq@gmail.com

5 Escuela Nacional de Estudios Superiores, Unidad Morelia, Universidad Nacional Autónoma de México, Antigua Carretera a Pátzcuaro No. 8701, Col. Ex. Hacienda de San José de la Huerta, Morelia 58190, Mexico; arlette_santiago@enesmorelia.unam.mx

6 Instituto de Investigaciones en Materiales, Unidad Morelia, Universidad Nacional Autónoma de México, Antigua Carretera a Pátzcuaro No. 8701, Col. Ex. Hacienda de San José de la Huerta, Morelia 58190, Mexico; ialfonso@iim.unam.mx

7 Facultad de Química, Universidad Autónoma del Carmen, Calle 56 No. 4 Av. Concordia, Ciudad del Carmen, Campeche 24180, Mexico; caguilar@pampano.unacar.mx

8 Instituto de Ciencias aplicadas y Tecnología, UNAM, Circuito Exterior, S/N, Ciudad Universitaria, A.P. 70-186, Delegación Coyoacán, Ciudad de México 04510, Mexico; america.vazquez@icat.unam.mx

* Correspondence: mabatal@pampano.unacar.mx; Tel.: +52-938-3811017 (ext. 1702)

Abstract: This study explored the effects of solution pH, biosorbent dose, contact time, and temperature on the Pb(II) biosorption process of natural and chemically treated leaves of *A. compressa* K. (Raw-AC and AC-OH, respectively). The results show that the surface characteristics of Raw-AC changed following alkali treatment. FT-IR analysis showed the presence of various functional groups on the surface of the biosorbent, which were binding sites for the Pb(II) biosorption. The nonlinear pseudo-second-order kinetic model was found to be the best fitted to the experimental kinetic data. Adsorption equilibrium data at pH = 2–6, biosorbents dose from 5 to 20 mg/L, and temperature from 300.15 to 333.15 K were adjusted to the Langmuir, Freundlich, and Dubinin–Radushkevich (D-R) isotherm models. The results show that the adsorption capacity was enhanced with the increase in the solution pH and diminished with the increase in the temperature and biosorbent dose. It was also found that AC-OH is more effective than Raw-AC in removing Pb(II) from aqueous solutions. This was also confirmed using artificial neural networks and genetic algorithms, where it was demonstrated that the improvement was around 57.7%. The nonlinear Langmuir isotherm model was the best fitted, and the maximum adsorption capacities of Raw-AC and AC-OH were 96 mg/g and 170 mg/g, respectively. The removal efficiency of Pb(II) was maintained approximately after three adsorption and desorption cycles using 0.5 M HCl as an eluent. This research delved into the impact of solution pH, biosorbent characteristics, and operational parameters on Pb(II) biosorption, offering valuable insights for engineering education by illustrating the practical application of fundamental chemical and kinetic principles to enhance the design and optimization of sustainable water treatment systems.

Keywords: *Ardisia compressa* K.; biosorption; heavy metals; artificial neural network; educational Innovation

1. Introduction

Over recent decades, heavy metals have been monitored in sediments, soils, and water systems in different regions of the world in order to study their risks to wildlife and human health [1–3]. In these studies, the reported concentrations of various toxic metals, such as chromium (Cr), mercury (Hg), lead (Pb), nickel (Ni), and cadmium (Cd), exceed the permissible limits according to the United States Environmental Protection Agency (US-EPA) and World Health Organization (WHO) [1]. In the previous investigations, urban sewage and effluents discharged from different manufacturers like pigment, battery, and pesticides or mining activities are considered the principal sources of these hazardous metals in the monitored sites [1]. The presence of toxic metals, even at low concentrations in soils and water, can cause severe animal and human health risks, even cancer and other diseases, such as respiratory irritation, bronchial spasms, and coughing [4–6]. Therefore, intense investigations have developed various industrial and urban wastewater treatment techniques, including coagulation, chemical precipitation, reverse osmosis, ion exchange, flocculation, and membrane filtration, that are used before the waste products are discharged into the aquatic system [7]. Among these methods, adsorption has become the most used technique for removing organic or inorganic pollutants due to its simplicity, high efficiency, and low cost [8]. Numerous adsorbents have been developed for Pb(II) ion removal, such as zeolites [9], kaolinite clay [10], carbon-based materials [11], agricultural waste [12], and polymeric materials [13]. The cost and availability of the adsorbent materials are crucial to the efficiency of the adsorption technique. In this regard, many agriculture and industrial waste materials have been widely used to remove heavy metals from wastewater due to their low cost, availability, and the possibility for generation over various adsorption/desorption cycles [14–17]. A number of agro-wastes were studied for the adsorption of Pb (II) from aqueous solution, such as *Azadirachta indica* leaves [18], *Tamarindus indica* seeds [19], and rape straw [20], as a result of their chemical components namely, steroids, saponins, flavonoids, alkaloids, tannins, and amino acid [21]. Recent investigations reported that chemical modification increased the number of binding ligand groups on the biosorbent's surface. Therefore, the adsorption capacity of biosorbents is expected to be enhanced because the heavy metal ion uptake is favored by increasing active sites on the biosorbent's surface [22].

A. compressa K. belongs to the Myrsinaceae family, which is found in tropical and subtropical regions and has been used as food and folk medicine in different parts of the world [23]. In Mexico, *A. compressa K.* is found in different states, such as Aguascalientes, Chiapas, Chihuahua, Guerrero, Hidalgo, Jalisco, Michoacán, Nayarit, Oaxaca, Puebla, Tlaxcala, San Luis Potosí, Sinaloa, Tamaulipas, and Veracruz [24]. Notably, in the North of Puebla, it is known as *Acachul*, and recent investigations reported by Vázquez-Sánchez A.Y. et al. [24] and by Joaquín-Cruz, Elvia et al. [23] showed that *A. compressa K.* fruits contain various phytochemicals like anthocyanins, polyphenols, flavonoids, diterpenes, gallotannins, and chlorogenic acids. In contrast, Chandra and Mejia [25] identified the presence of phenolic constituents, such as catechin, epicatechin gallate, gallic acid, kaempferol, and ardisin, in the aqueous extracts of the leaves of *A. compressa K.* Therefore, *A. compressa K.* biomass can be considered as an alternative potential biosorbent for heavy metal removal from water due to its interesting properties and its availability.

Another novel aspect of this research is to resolve such a multivariate problem to find the optimal combination of all the tuning variables for the maximization of the removal fraction. This is attained through the application of artificial neural networks for the regression model and the genetic algorithm for optimization purposes.

To the best of our knowledge, *A. compressa* K. biomass has not been used as a biosorbent. Thus, this work contributes to studying the biosorption capacity of natural and treated-with-NaOH *A. compressa* K. leaves regarding removing Pb(II) from aqueous solutions varying solution pH, contact time, adsorbent dose, and temperature. In addition, the adsorption–desorption cycle experiments studied the reuse of raw and treated biosorbents.

2. Results and Discussion

2.1. Physicochemical Characterization

Table 1 presents the physicochemical parameters of *A. compressa* K. leaves in a sense to determine their nutritional composition. The results are in agreement with various reports of the proximate chemical composition of the other biosorbents, such as *Leucaena leucocephala* leaves [26,27] and *Nostoc commune* [28]. For example, if we compare our results with those reported for *Leucaena leucocephala* leaves, the ash content was higher than the 5.35% reported by Thamaga et al. (2021) [26] but lower than the 6.62% reported by G.W. Garcia et al. (2013) [27]; likewise, the protein contents were lower than 19.53% and 22.03% [26,27]. These variations could be attributed to differences in the ages of trees, agro-climatic conditions, and maturity stages [29]. Also, based on the results and in comparison, *N. commune* samples had a high moisture content (96.52 ± 0.11%) and high amounts of non-nitrogen-containing components (69.56 ± 0.09%) compared with *A. compressa* K. leaves. Furthermore, the values of the crude fat (0.24 ± 0.02%), crude fiber (3.42 ± 0.06%), and ash (5.25 ± 0.08%) were lower than *A. compressa* K. leaves [30].

Table 1. Physicochemical parameters of *Ardisia compressa* K. leaves.

Parameter	Percent Composition (% *w*/*w*)
Total soluble solids (%)	86.49 ± 0.12
Moisture (%)	13.51 ± 0.12
Ash (%)	6.74 ± 0.01
Ethereal extract (%)	3.60 ± 0.29
Raw fiber (%)	16.51 ± 0.34
Protein (%)	5.83 ± 0.12
Nitrogen-free extract (%)	53.80 ± 0.36

All values are reported in % on a dry weight basis.

2.2. Characterization

2.2.1. Zero-Point Charge

Figure 1 shows the initial pH ($pH_{initial}$) variation vs. the final pH (pH_{final}) of Raw-AC and AC-OH biosorbents. As can be observed from Figure 1, the intersection of the diagonal line and the $pH_{initial}$ vs. pH_{final} curves indicates that the pH_{PZC} values of Raw-AC and AC-OH were 4.8 and 6.4, respectively. The increase in the pH_{PZC} after the alkali treatment of Raw-AC confirmed the modification in its surface charge, which might improve the biosorption of metal ions since when $pH > pH_{PZC}$, the surface becomes negative, whereas it is positive when $pH < pH_{PZC}$. Similar results were obtained and are presented in Table 2, where the pH_{PZC} values of some raw biosorbents increased with alkaline treatment.

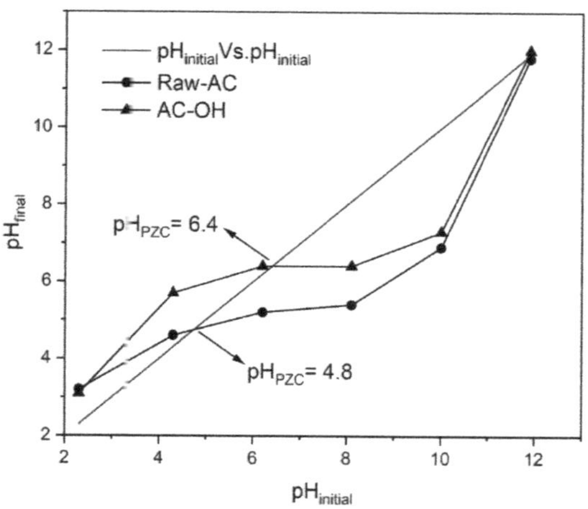

Figure 1. $pH_{initial}$ vs. pH_{final} of Raw-AC and AC-OH biosorbents.

Table 2. Comparison of pH_{PZC} values of untreated and alkaline treated biosorbents.

Biosorbent	pH_{PZC} (Raw Biosorbent)	pH_{PZC} (Alkali-Treated Biosorbent)	Ref.
Prunus armeniaca L. *shells*	4.9	5.7	[31]
Leucaena leucephala leaves	6.7	7.2	[32]
Cupressus sempervirens	6.1	6.7	[33]
Carob shells	5.4	6.52	[34]
Nostoc commune	1.3	2.5	[28]
Ardisia compressa K.	4.8	6.4	This study

2.2.2. Scanning Electron Microscopy

The chemical composition before and after contact with the Pb(II) solution of the RAW-AC and AC-OH samples was estimated using energy-dispersive X-ray (EDX) spectroscopy. Both samples show mainly the presence of carbon (C), oxygen (O), calcium (Ca), and aluminum (Al). The EDX analysis results show that the amount of Pb was 4.4 Wt% for the RAW-AC sample, whereas for the AC-OH sample, the Pb adsorbed was 7.5% Wt%. These results and their corresponding SEM images of RAW-AC and AC-OH before and after contact with Pb (II) are presented in Figures 2 and 3, respectively. For both samples, the presence of Pb (II) is clear in the zoomed-in image, as represented by red dots.

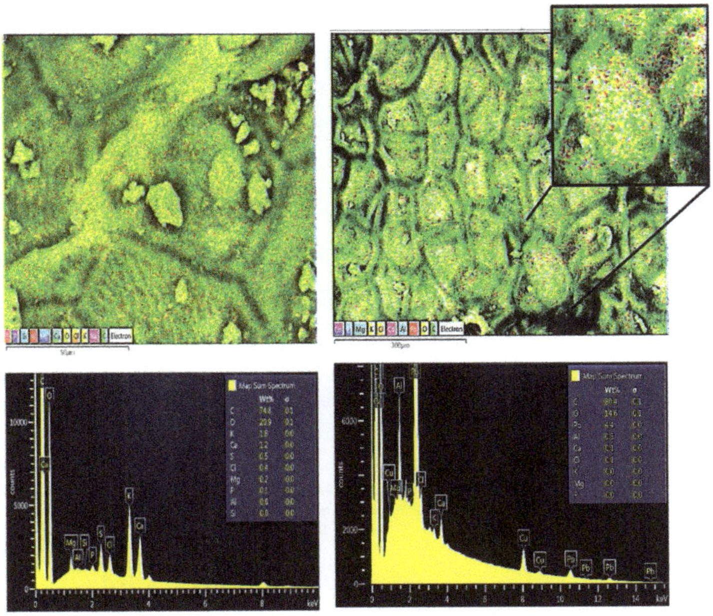

Figure 2. SEM images and EDS analysis of RAW-AC (**left**) and RAW-AC-Pb (**right**); in the zoomed-in image, the red dots represent the Pb(II) on the surface of the sample.

2.2.3. FTIR Analysis

Fourier transform infrared (FTIR) analysis of the biosorbent samples was performed to confirm the presence of functional groups. Figure 4 shows the bands corresponding to the identified functional groups in the biosorbents. The broad band, observed from 3310 to 3280 cm^{-1} for both the before-biosorption and after-biosorption samples, indicates the presence of -OH and -N-H groups [35]. The bands at 2920, 2900, and 2850 cm^{-1} represent the existence of methyl and methylene groups. The bands at 1730 and 1610 cm^{-1} represent the characteristics of the carbonyl group (C=O) stretching from aldehydes and ketones, which could be conjugated or non-conjugated to aromatic rings [36]. The bands at

1450 cm^{-1} show CH_2 and CH_3 groups, while the 1160 and 1220 cm^{-1} bands illustrate C-N [37]. The band at 1610 cm^{-1} was shifted to 1590 cm^{-1} after the Pb(II) uptake, suggesting the interaction of the carbonyl group with Pb(II) during the biosorption process. The *A. compressa* K. has a band at 1030 cm^{-1} and was assigned to the stretch of the C-O group of ether or ester. After the Pb(II) uptake, this band was shifted to 1020 cm^{-1}, indicating the interaction of biomass with Pb(II) [38]. As expected, the surface of the biosorbents contains different functional groups, such as carboxyl, hydroxyl, and amino groups, which could be a potential adsorption site for biosorption.

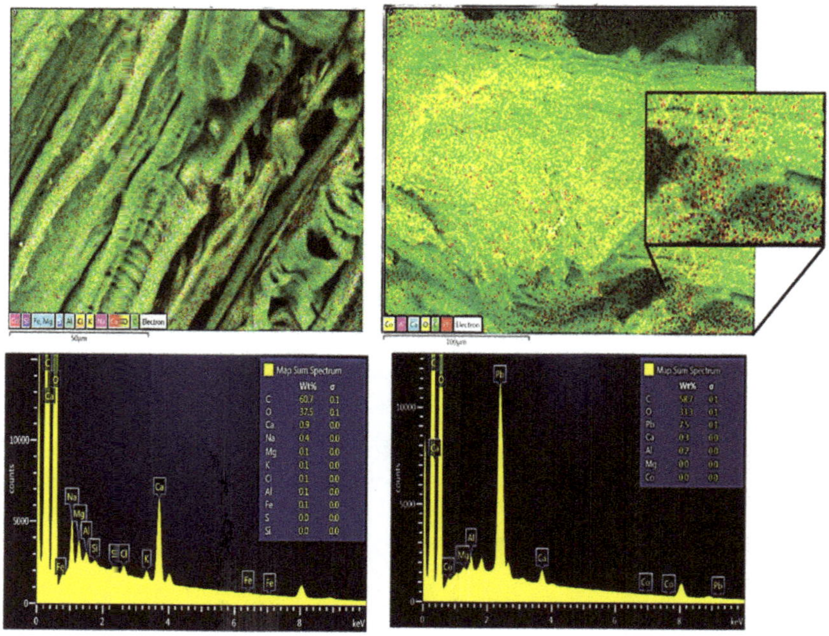

Figure 3. SEM images and EDS analysis of AC-OH (**left**) and AC-OH-Pb (**right**); in the zoomed-in image, the red dots represent the Pb(II) on the surface of the sample.

2.3. Biosorption Results

2.3.1. Effect of Solution pH

The effect of the solution pH on the Pb(II) sorption capacity of Raw-AC and AC-OH was studied using the isotherm experiments. Figure 5 shows the isotherm adsorption of Pb(II) for solution pHs at 2, 4, and 6. As observed, the Pb(II) sorption capacity of Raw-AC and AC-OH increased with the increase in adsorbate solution pH. This dependence can be associated with the metal species present at different pH solutions and the biosorbent's surface charge [39].

The lower removal of Pb(II) at acidic pH (pH = 2 and 4) can be attributed to two reasons: (a) the competition between H^+ present in the solution and Pb^{2+} ions that are the predominant species in this pH region, and (b) the electrostatic repulsion between the Pb^{2+} and the positively charged surface of Raw-AC and AC-OH biosorbents because the solution pH was lower than pH_{PZC}(Raw-AC) = 4.8 and pH_{PZC}(AC-OH) = 6.4. At pH = 6, the capacity of adsorption of biosorbents was superior when compared with pH = 2 and 4; this can be explained by the decrease in competition between protons and cations of Pb on the surface of the biosorbent due to the lower number of protons in the solutions. Furthermore, at pH = 6, the surfaces of biosorbents became negative, and the biosorption of Pb ions was favored over the ionic states of the carboxyl, hydroxyl, and amino groups [40].

Table 3 gives the parameter values of the Langmuir, Freundlich, and Dubinin–Radushkevich isotherm models using their non-linear forms (Equations (4), (5), and (6), respectively). Comparing the correlation coefficients (R^2) obtained from the nonlinearized Langmuir, Freundlich, and Dubinin–Radushkevich isotherm models, the Langmuir model was the best fitted to the experimental equilibrium data.

Table 3. Nonlinear Langmuir, Freundlich, and Dubinin–Radushkevich isotherm parameters for Pb(II) biosorption on Raw-AC and AC-OH biomass at solution pH = 2, 4, and 6.

pH	Model	Parameters	Raw-AC	AC-OH
2	Langmuir	Q_{max} (mg/g)	26.4	43.7
		$K_L \cdot 10^{-2}$ (L/mg)	2.001	59.431
		R^2	0.981	0.971
	Freundlich	K_F (mg/g)(L/mg)$^{1/n}$	3.580	16.226
		n	3.200	5.404
		R^2	0.960	0.924
	Dubinin–Radushkevich (D-R)	Q_{max} (mg/g)	21.948	41.965
		β (mol^2/kJ^2)	93.593	0.242
		E (kJ/mol)	0.073	1.437
		R^2	0.897	0.917
4	Langmuir	Q_{max} (mg/g)	48.1	69.41
		$K_L \cdot 10^{-2}$ (L/mg)	0.882	56.634
		R^2	0.992	0.984
	Freundlich	K_F (mg/g)(L/mg)$^{1/n}$	2.229	23.975
		n	2.084	4.979
		R^2	0.970	0.913
	Dubinin–Radushkevich (D-R)	Q_{max} (mg/g)	34.183	62.050
		β (mol^2/kJ^2)	413.006	0.100
		E (kJ/mol)	0.035	2.236
		R^2	0.932	0.867
6	Langmuir	Q_{max} (mg/g)	96.4	171.0
		$K_L \cdot 10^{-2}$ (L/mg)	0.781	9.076
		R^2	0.997	0.990
	Freundlich	K_F (mg/g)(L/mg)$^{1/n}$	2.654	22.311
		n	1.710	2.043
		R^2	0.992	0.976
	Dubinin–Radushkevich (D-R)	Q_{max} (mg/g)	55.216	120.205
		β (mol^2/kJ^2)	239.383	2.729
		E (kJ/mol)	0.046	0.428
		R^2	0.910	0.913

For Raw-AC, at pH = 2, R^2(Langmuir) = 0.981 > R^2(Freundlich) = 0.960 > R^2(Dubinin–Radushkevich) = 0.897; at pH = 4, R^2(Langmuir) = 0.992 > R^2(Freundlich) = 0.970 > R^2 (Dubinin–Radushkevich) = 0.932; and at pH = 6, R^2(Langmuir) = 0.997 > R^2(Freundlich) = 0.972 > R^2(Dubinin–Radushkevich) = 0.910.

For AC-OH, at pH = 2, R^2(Langmuir) = 0.971 > R^2(Freundlich) = 0.924 > R^2(Dubinin–Radushkevich) = 0.917; at pH = 4, R^2(Langmuir) = 0.984 > R^2(Freundlich) = 0.913 > R^2 (Dubinin–Radushkevich) = 0.867; and at pH = 6, R^2(Langmuir) = 0.990 > R^2(Freundlich) = 0.976, > R^2(Dubinin–Radushkevich) = 0.913.

At pH = 2, 4, and 6, the maximum sorption capacities of Raw-AC determined from the Langmuir model (Q_m) were 26.4, 48.1, and 96.4 mg/g, whereas, for AC-OH, the Q_m values were 43.7, 49.6, and 170.9 mg/g, respectively. As shown in Table 3, the values of K_L for both biosorbents decreased with the increase in solution pH (from $2.001.10^{-2}$ to $0.781.10^{-2}$ L/mg for Raw-AC and from 59.431.10–2 to 9.076 10^{-2} L/mg for AC-OH), indicating that the adsorption of Pb(II) was not favored in an acidic solution. Similar results were reported using other biosorbents [39,40].

The values of the average relative error (ARE), the sum of square error (SSE), normalized standard deviation Δq (%), chi-squared test (χ^2), the sum of absolute error (EABS), and the root mean square error (RMSE) are presented in Table 4.

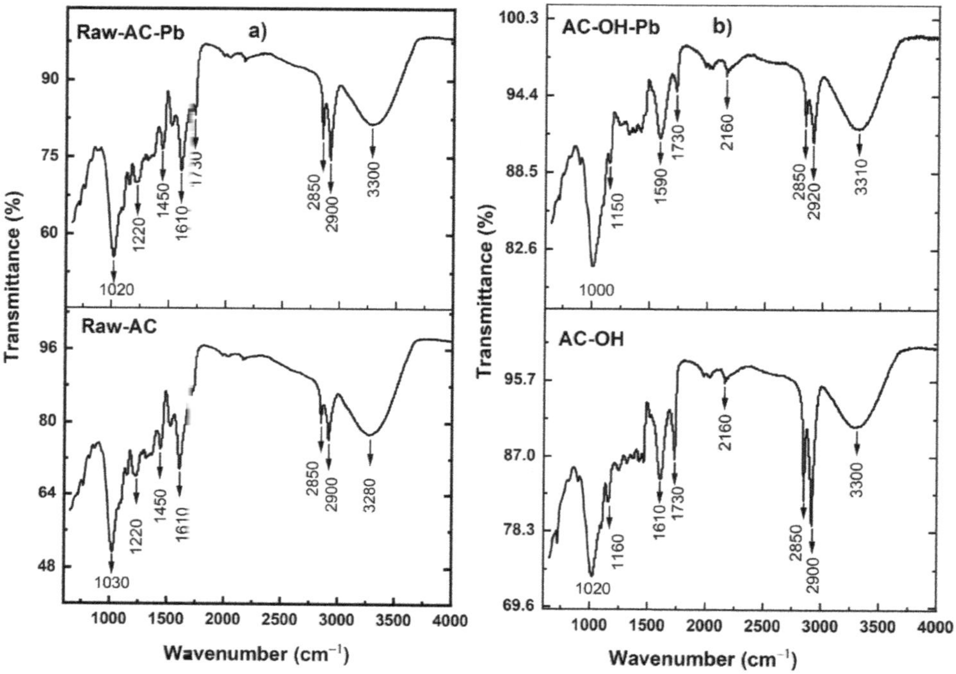

Figure 4. FTIR spectra of (**a**) Raw-AC and Raw-AC-Pb and (**b**) AC-OH and AC-OH-Pb.

Table 4. Error function values of the Langmuir, Freundlich, and Dubinin–Radushkevich isotherm models at solution pH = 2, 4, and 6.

Biosorbent	pH	Model	Error Functions					
			APE	SSE	Δq(%)	χ^2	EABS	RMSE
Raw-AC	2	Langmuir	9.934	4.703	19.201	1.790	5.898	0.970
		Freundlich	16.261	21.035	19.881	1.581	10.604	2.051
		D-R	36.483	51.896	57.739	353.992	15.899	3.222
	4	Langmuir	12.051	8.054	17.513	1.223	5.875	1.269
		Freundlich	19.734	36.164	23.840	2.180	13.922	2.689
		D-R	47.706	81.307	70.448	509.599	21.196	4.033
	6	Langmuir	16.786	9.119	30.325	3.133	6.721	1.510
		Freundlich	17.545	25.427	29.171	1.516	12.695	2.255
		D-R	53.582	272.070	72.248	903.793	41.273	7.377
AC-OH	2	Langmuir	6.402	54.695	8.453	3.009	14.617	3.307
		Freundlich	17.359	142.978	24.869	7.253	28.274	5.347
		D-R	30.443	156.436	47.782	911.672	30.073	5.593
	4	Langmuir	24.359	87.296	30.821	10.184	19.823	4.178
		Freundlich	68.640	469.134	43.902	16.675	52.871	9.686
		D-R	38.773	719.036	53.671	51.392	54.909	11.992
	6	Langmuir	16.701	169.569	21.784	4.109	25.363	5.824
		Freundlich	81.765	402.405	131.184	16.084	48.817	8.971
		D-R	55.164	1427.996	71.002	427.180	88.157	16.900

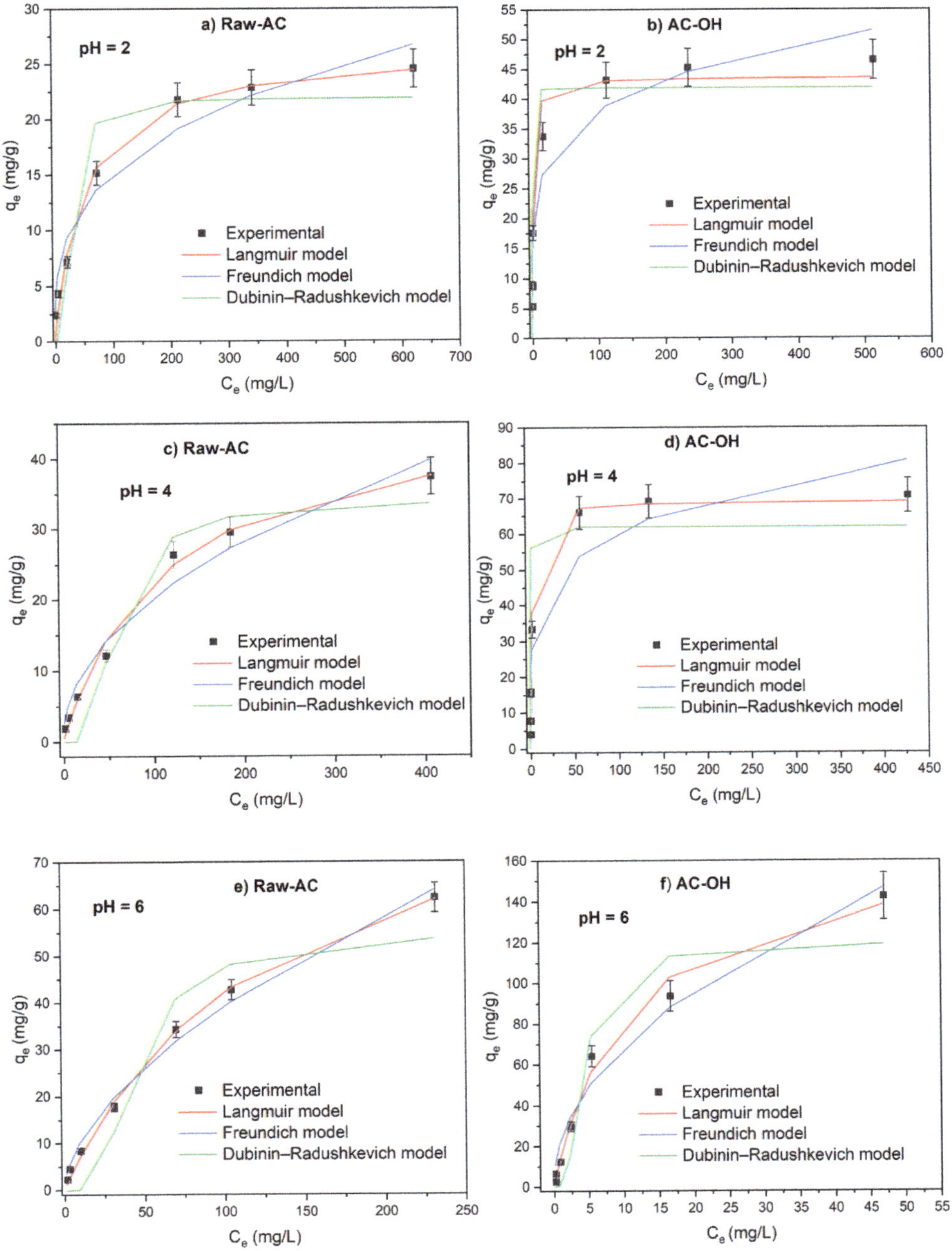

Figure 5. Adsorption isotherms of Pb(II) on Raw-AC and AC-OH at pH = 2 (**a**,**b**), pH = 4 (**c**,**d**), and pH = 6 (**e**,**f**).

Comparing the values calculated for both isotherm models, it can be observed that the lower values of all error functions were obtained for the nonlinearized Langmuir model, confirming the same result predicted using correlation coefficient nonlinear analysis.

2.3.2. Adsorbent Mass Effect

The effect of biosorbent mass on the adsorption of Pb(II) was explored using isotherm experiments. Figure 6 shows that the Pb(II) adsorption isotherms for doses of Raw-AC and AC-OH that varied from 5 to 20 g/L. The experimental data graphs were plotted alongside nonlinear Langmuir and Freundlich isotherm models.

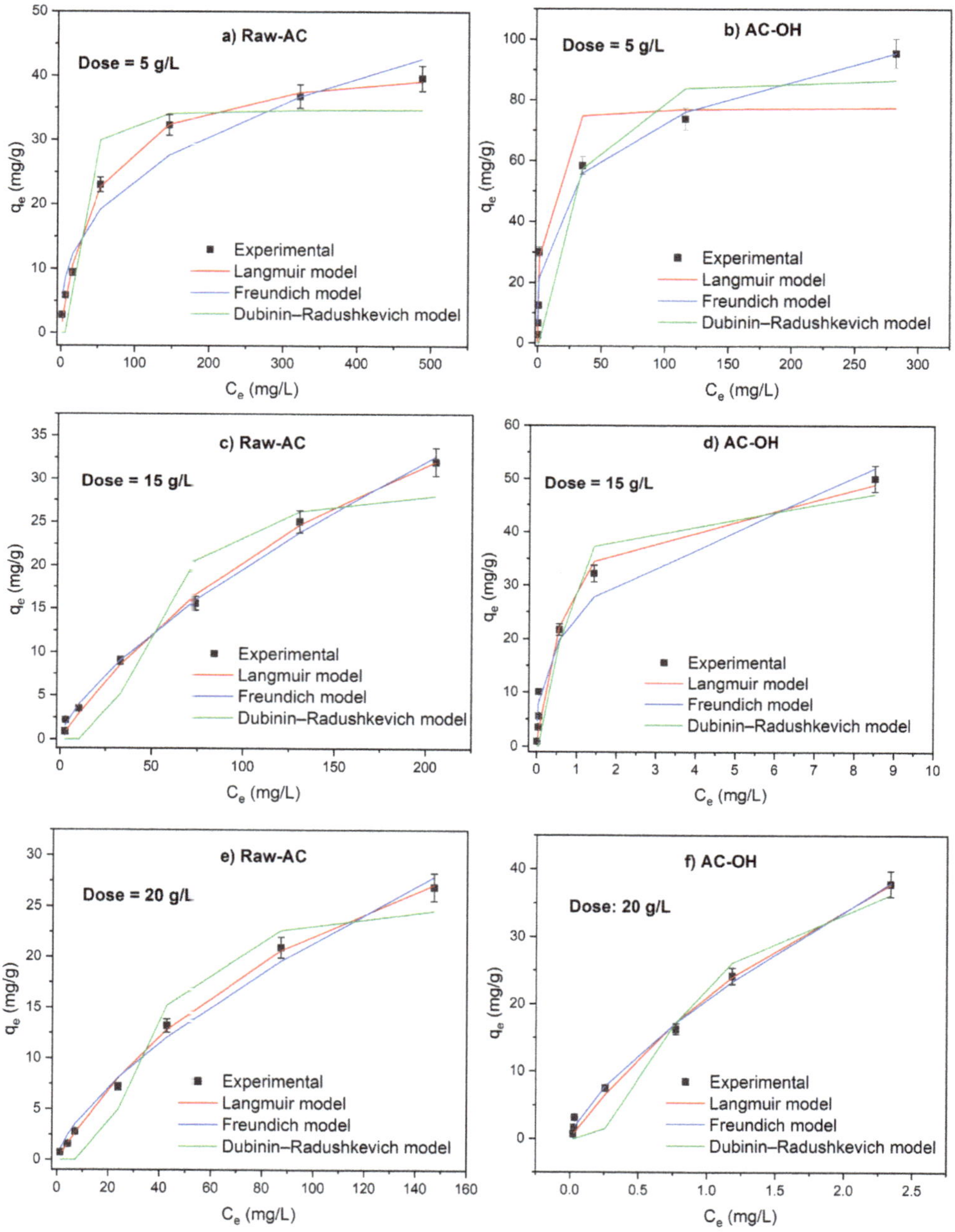

Figure 6. Adsorption isotherms of Pb(II) on Raw-AC and AC-OH at doses of 5 g/L (**a**,**b**), 15 g/L (**c**,**d**), and 20 g/L (**e**,**f**).

The removal percentage of Pb(II) was found to increase by increasing the biosorbent dose and decrease with increasing of initial concentration of Pb(II). For example, at C_i Pb(II) = 500 mg/L, the removal percentages of Pb(II) by Raw-AC and AC-OH were found to be 52.59–89.47%, 76.04–99.83%, and 86.07–99.76% at a doses of 5, 15, and 20 g/L, respectively. This result can be explained by the increase in the number of adsorption sites with increasing biosorbent mass [31].

Experimental equilibrium adsorption data were accessed using the nonlinear forms of Langmuir, Freundlich, and Dubinin–Radushkevich isotherm models. Table 5 shows the parameters of the isotherm models and their respective correlation coefficients (R^2).

Table 5. Nonlinear Langmuir, Freundlich, and Dubinin–Radushkevich isotherm parameters for Pb(II) biosorption on Raw-AC and AC-OH biomass at dose = 5 g/L, 15 g/L, and 20 g/L.

Dose (g/L)	Model	Parameters	Raw-AC	AC-OH
5	Langmuir	Q_{max} (mg/g)	53.7	73.7
		$K_L \cdot 10^{-2}$ (L/mg)	2.105	0.773
		R^2	0.996	0.957
	Freundlich	K_F (mg/g)(L/mg)$^{1/n}$	4.905	23.916
		n	2.494	4.414
		R^2	0.953	0.953
	Dubinin–Radushkevich (D-R)	Q_{max} (mg/g)	34.771	87.074
		β (mol^2/kJ^2)	68.942	81.735
		E (kJ/mol)	0.085	0.078
		R^2	0.908	0.836
15	Langmuir	Q_{max} (mg/g)	53.5	53.3
		$K_L \cdot 10^{-2}$ (L/mg)	0.613	128.915
		R^2	0.996	0.966
	Freundlich	K_F (mg/g)(L/mg)$^{1/n}$	0.952	24.446
		n	1.539	2.829
		R^2	0.986	0.964
	Dubinin–Radushkevich (D-R)	Q_{max} (mg/g)	29.303	47.476
		β (mol^2/kJ^2)	310.592	0.136
		E (kJ/mol)	0.040	1.916
		R^2	0.914	0.905
20	Langmuir	Q_{max} (mg/g)	50.1	90.6
		$K_L \cdot 10^{-2}$ (L/mg)	0.801	30.554
		R^2	0.998	0.993
	Freundlich	K_F (mg/g)(L/mg)$^{1/n}$	0.941	20.674
		n	1.472	1.395
		R^2	0.989	0.997
	Dubinin–Radushkevich (D-R)	Q_{max} (mg/g)	25.706	42.859
		β (mol^2/kJ^2)	157.144	0.212
		E (kJ/mol)	0.056	1.537
		R^2	0.955	0.951

Based on the correlation coefficient shown, for all Raw-AC dose values, the nonlinear Langmuir isotherm model showed a better fit to the adsorption data than the Freundlich and Dubinin–Radushkevich isotherm models. The maximum adsorption capacities (q_{max}) were 53.7 mg/g, 53.5 mg/g, and 50.1 mg/g for 5 g/L, 15 g/L, and 20 g/L of Raw-AC, respectively. The decrease in the adsorption amount with increasing biosorbent concentration has been reported in various investigations, which can be attributed to the declining total biosorbent surface area caused by the aggregation of sorption sites [41].

At biosorbent concentrations of 5 g/L and 15 g/L, experimental adsorption isotherms of AC-OH were better fitted to the Langmuir and Dubinin–Radushkevich models, whereas at a higher biosorbent concentration (20 g/L), the biosorption data were better fitted using the Freundlich and Dubinin–Radushkevich models.

2.3.3. Effect of Contact Time

Figure 7a,b show the uptake of Pb(II) versus contact time with Raw-AC and AC-OH, respectively. It was observed that for all initial Pb(II) concentrations ranging from 25 mg/L to 250 mg/L, between 80 and 85% of Pb(II) was removed in the first 10 min of contact time. Following this time, the Pb(II) removal rate became negligible, and equilibrium was achieved in 30 min for the biosorption of Pb(II) by both biosorbents. The rapid uptake of lead ions in the first stage, which is commonly observed for biosorption of metal ions, is due to physicochemical interactions between the metal solution and available functional groups on the biomass surface [42]. As can be seen from Figure 7a,b, the equilibrium time of the biosorption process was 30 min, and after this time, the insignificant removal rate of Pb(II) could be attributed to the saturation of the available sites on the Raw-AC and AC-OH biosorbents surfaces [43].

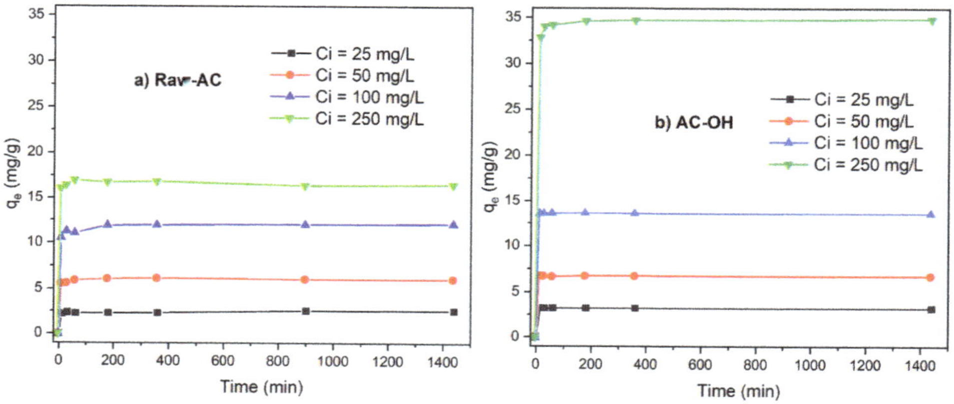

Figure 7. q_e (mg/g) vs. contact time t (min) of (**a**) Raw-AC and (**b**) AC-OH.

Table 6 shows the values of kinetic parameters of Pb(II) biosorption on Raw-AC and AC-OH biosorbents obtained from the nonlinear pseudo-first- (PFO) and pseudo-second-order (PSO) models (non-linear equations of PFO and PSO are shown in Supplementary Material).

Table 6. Parameter values calculated using the nonlinear forms of the pseudo-first-order and pseudo-second-order kinetic models for Pb(II) biosorption on Raw-AC and AC-OH biomass at C_i = 25 mg/L, 50 mg/L, 100 mg/L, and 250 mg/L.

Biosorbent	Ci (mg/L)	Pseudo-First-Order			Pseudo-Second-Order		
		$q_{e,cal}$ (mg/g)	k_1 (1/min)	R^2	$q_{e,cal}$ (mg/g)	k_2 (g/mg min)	R^2
Raw-AC	25	2.4	0.341	0.976	2.4	0.970	0.994
	50	6.0	0.118	0.985	6.2	0.038	0.987
	100	11.7	0.228	0.941	11.9	0.058	0.973
	250	16.7	0.332	0.990	16.7	0.148	0.991
AC-OH	25	3.2	0.409	0.999	3.2	3.175	0.999
	50	6.7	0.298	0.999	6.8	0.728	0.999
	100	13.6	0.232	0.999	13.7	0.153	0.999
	250	34.5	0.200	0.997	34.8	0.032	1.000

Comparing the values of the determination coefficient (R^2) and the $q_{e,calc}$ obtained from both kinetic models, it can be observed that for all initial concentrations (Ci(Pb(II) = 25–250 mg/L), the nonlinear PSO model fit better than the PFO model. Furthermore, the PSO rate constants (k_2) were found to decrease from 0.084–0.484 to 0.024–0.148 (g mg min) for Raw-AC and from 0.488–3.175 to 0.022–0.032 (g/mg min) for AC-OH, for an increase in the concentration of Pb(II) from 25 to 250 mg/L. Biomass treated with NaOH showed a higher uptake of Pb(II) and biosorption rate k_2 compared with the raw biomass. The enhancement of biosorption capacity of treated *Ardisia compressa* K. leaves was also observed in other biosorbents [32,44].

As shown in Table 6, the regression coefficients obtained from the nonlinear form of the PSO and PSO models were higher than 0.941; furthermore, the $q_{e,cal}$ obtained by both kinetic models were similar to $q_{e,exp}$. Therefore, the comparison of R^2 and $q_{e,cal}$ resulted in inadequate criteria for defining the best-fitting kinetic model. However, the values calculated by the six function errors for all initial concentrations of Pb(II) resulted in lower PSO than the PFO (Table 7), confirming that the PSO was the best model to fit the kinetic data. This result suggests that the removal of Pb(II) was controlled by chemisorption, which involved the strong surface complexation of metal ions with the functional groups on the surface of biosorbents [31].

Table 7. Error functions of the PFO and PSO kinetic models.

Biosorbent	Ci (mg/L)	Model	Error Functions					
			ARE	SSE (10^{-2})	Δq(%) (10^{-2})	χ^2 (10^{-2})	EABS	RMSE
Raw-AC	25	PFO	1.486	1.531	2.099	0.635	0.251	0.055
		PSO	1.285	1.065	1.728	0.444	0.218	0.046
	50	PFO	1.758	9.475	2.218	1.633	0.698	0.138
		PSO	1.725	8.188	2.166	1.510	0.663	0.128
	100	PFO	2.790	99.482	3.577	8.505	2.252	0.446
		PSO	1.712	44.854	2.446	3.874	1.360	0.300
	250	PFO	1.051	28.115	1.297	1.688	1.226	0.237
		PSO	0.933	26.359	1.253	1.583	1.089	0.230
AC-OH	25	PFO	0.310	0. 093	0.424	0.029	0.060	0.015
		PSO	0.298	0. 064	0.352	0.019	0.057	0.013
	50	PFO	0.318	0.552	0.495	0.082	0.128	0.037
		PSO	0.295	0.459	0.452	0. 068	0.119	0.034
	100	PFO	0.136	0.446	0.220	0.033	0.111	0.033
		PSO	0.341	1.560	0.414	0.116	0.276	0.062
	250	PFO	0.711	46.737	0.889	1.355	1.467	0.342
		PSO	0.263	6.943	0.345	0.204	0.539	0.132

2.3.4. Temperature Effect on Sorption Capacity

The study of the temperature effect on the sorption of Pb(II) was evaluated using adsorption isotherm experiments by varying the temperature from 300.15 to 330.15 K. Figure 8 shows the adsorption isotherms for Raw-AC and AC-OH at T = 300.15, 315.15, and 330.15 K. It can be observed that the adsorption capacity of Pb(II) of Raw-AC and AC-OH was higher at room temperature compared with that at 315.15 and 330.15 K, confirming that the adsorption was an exothermic process. The decrease in adsorption capacity with temperature can be explained by the breaking of adsorption forces between the metal ions and active sites on the biosorbents caused by the excess of energy supplied to the system, which promoted the desorption of metal ions from surface biosorbent [45,46]. The exothermic adsorption of Pb(II) ions using various types of sorbents, such as *Azadirachta indica* [47], montmorillonite [45], rice husk [48], and activated tea waste [49], was previously reported.

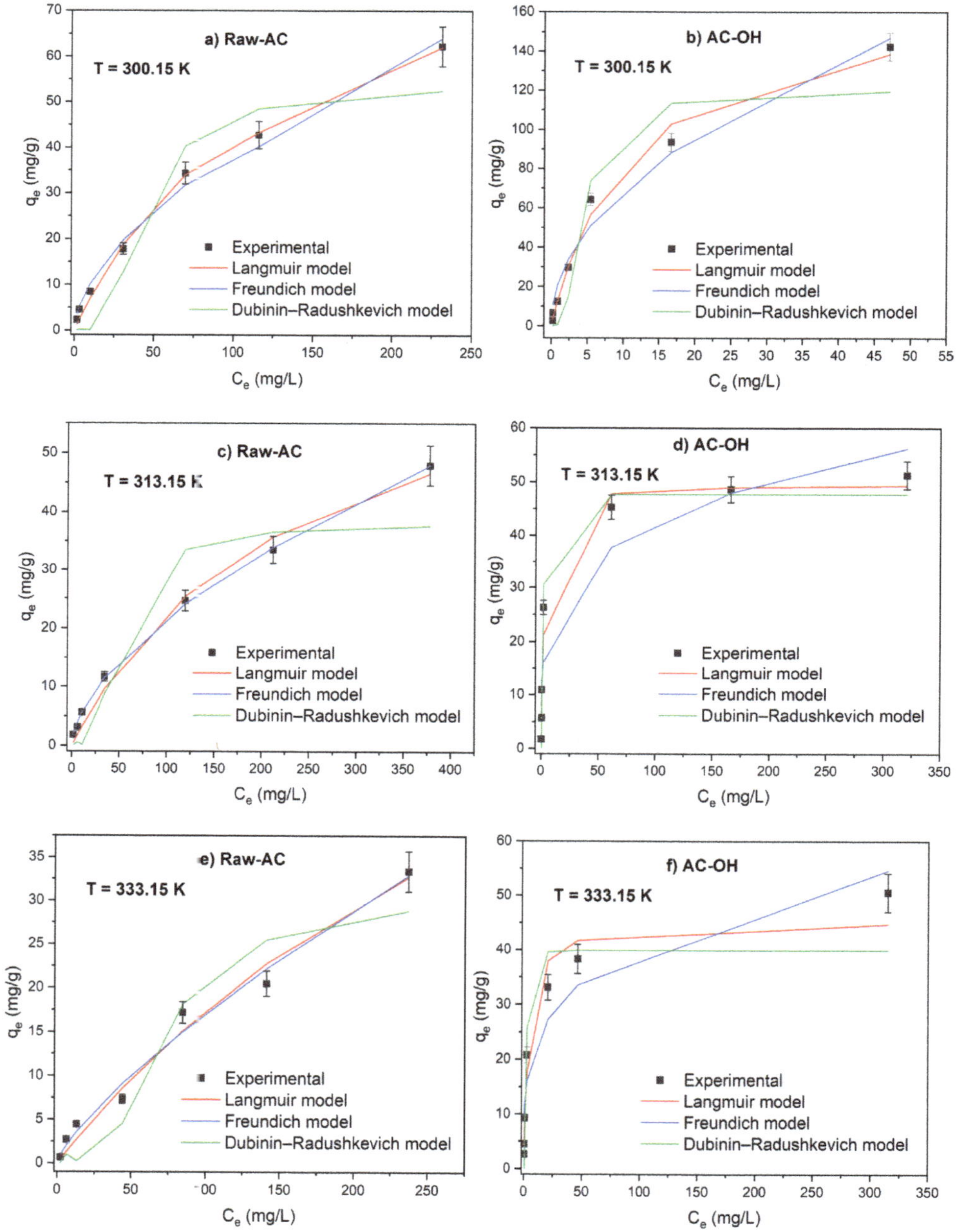

Figure 8. Adsorption isotherms of Pb(II) on Raw-AC and AC-OH at temperatures 300.15 (**a**,**b**), 313.15 (**c**,**d**), and 330.15 K (**e**,**f**).

The isotherm parameters were calculated using the nonlinear equations of Langmuir, Freundlich, and Dubinin–Radushkevich isotherm models.

As shown in Table 8, the adsorption of Pb(II) on AC-OH on the surface of different temperatures studied showed a best fitting to the Langmuir model, as indicated by the high correlation coefficient (R^2 = 0.959–0.990) when the temperature increased from 300.15 K to 330.15 K, the Langmuir constant Q_m of AC-OH, which indicates the monolayer saturation at equilibrium, decreased from 171.0 to 45.3 mg/g. In the case of Raw-AC, the adsorption of Pb(II) at ambient temperature showed a good fit to the Langmuir model, whereas at T = 313.15 and 330.15 K, it was best fitted to the Freundlich model and the values of K_F and n parameters decreased with the temperature rise. At all temperatures studied, the values of n parameters were between 1 and 10, indicating that AC-OH and Raw-AC favorably adsorb Pb(II) ions.

Table 8. Langmuir, Freundlich, and Dubinin–Radushkevich (D-R) isotherm parameters using nonlinear regression analysis for Pb(II) biosorption on Raw-AC and AC-OH biomass at different temperatures.

Temperature (K)	Model	Parameters	Raw-AC	AC-OH
300.15	Langmuir	Q_{max} (mg/g)	96.4	171
		$K_L \cdot 10^{-2}$ (L/mg)	0.781	9.076
		R^2	0.997	0.990
	Freundlich	K_F (mg/g)(L/mg)$^{1/n}$	2.654	22.311
		n	1.71	2.043
		R^2	0.992	0.976
	Dubinin–Radushkevich (D-R)	Q_{max} (mg/g)	53.840	120.205
		β (mol^2/kJ^2)	227.720	2.729
		E (kJ/mol)	0.047	0.428
		R^2	0.899	0.913
313.15	Langmuir	Q_{max} (mg/g)	76.1	49.6
		$K_L \cdot 10^{-2}$ (L/mg)	0.417	42.48
		R^2	0.988	0.971
	Freundlich	K_F (mg/g)(L/mg)$^{1/n}$	1.375	14.096
		n	1.671	4.173
		R^2	0.999	0.884
	Dubinin–Radushkevich (D-R)	Q_{max} (mg/g)	38.081	47.671
		β (mol^2/kJ^2)	264.046	0.324
		E (kJ/mol)	0.044	1.243
		R^2	0.865	0.970
330.15	Langmuir	Q_{max} (mg/g)	92.4	45.3
		$K_L \cdot 10^{-2}$ (L/mg)	0.231	24.844
		R^2	0.981	0.959
	Freundlich	K_F (mg/g)(L/mg)$^{1/n}$	0.508	12.635
		n	1.311	3.329
		R^2	0.985	0.904
	Dubinin–Radushkevich (D-R)	Q_{max} (mg/g)	30.879	39.923
		β (mol^2/kJ^2)	494.126	0.503
		E (kJ/mol)	0.032	0.997
		R^2	0.904	0.887

Error functions for all temperatures studied were calculated. Table 9 shows that for AC-OH, the Langmuir isotherms exhibit lower values of error functions for all temperatures studied and is a better fit compared with the Freundlich isotherm.

Table 9. Error function values of the Langmuir, Freundlich, and Dubinin–Radushkevich equilibrium models at 300.15, 313.15, and 330.15 K.

Biosorbent	T (K)	Model	Error Functions					
			APE	SSE	Δq(%)	χ^2	EABS	RMSE
Raw-AC	300.15	Langmuir	19.583	9.119	0.303	3.133	6.721	1.350
		Freundlich	20.027	20.996	0.291	1.405	11.534	2.049
		D-R	52.451	298.240	70.076	608.200	42.590	7.723
	313.15	Langmuir	26.940	20.969	0.397	7.404	11.441	2.048
		Freundlich	5.468	1.262	0.114	0.211	2.160	0.502
		D-R	52.210	244.613	67.720	210.523	35.096	6.994
	330.15	Langmuir	23.035	16.390	0.305	3.404	9.584	1.811
		Freundlich	18.858	12.375	0.242	1.310	7.947	1.573
		D-R	47.879	75.587	63.686	94.235	19.872	3.888
AC-OH	300.15	Langmuir	16.701	169.569	0.218	4.109	25.363	5.824
		Freundlich	81.765	402.405	1.312	16.084	48.817	8.971
		D-R	55.163	1427.996	71.002	427.158	88.157	16.900
	313.15	Langmuir	54.847	80.379	1.131	7.002	20.212	4.009
		Freundlich	102.467	317.822	2.177	19.917	40.802	7.973
		D-R	37.865	97.503	57.454	78.557	23.568	4.416
	330.15	Langmuir	18.713	84.480	0.244	3.088	20.839	4.110
		Freundlich	71.759	196.337	1.261	13.284	35.860	6.266
		D-R	43.927	231.236	59.709	10.934	35.336	6.801

2.4. Adsorption and Desorption Study

Figures 9a and 9b show the adsorption and desorption cycles of Pb(II) on Raw-AC and AC-OH, respectively. It can be seen that after the three cycles, the percentage of Pb(II) removal on Raw-AC and AC-OH decreased from 80.82% to 45.70% and from 93.80% to 47.05%, respectively. Similarly, the Pb(II) desorption decreased from 67.75 to 45.70% and from 46.94% to 32.25% for Raw-AC and AC-OH, respectively. The decrease in the performance of adsorption and desorption of Raw-AC and AC-OH biosorbents was previously reported in other biosorbents, which was attributed to the possible remanent of metal ions on the surface of the biosorbent after the desorption step and the partial destruction of active functional groups by HCl after each regeneration cycle [22]. Despite the reduction in the percentage of Pb(II) removal after three cycles, both biosorbents have still shown efficient removal and desorption of Pb(II) after three times of recycling. As such, Raw-AC and AC-OH can be considered suitable recyclable biosorbents.

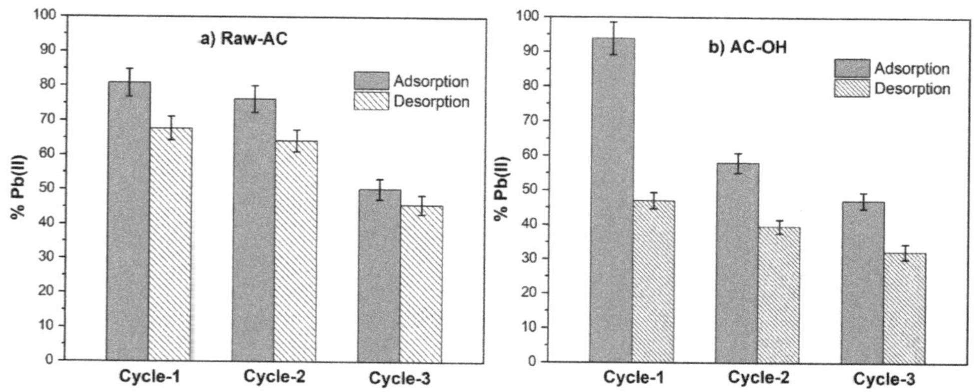

Figure 9. Adsorption and desorption cycles of Pb(II) with (**a**) Raw-AC and (**b**) AC-OH biosorbents.

2.5. Comparison of Natural and Alkali Ardisia Compressa K. with Other Biosorbents

In this study, the maximum Pb(II) biosorption capacity, according to Langmuir's model, was 96.4 and 170.9 mg/g using Raw-AC and AC-OH, respectively. As shown in Table 10, the results obtained in this work are within the usual values, and in some cases, they are between the highest maximum biosorption capacities for other similar biosorbents reported in the scientific literature, indicating the potential for its application in the removal of lead, and perhaps other heavy metals too, from contaminated effluents.

Table 10. Comparison of Pb(II) adsorption capacity of different sorbents.

Biosorbent	Dose (g/L)	Ci (mg/L)	pH	Kinetic Model	Mechanism	$q_{e,max}$ ($mg \cdot g^{-1}$)	Ref.
Moringa oleifera tree leaves	4	20–200	5	PSO	Chemisorption	34.6	[42]
Nostoc commune	0.5	127- 442	4.5–5.5	PSO	Chemisorption	384.6	[28]
Cystoseira stricta	1	0.5–1.0	3	-	-	65.0	[50]
Leucaena leucocephala leaves	10	10–500	6	PSO	Chemisorption	25.5	[51]
Ceratophyllum demursum	0.1	1.46–1.73	7	-	Physical, monolayer	44.8	[52]
Punica geranatum leaves	0.1	1.46–1.73	7	-	Physical, monolayer	31.78	[53]
Mirabilis jalapa leaves	0.1	50–125	4.5	-	Multimolecular layers	38.5	[54]
Laurus nobilis leaves waste	4	5–10	6	PFO	Chemisorption	96.1	[55]
Cladophora fascicularis	2	225.4–838.3	5	PSO	Chemisorption	227.7	[56]
Nickel ferrite nanoparticle	-	0.01–0.05	5	PSO	-	19.2	[57]
MoS_2/thiol-functionalized multiwalled carbon nanotube	2	20–150	6	PSO	Multilayer	90.0	[58]
Nano Zero-Valent Iron particles	0.1	10–1000	6	PSO	Chemisorption	140.8	[59]
Cu-Based Composite Track-Etched Membranes	0.0033–0.0052	50	5	PSO	Chemisorption	0.44–0.56	[60]
Raw-AC	10	25–1000	6	PSO	Chemisorption	96.4	In this study
AC-OH	10	25–1000	6	PSO	Chemisorption	170.9	In this study

2.6. Proposed Mechanism of Interaction of Pb^{2+} with Raw-AC and AC-OH

The mechanism of interaction of Pb^{2+} with Raw-AC and AC-OH could be represented as shown in Figure 10 below:

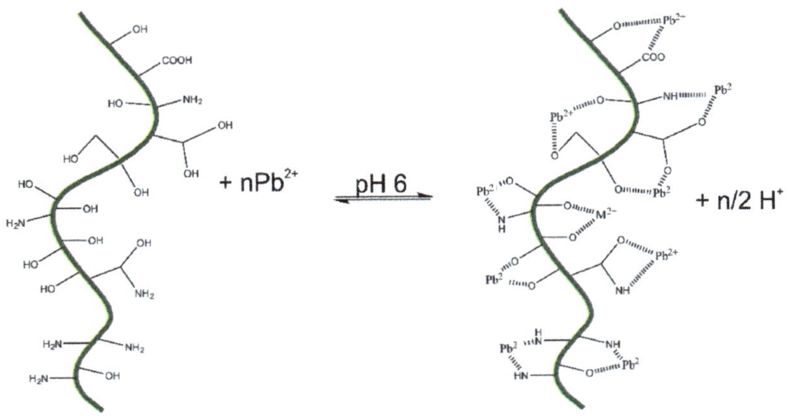

Figure 10. Interaction mechanism of the uptake of Pb^{2+} with raw-AC and AC-OH.

According to the FTIR results, both biomasses (Raw-AC and AC-OH) present, OH-,-COOH, and NH_2 could be complexed with Pb^{2+} species, or some ion exchange could occur. Adjusting the pH to 4–6, the COOH groups of carboxylic acid release H^+ into the solution.

This pH adjustment helps the complexation or electrostatic attraction of Pb^{2+} with both Raw-AC and AC-OH biomasses.

2.7. Data-Driven Optimization

An optimization study was conducted for the two materials (Raw-AC and AC-OH) and its objective was to find which type of material, along with its multivariate combination of initial concentration of Pb, mass, temperature, pH of the solution, and contact time, could provide the maximum q_e. However, optimization using the experimentation data required a couple of steps which are mentioned herewith:

1. In the first step: the original data was arranged in the form of a database (see Table 11), where all the tuning variables and the performance indicators were placed together. The database contained a total of 178 experimental points. Bringing the data in such a format would help in the development of an empirical model [61]. It was necessary to develop an empirical model because the optimization algorithms required a direct correlation between the tuning variables and the performance indicator. Thus, it was also necessary to understand the data through the heatmap of the Pearson coefficient of correlation (see Table 12) [62]. It was noted that the initial concentration of Pb was the variable with the highest level of sensitivity for q_e, and it was also noted that material 2 had a relatively higher tendency for higher values of q_e. The temperature and the pH of the solution were the least sensitive.

Table 11. Format of the database for the development of the empirical model. Note: material 1 refers to the combination of Raw-AC and material 2 refers to the combination of AC-OH.

Material	Initial Concentration of Pb (mg/L	Mass (g)	Temperature (K)	pH of the Solution	Contact Time (min)	q_e (mg/g)
1	24.414	0.1	298.15	2	1440	2.41213
1	48.204	0.1	298.15	2	1440	4.33674
1	92.492	0.1	298.15	2	1440	7.1832
1	412.92	0.1	298.15	6	1440	34.3227
1	531.76	0.1	298.15	6	1440	41.574
1	854.16	0.1	298.15	6	1440	62.287
2	26.958	0.05	298.15	2	1440	5.36626
2	44.613	0.05	298.15	2	1440	8.84096
2	89.1	0.05	298.15	2	1440	17.61704
2	333.02	0.05	298.15	6	1440	63.7672
2	757.88	0.05	298.15	6	1440	127.848
2	1012.98	0.05	298.15	6	1440	147.475

Table 12. Understanding the correlation coefficient between the tuning and performance indicators through Pearson's coefficient of correlation.

	Material	Initial Concentration of Pb	Mass	Temperature	pH of the Solution	Contact Time	q_e
Material	1.00	−0.01	−0.24	0.01	−0.01	0.01	0.27
Initial concentration of Pb	−0.01	1.00	0.14	−0.02	−0.12	0.27	0.79
Mass	−0.24	0.14	1.00	−0.06	0.06	0.38	−0.18
Temperature	0.01	−0.02	−0.06	1.00	0.16	0.22	−0.03
pH of the solution	−0.01	−0.12	0.06	0.16	1.00	−0.23	−0.03
Contact time	0.01	0.27	0.38	0.22	−0.23	1.00	0.22
q_e	0.27	0.79	−0.18	−0.03	−0.03	0.22	1.00

2. In the second step, an empirical model was developed for the selected database. Such a model was developed through the application of artificial neural networks for regression models. The MATLAB environment using the nftool box was implemented.

All the input variables were placed in the input layer, and the performance indicator was placed in the performance indicator. A backpropagation method called Bayesian regularization backpropagation was implemented, which was suitable for the noisy database as the origin was experimentation [63]. The training, testing, and validation percentages were set to be 70, 15, and 15, respectively. The number of hidden neurons was iterated and a total of eight neurons came out to be optimal for the minimum mean square error. The regression fit and the error histogram are displayed in Figure 11. During the training, testing, and total phase, the values of R came out to be 0.99276, 0.977, and 0.9905, respectively, which indicates a high goodness of fit. The error histogram also fulfills the normality assumption of errors.

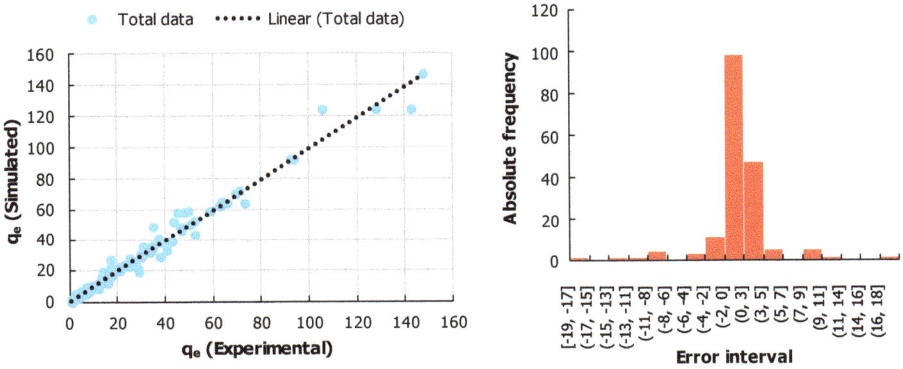

Figure 11. (**Left**) Results of empirical modeling through a comparison between experimental and simulated q_e. (**Right**) Histogram of the errors between the experimental and simulated q_e values.

3. Once the empirical model was developed, in the last step, an optimization study could be conducted. The optimization study was conducted to maximize qe for the given range of each of the tuning variables. The genetic algorithm was implemented in the MATLAB environment through its Optimization toolbox. All the generic configuration of the algorithm was taken to be default [64]. The optimization was conducted two times, one for each material, and the results are shown in Table 13. It is noted that the combination of tuning variables for each type of material was quite similar. For example, for Raw-AC, the optimal combination of temperature, pH of the solution, and contact time came out to be the same, which was 298.15 K, 6, and 1440 min, respectively. The only difference was noted in the cases of initial concentration of Pb and mass, which were 854.16 mg/L and 0.1 g for Raw-AC and 1012.98 mg/L and 0.05 g for AC-OH, respectively. It is quite interesting that even though a couple of variables were the same, the value of optimal q_e was quite different. For the combination of Raw-AC and AC-OH, the optimal q_e values came out to be 62.287 mg/g and 147.475 mg/g, respectively. It is remarkable how doing such a treatment can improve the results by 57.76%. It is also to be stressed here that the optimal point evaluated through the computational method was placed back into the experimental conditions and conformity between the optimal point through the computational method and the experimentation was met.

Table 13. Optimality of Raw-AC and AC-OH through conformity between the computational method and experimental procedure.

Biomass	Initial Concentration of Pb (mg/L)	Mass (g)	Temperature (K)	pH of the Solution	Contact Time (min)	Optimal q_e (mg/g)
Raw-AC	854.16	0.1	298.15	6	1440	62.287
AC-OH	1012.98	0.05	298.15	6	1440	147.475

3. Materials and Methods

3.1. Pb standard Solutions

Stock solutions (1000 mg/L) of Pb(II) were prepared by dissolving 0.337 g of $PbCl_2$ (99%, Sigma Aldrich, St. Louis, MO, USA) in deionized water. Solutions with 25 to 700 mg/L concentrations were prepared by diluting the stock solutions using deionized water. Pb(II) calibration curves were obtained by diluting 1000 mg/L standard solutions of Pb in 2% *v*/*v* HNO_3 (Fluka Analytical, St. Louis, MO, USA).

3.2. Biosorbents

Leaves of *A. compressa* K. were manually collected from the experimental garden at Universidad Tecnológica de Xicotepec de Juárez, located in Xicotepec de Juárez, Puebla (−97.96 20° 15′ 33″ N, 97° 57′ 36″ W). Fresh leaves of *A. compressa* K. were washed, disinfected, and air-dried. The dry material was ground in a blender until a homogeneous fine powder (particle size between 0.42 and 0.80 mm) was obtained. The powder was preserved until use at room temperature in hermetically sealed plastic bags in the absence of light and humidity.

Alkali treatment of raw *A. compressa* K. (Raw-AC) was done as follows: 10 g of pulverized Raw-AC was added to 250 mL of 1.0 M NaOH in an Erlenmeyer flask (500 mL) and shaken at 140 rpm for 2 h. Then, the sample was filtered under a vacuum and rinsed several times using deionized water until the wash solution pH reached approximately 6.5. The treated biomass labeled AC-OH was air-dried and stored in glass bottles. Figure 12 illustrates the preparation procedure of the biosorbents.

Figure 12. Schematic representation of the preparation of the biosorbents.

3.3. Physicochemical Analysis

Association of Analytical Communities (AOAC (2007)) methods were used for the chemical characterization of *A. compressa* leaves [65]: ash, moisture, raw fiber, and ethereal extract were found using the Soxhlet method; protein was found using the Kjeldahl method and nitrogen-free extract (NFE) was calculated using the percentage differences. The total soluble solids (TSSs) were identified using a digital refractometer (PR-101ATAGO PALETTE) according to the method previously described [65]. All determinations were done in triplicate.

3.4. Characterization Techniques

The pH_{PZC} was obtained as previously described [32]. Samples of Raw-AC and AC-OH, before and after contact with metal ions, were characterized using a Fourier transform infrared spectrometer (Nicolet Nexus 670) to identify the functional groups on the biomass surface. Surface characteristics of Raw-AC and AC-OH biosorbents before and after the uptake of Pb(II) were studied using scanning electron microscopy (JEOL JSM-IT300). An energy-dispersive X-ray spectrometer was used to determine the approximate chemical composition on the surface of Raw-AC and AC-OH biosorbents.

3.5. Biosorption Study

In this work, the biosorption process was carried out using a batch technique. For the kinetic study, 0.1 g of Raw-AC and AC-OH was mixed with 10 mL of Pb(II) solutions at initial concentrations from 25 mg/L to 250 mg/L. The samples were shaken at 140 rpm in an orbital shaker (CRP-0228, Scientific) for times ranging from 15 min to 1440 min. After each contact time, the samples were centrifuged (CRM Globe, Certificient, Chicago, IL, USA) for 5 min, and the Pb(II) contents in the aqueous solutions were determined using flame atomic absorption spectrometry (iCe Serie TermoScientific).

After each contact time, the amount of Pb(II) adsorbed on the surface of Raw-AC and AC-OH was calculated using Equation (1):

$$q_t = \frac{C_i - C_t}{m} \cdot V \quad (1)$$

where C_i and C_t are, respectively, the initial and final concentrations of Pb(II) (mg/L); V (L) is the volume of solution; and m (g) is the mass of biosorbents.

In this study, the effects of the solution pH, dose of the biosorbents, and temperature were carried out by varying the initial lead concentration from 25 to 1000 mg/L. For the kinetic study, the temperature and solution pH were fixed, whereas the initial lead concentration was varied from 25 to 250 mg/L and the contact time was changed from 5 to 1440 min.

Table 14 gives the experimental conditions used to investigate the effects of contact times, solution pH, temperature, and dose of biosorbents on the removal of Pb(II).

Table 14. Values of contact times, solution pH, temperature, and adsorbent dose parameters used for the experimental process.

Experiments	Experimental Parameters				
	pH	**Time (min)**	**Ci (mg/L)**	**Dose (g/L)**	**Temperature (K)**
Solution pH	2–6	1440	25–1000	10	300.15
Dose	6	1440	25–1000	5–20	300.15
Contact times	6	5–1440	25–250	10	300.15
Temperature	6	1440	25–1000	10	300.15–330.15

All experiments were performed in duplicate and the results were expressed as averaged values. The Supplementary Materials provide the kinetic and isotherm models, as well as the error functions analysis used in this study.

3.6. Kinetic Models

This study employed the nonlinear equations of pseudo-first- and pseudo-second-order models (Equation (2) and Equation (3), respectively) to investigate the kinetics of Pb(II) biosorption onto Raw-AC and AC-OH:

$$q_t = q_e(1 - \exp(-k_1 t)) \quad (2)$$

$$q_t = \frac{k_2 {q_e}^2 t}{1 + k_2 q_e t} \quad (3)$$

where q_e is the theoretical adsorption capacity (mg/g), and k_1 (1/min) and k_2 (g/mg min) are the pseudo-first- and pseudo-second-order rate constants, respectively.

3.7. Isotherm Models

Experimental data were examined using Langmuir, Freundlich, and Dubinin-Radushkevich (D-R) isotherm models.

Equation (4) describes the non-linear form of the Langmuir isotherm model:

$$q_e = \frac{q_m K_L C_e}{1 + K_L C_e} \tag{4}$$

where q_m (mg/g) is the maximum sorption capacity, and K_L (L/mg) is the Langmuir equilibrium constant, which indicates the affinity of the sorbate for the solute [55].

The Langmuir model assumes that adsorption takes place for monolayer adsorption of the adsorbate from a liquid solution on a surface of biosorbent containing a finite number of identical sites and all adsorption sites have equal adsorption energies [55,66].

The Freundlich model is an empirical isotherm that can be used for non-ideal adsorption, which is mostly used to understand the adsorption of metal ions on a heterogeneous surface with multilayer adsorption. It can also define an exponential distribution of active sites and their energy [54]. The non-linear form of the Freundlich isotherm model is expressed as Equation (5):

$$q_e = K_F {C_e}^{1/n} \tag{5}$$

where K_F (mg/g)(L/mg)$^{1/n}$ and n are the Freundlich constants associated with the adsorption capacity and adsorption intensity of the adsorbent, respectively. C_e (mg/L) is the adsorbate equilibrium concentration. When $n = 1$, the isotherm is linear and indicates that all sites on the adsorbent have equal affinity for the adsorbates. Values of $n > 1$ indicate the affinities decreased with increasing adsorption density [54]. The greater the value of n, the more favorable the adsorption and the more heterogeneous the surface of the particles will be [67].

The D-R isotherm is an empirical model that is generally employed to find whether the adsorption involved in the experimental data is physical or chemical [68]. The non-linear form of the D-R isotherm model is shown in the following Equation (6):

$$q_e = q_m e^{-\beta \epsilon^2} \tag{6}$$

where β (mol^2/kJ^2) is a constant related to adsorption energy, $\varepsilon = RTln\left(1 + \frac{1}{C_e}\right)$ is the adsorption potential (kJ mol^{-1}), R (8.314 J mol^{-1} K^{-1}) is the universal gas constant, T (K) is the absolute temperature, and $E = \frac{1}{\sqrt{2\beta}}$ is the adsorption energy (kJ/mol).

The numerical value from the free energy of the molecule (E) gives an indication of the nature of the interaction forces between lead ions and the active sites on the composite surface. When the value of the adsorption energy was less than 8 kJ/mol, it indicates that the adsorption occurred physically; if it was between 8 and 16 kJ/mol, it suggests that the adsorption process was carried out chemically, driven by the ion exchange mechanism; and chemisorption was observed when E values are between 20 and 40 kJ/mol [69].

3.8. Adsorption–Desorption Cycles

Experiments of adsorption and desorption cycles were conducted: 10 g/L of Raw-AC and AC-OH were mixed with 1000 mg/L of Pb(II) for 1440 min at room temperature and solution pH at 6. The samples were filtered, and the adsorption efficiency was obtained utilizing Equation (7):

$$Adsorption\ efficiency\ \% = \frac{C_{ads}}{C_{initial}} \times 100 \tag{7}$$

After the adsorption process, the biosorbents were dried and weighed; then, the desorption experiments were conducted by mixing the biosorbents at ambient temperature for 2 h with 0.5 M HCl. The Pb(II) concentrations in each solution were measured, and the desorption efficiency was obtained utilizing Equation (8):

$$Desorption\ efficiency\ \% = \frac{C_{des}}{C_{ads}} \times 100 \quad (8)$$

where C_{des} is the concentration of Pb(II) desorbed in mg/g and C_{ads} is the concentration of Pb(II) adsorbed in mg/g.

After desorption, the samples were rinsed several times with deionized water to remove the eluent agent from the biosorbents. The samples were then dried and weighed for further reuse. In this study, three adsorption and desorption cycles were repeated.

3.9. *Framework of Empirical Model and Optimization Process*

Owing to the complexity of the involvement of so many tuning variables, it is not easy to determine the optimal set conditions for this problem. For this purpose, an artificial intelligence method consisting of the artificial neural network was adapted for the empirical modeling, and later on, a genetic algorithm was applied for optimization purposes, which would eventually give out the best combination of all the multivariate tuning variables [70].

4. Conclusions

In conclusion, this study demonstrates that *Ardisia compressa* K. leaves, particularly those treated with a solution of 1M NaOH (AC-OH), exhibit a higher sorption capacity for Pb(II) from aqueous solutions compared with natural *Ardisia compressa* K. leaves (Raw-AC). The surface modification of the biosorbent after the alkali treatment was confirmed via pH_{PZC} and FTIR analysis, which revealed the presence of carboxyl, hydroxyl, and amino groups responsible for Pb(II) binding.

The removal of Pb(II) was found to be dependent on various factors, including solution pH, contact time, biosorbent dose, and water temperature. The optimum experimental conditions for maximum Pb(II) removal by both biosorbents were at pH 6 and T = 300.15 K.

The kinetic data analysis indicated that the pseudo-second-order kinetic model best described Pb(II) biosorption on both Raw-AC and AC-OH, which suggests that chemisorption controlled the rate of reaction. The Langmuir isotherm model described the adsorption isotherms for both biosorbents at different solution pH values well, suggesting monolayer sorption on a homogeneous surface took place. Furthermore, the study introduced the application of artificial neural networks and genetic algorithms for regression modeling and optimization purposes.

The model showed high reliability, and the optimization study revealed a significant improvement of around 57.76% in the indicator "q_e" for AC-OH. The optimal conditions for AC-OH were determined as follows: initial concentration of Pb: 1012.98 mg/L, mass: 0.05 g, temperature: 298.15 K, pH of the solution: 6, and contact time: 1440 min, resulting in an optimal q_e of 147.475 mg/g. In summary, the results indicate that *Ardisia compressa* K. leaves can serve as a promising, cost-effective, and efficient biosorbent for removing Pb(II) ions from aqueous solutions. The study also highlights the potential of artificial neural networks and genetic algorithms for optimizing biosorption processes. These findings contribute to the development of environmentally friendly and sustainable methods for heavy metal removal from water sources.

A significant finding of this study pertains to the coefficient of determination, which indicated that the Dubinin–Radushkevich isotherm model exhibited lower values compared with the Langmuir and Freundlich, suggesting a need for further investigation into the applicability and parameter-based assessment of the Dubinin–Radushkevich model to elucidate sorption mechanisms.

The insights from this research hold valuable implications for engineering education, as it showcases how modifying biosorbents, optimizing process conditions, employing kinetic and isotherm models, and utilizing advanced techniques like artificial neural networks and genetic algorithms can be integrated into engineering curricula to foster a comprehensive understanding of designing effective and eco-friendly solutions for heavy metal removal in water treatment systems. The social insights derived from this research underscore the potential impact of innovative engineering approaches on addressing environmental challenges, demonstrating the significance of interdisciplinary collaboration between engineering and environmental sciences to develop sustainable solutions for water pollution, thereby imparting the importance of considering broader societal and ecological implications of technological advancements in engineering education.

Supplementary Materials: The following supporting information can be downloaded from https://www.mdpi.com/article/10.3390/molecules28176387/s1. Error functions analysis.

Author Contributions: A.Y.V.-S., M.A. and A.A.S. conceived and planned the experiments; A.A.S., C.A. and I.A. contributed to the characterization of the samples. M.A., R.T. and A.R.V.-O. wrote the manuscript with input from all authors. E.C.L., M.A. and I.A. contributed to the interpretation of the results. All authors provided critical feedback and helped to shape the research, analysis, and manuscript. All authors have read and agreed to the published version of the manuscript.

Funding: This research received no external funding.

Institutional Review Board Statement: Not applicable.

Informed Consent Statement: Not applicable.

Data Availability Statement: The data presented in this study are available on request from the corresponding author.

Acknowledgments: The authors are grateful to Adriana Tejeda Cruz and Eliezer Hernández Me-cinas for their assistance in the X-ray diffraction and FTIR measurements.

Conflicts of Interest: The authors declare no conflict of interest.

Sample Availability: Not applicable.

References

1. Kinuthia, G.K.; Ngure, V.; Beti, D.; Lugalia, R.; Wangila, A.; Kamau, L. Levels of heavy metals in wastewater and soil samples from open drainage channels in Nairobi, Kenya: Community health implication. *Sci. Rep.* **2020**, *10*, 8434. [CrossRef]
2. Ben Omar, M.; Mendiguchía, C.; Er-Raioui, H.; Marhraoui, M.; Lafraoui, G.; Oulad-Abdellah, M.K.; García-Vargas, M.; Moreno, C. Distribution of heavy metals in marine sediments of Tetouan coast (North of Morocco): Natural and anthropogenic sources. *Environ. Earth Sci.* **2015**, *74*, 4171–4185. [CrossRef]
3. Mendiguchía, C.; Moreno, C.; García-Vargas, M. Evaluation of natural and anthropogenic influences on the Guadalquivir River (Spain) by dissolved heavy metals and nutrients. *Chemosphere* **2007**, *69*, 1509–1517. [CrossRef] [PubMed]
4. Sandeep, G.; Vijayalatha, K.R.; Anitha, T. Heavy metals and its impact in vegetable crops. *Int. J. Chem. Stud.* **2019**, *7*, 1612–1621.
5. Vajargah, M.F. A Review on the Effects of Heavy Metals on Aquatic Animals. *J. Biomed. Res. Environ. Sci.* **2021**, *2*, 865–869. [CrossRef]
6. Munir, N.; Jahangeer, M.; Bouyahya, A.; El Omari, N.; Ghchime, R.; Balahbib, A.; Aboulaghras, S.; Mahmood, Z.; Akram, M.; Shah, S.M.A.; et al. Heavy metal contamination of natural foods is a serious health issue: A review. *Sustainability* **2022**, *14*, 161. [CrossRef]
7. Vievard, J.; Alem, A.; Pantet, A.; Ahfir, N.-D.; Arellano-Sánchez, M.G.; Devouge-Boyer, C.; Mignot, M. Bio-Based Adsorption as Ecofriendly Method for Wastewater Decontamination: A Review. *Toxics* **2023**, *11*, 404. [CrossRef]
8. Rajendran, S.; Priya, A.K.; Kumar, P.S.; Hoang, T.K.; Sekar, K.; Chong, K.Y.; Khoo, K.S.; Ng, H.S.; Show, P.L. A critical and recent developments on adsorption technique for removal of heavy metals from wastewater—A review. *Chemosphere* **2022**, *303*, 135146. [CrossRef]
9. Pandey, P.K.; Sharma, S.K.; Sambi, S.S. Removal of lead(II) from waste water on zeolite-NaX. *J. Environ. Chem. Eng.* **2015**, *3*, 2604–2610. [CrossRef]
10. Uddin, M.K. A review on the adsorption of heavy metals by clay minerals, with special focus on the past decade. *Chem. Eng. J.* **2017**, *308*, 438–462. [CrossRef]
11. Nordin, N.A.; Rahman, N.A.; Abdullah, A.H. Effective Removal of Pb(II) Ions by Electrospun PAN/Sago Lignin-Based Activated Carbon Nanofibers. *Molecules* **2020**, *25*, 3081. [CrossRef] [PubMed]
12. Nguyen, T.; Ngo, H.; Guo, W.; Zhang, J.; Liang, S.; Yue, Q.; Li, Q. Applicability of agricultural waste and by-products for adsorptive removal of heavy metals from wastewater. *Bioresour. Technol.* **2013**, *148*, 574–585. [CrossRef]
13. Ge, Y.; Qin, L.; Li, Z. Lignin microspheres: An effective and recyclable natural polymer-based adsorbent for lead ion removal. *Mater. Des.* **2016**, *95*, 141–147. [CrossRef]

14. Chakraborty, R.; Asthana, A.; Singh, A.K.; Jain, B.; Susan, A.B.H. Adsorption of heavy metal ions by various low-cost adsorbents: A review. *Int. J. Environ. Anal. Chem.* **2022**, *102*, 342–379. [CrossRef]
15. Hegazi, H.A. Removal of heavy metals from wastewater using agricultural and industrial wastes as adsorbents. *HBRC J.* **2013**, *9*, 276–282. [CrossRef]
16. Joseph, L.; Jun, B.M.; Flora, J.R.V.; Park, C.M.; Yoon, Y. Removal of heavy metals from water sources in the developing world using low-cost materials: A review. *Chemosphere* **2019**, *229*, 142–159. [CrossRef]
17. Demirbas, A. Heavy metal adsorption onto agro-based waste materials: A review. *J. Hazard. Mater.* **2008**, *157*, 220–229. [CrossRef]
18. Elkhaleefa, A.; Ali, I.H.; Brima, E.I.; Shigidi, I.; Elhag, A.B.; Karama, B. Evaluation of the adsorption efficiency on the removal of lead(II) ions from aqueous solutions using *Azadirachta indica* leaves as an adsorbent. *Processes* **2021**, *9*, 559. [CrossRef]
19. Jayasree, R.; Kumar, P.S.; Saravanan, A.; Hemavathy, R.; Yaashikaa, P.; Arthi, P.; Shreshta, J.; Jeevanantham, S.; Karishma, S.; Arasu, M.V.; et al. Sequestration of toxic Pb(II) ions using ultrasonic modified agro waste: Adsorption mechanism and modelling study. *Chemosphere* **2021**, *285*, 131502. [CrossRef]
20. Zhang, Z.; Wang, T.; Zhang, H.; Liu, Y.; Xing, B. Adsorption of Pb(II) and Cd(II) by magnetic activated carbon and its mechanism. *Sci. Total Environ.* **2021**, *757*, 143910. [CrossRef]
21. Saleh Al-Hashemi, Z.S.; Hossain, M.A. Biological activities of different neem leaf crude extracts used locally in Ayurvedic medicine. *Pac. Sci. Rev. A Nat. Sci. Eng.* **2016**, *18*, 128–131. [CrossRef]
22. Tran, H.N.; Chao, H.P. Adsorption and desorption of potentially toxic metals on modified biosorbents through new green grafting process. *Environ. Sci. Pollut. Res.* **2018**, *25*, 12808–12820. [CrossRef] [PubMed]
23. Joaquín-Cruz, E.; Dueñas, M.; García-Cruz, L.; Salinas-Moreno, Y.; Santos-Buelga, C.; García-Salinas, C. Anthocyanin and phenolic characterization, chemical composition and antioxidant activity of chagalapoli (*Ardisia compressa* K.) fruit: A tropical source of natural pigments. *Food Res. Int.* **2015**, *70*, 151–157. [CrossRef]
24. Vázquez-Sánchez, A.Y.; Corfield, R.; Sosa, N.; Salvatori, D.; Schebor, C. Physicochemical, functional, and sensory characterization of apple leathers enriched with acáchul (*Ardisia compressa* Kunth) powder. *LWT* **2021**, *146*, 111472. [CrossRef]
25. Chandra, S.; De Mejia, E.G. Polyphenolic compounds, antioxidant capacity, and quinone reductase activity of an aqueous extract of *Ardisia compressa* in comparison to mate (*Ilex paraguariensis*) and green (*Camellia sinensis*) teas. *J. Agric. Food Chem.* **2004**, *52*, 3583–3589. [CrossRef] [PubMed]
26. Thamaga, M.W.; Mokoboki, H.K.; Sebola, N.A.; Ravhuhali, K.E. Apparent digestibility and nutritional composition of *Leucaena leucocephala* (Lam) leaf meal incorporated in the diets of Black Australorp and Potchefstroom Koekoek chicken breeds. *Trop. Anim. Health Prod.* **2021**, *53*, 458. [CrossRef] [PubMed]
27. Garcia, G.W.; Ferguson, T.U.; Neckles, F.A.; Archibald, K.A.E. The nutritive value and forage productivity of *Leucaena leucocephala*. *Anim. Feed. Sci. Technol.* **1996**, *60*, 29–41. [CrossRef]
28. Lavado-Meza, C.; De la Cruz-Cerrón, L.; Lavado-Puente, C.; Angeles-Suazo, J.; Dávalos-Prado, J.Z. Efficient Lead Pb(II) Removal with Chemically Modified Nostoc commune Biomass. *Molecules* **2023**, *28*, 268. [CrossRef] [PubMed]
29. Ayssiwede, S.B.; Dieng, A.; Chrysostome, C.; Ossebi, W.; Hornick, J.L.; Missohou, A. Digestibility, metabolic utilization and nutritional value of *Leuceana leucocephala* (Lam.) leaf meal incorporated in indigenous Senegal chickens diets. *Adv. Anim. Biosci.* **2010**, *1*, 425–426. [CrossRef]
30. Torres-Maza, A.; Yupanqui-Bacilio, C.; Castro, V.; Aguirre, E.; Villanueva, E.; Rodríguez, G. Comparison of the hydrocolloids *Nostoc commune* and *Nostoc sphaericum*: Drying, spectroscopy, rheology and application in nectar. *Sci. Agropecu.* **2020**, *11*, 583–589. [CrossRef]
31. Šoštarić, T.D.; Petrović, M.S.; Pastor, F.T.; Lončarević, D.R.; Petrović, J.T.; Milojković, J.V.; Stojanović, M.D. Study of heavy metals biosorption on native and alkali-treated apricot shells and its application in wastewater treatment. *J. Mol. Liq.* **2018**, *259*, 340–349. [CrossRef]
32. Abatal, M.; Lima, E.C.; Anastopoulos, I.; Giannakoudakis, D.A.; Vargas, J.; Aguilar, C.; Olguín, M.; Anguebes-Fransechi, F. Effect of alkali treatment on the removal of Co(II) ions by *Leucaena leucephala* biomass. *J. Mol. Liq.* **2022**, *367*, 120419. [CrossRef]
33. Fernandez, M.E.; Nunell, G.V.; Bonelli, P.R.; Cukierman, A.L. Batch and dynamic biosorption of basic dyes from binary solutions by alkaline-treated cypress cone chips. *Bioresour. Technol.* **2012**, *106*, 55–62. [CrossRef] [PubMed]
34. Farnane, M.; Tounsadi, H.; Elmoubarki, R.; Mahjoubi, F.; Elhalil, A.; Saqrane, S.; Abdennouri, M.; Qourzal, S.; Barka, N. Alkaline treated carob shells as sustainable biosorbent for clean recovery of heavy metals: Kinetics, equilibrium, ions interference and process optimisation. *Ecol. Eng.* **2017**, *101*, 9–20. [CrossRef]
35. Bayramoğlu, G.; Çelik, G.; Arica, M.Y. Biosorption of Reactive Blue 4 dye by native and treated fungus *Phanerocheate chrysosporium*: Batch and continuous flow system studies. *J. Hazard. Mater.* **2006**, *137*, 1689–1697. [CrossRef] [PubMed]
36. Chen, G.Q.; Zeng, G.M.; Tu, X.; Niu, C.G.; Huang, G.H.; Jiang, W. Application of a by-product of Lentinus edodes to the bioremediation of chromate contaminated water. *J. Hazard. Mater.* **2006**, *135*, 249–255. [CrossRef] [PubMed]
37. Rezaei, M.; Pourang, N.; Moradi, A.M. Removal of lead from aqueous solutions using three biosorbents of aquatic origin with the emphasis on the affective factors. *Sci. Rep.* **2022**, *12*, 751. [CrossRef] [PubMed]
38. Bhattacharya, P.; Mallick, K.; Ghosh, S.; Banerjee, P.; Mukhopadhyay, A.; Bandyopadhyay, S. Algal biomass as potential biosorbent for reduction of organic load in gray water and subsequent reuse: Effect on seed germination and enzyme activity. *Bioremediat. J.* **2014**, *18*, 56–70. [CrossRef]

39. Faghihian, H.; Peyvandi, S. Adsorption isotherm for uranyl biosorption by *Saccharomyces cerevisiae* biomass. *J. Radioanal. Nucl. Chem.* **2012**, *293*, 463–468. [CrossRef]
40. Aksu, Z.; Açıkel, Ü.; Kabasakal, E.; Tezer, S. Equilibrium modelling of individual and simultaneous biosorption of chromium(VI) and nickel(II) onto dried activated sludge. *Water Res.* **2002**, *36*, 3063–3073. [CrossRef]
41. Huang, Y.; Li, S.; Chen, J.; Zhang, X.; Chen, Y. Adsorption of Pb(II) on mesoporous activated carbons fabricated from water hyacinth using H_3PO_4 activation: Adsorption capacity, kinetic and isotherm studies. *Appl. Surf. Sci.* **2014**, *293*, 160–168. [CrossRef]
42. Reddy, D.H.K.; Seshaiah, K.; Reddy, A.; Rao, M.M.; Wang, M.C. Biosorption of Pb2+ from aqueous solutions by *Moringa oleifera* bark: Equilibrium and kinetic studies. *J. Hazard. Mater.* **2010**, *174*, 831–838. [CrossRef] [PubMed]
43. Yuvaraja, G.; Krishnaiah, N.; Subbaiah, M.V.; Krishnaiah, A. Biosorption of Pb(II) from aqueous solution by *Solanum melongena* leaf powder as a low-cost biosorbent prepared from agricultural waste. *Colloids Surf. B Biointerfaces* **2014**, *114*, 75–81. [CrossRef] [PubMed]
44. Patel, R.; Suresh, S. Kinetic and equilibrium studies on the biosorption of reactive black 5 dye by *Aspergillus foetidus*. *Bioresour. Technol.* **2008**, *99*, 51–58. [CrossRef] [PubMed]
45. Bhattacharyya, K.G.; Gupta, S.S. Adsorptive accumulation of Cd(II), Co(II), Cu(II), Pb(II), and Ni(II) from water on montmorillonite: Influence of acid activation. *J. Colloid. Interface Sci.* **2007**, *310*, 411–424. [CrossRef] [PubMed]
46. Alsohaimi, I.H.; Wabaidur, S.M.; Kumar, M.; Khan, M.A.; Alothman, Z.A.; Abdalla, M.A. Synthesis, characterization of PMDA/TMSPEDA hybrid nano-composite and its applications as an adsorbent for the removal of bivalent heavy metals ions. *Chem. Eng. J.* **2015**, *270*, 9–21. [CrossRef]
47. Bhattacharyya, K.G.; Sharma, A. Adsorption of Pb(II) from aqueous solution by *Azadirachta indica* (Neem) leaf powder. *J. Hazard. Mater.* **2004**, *113*, 97–109. [CrossRef] [PubMed]
48. Kumar, K.V.; Sivanesan, S. Prediction of optimum sorption isotherm: Comparison of linear and non-linear method. *J. Hazard. Mater.* **2005**, *126*, 198–201. [CrossRef]
49. Mondal, M.K. Removal of Pb(II) from aqueous solution by adsorption using activated tea waste. *Korean J. Chem. Eng.* **2010**, *27*, 144–151. [CrossRef]
50. Iddou, A.; Hadj Youcef, M.; Aziz, A.; Ouali, M.S. Biosorptive removal of lead (II) ions from aqueous solutions using *Cystoseira stricta* biomass: Study of the surface modification effect. *J. Saudi Chem. Soc.* **2011**, *15*, 83–88. [CrossRef]
51. Cimá-Mukul, C.A.; Abdellaoui, Y.; Abatal, M.; Vargas, J.; Santiago, A.A.; Barrón-Zambrano, J.A. Eco-Efficient Biosorbent Based on *Leucaena leucocephala* Residues for the Simultaneous Removal of Pb (II) and Cd (II) Ions from Water System: Sorption and Mechanism. *Bioinorg. Chem. Appl.* **2019**, *2019*, 2814047. [CrossRef]
52. Keskinkan, O.; Goksu, M.Z.L.; Basibuyuk, M.; Forster, C.F. Heavy metal adsorption properties of a submerged aquatic plant (*Ceratophyllum demersum*). *Bioresour. Technol.* **2004**, *92*, 197–200. [CrossRef] [PubMed]
53. Salehi, P.; Asghari, B.; Mohammadi, F. Biosorption of Ni(II), Cu(II) and Pb(II) by *Punica geranatum* from Aqueous Solutions. *J. Water Resour. Prot.* **2010**, *2*, 701–705. [CrossRef]
54. Begum, S.A.S.; Tharakeswar, Y.; Kalyan, Y.; Naidu, G.R. Biosorption of Cd (II), Cr (VI) & Pb (II) from Aqueous Solution Using *Mirabilis jalapa* as Adsorbent. *J. Encapsulation Adsorpt. Sci.* **2015**, *5*, 93–104. [CrossRef]
55. Gümüş, D. Biosorptive application of defatted *Laurus nobilis* leaves as a waste material for treatment of water contaminated with heavy metal. *Int. J. Phytoremediation* **2019**, *21*, 556–563. [CrossRef]
56. Deng, L.; Su, Y.; Su, H.; Wang, X.; Zhu, X. Sorption and desorption of lead (II) from wastewater by green algae *Cladophora fascicularis*. *J. Hazard. Mater.* **2007**, *143*, 220–225. [CrossRef] [PubMed]
57. Khoso, W.A.; Haleem, N.; Baig, M.A.; Jamal, Y. Synthesis, characterization and heavy metal removal efficiency of nickel ferrite nanoparticles (NFN's). *Sci Rep.* **2021**, *11*, 3790. [CrossRef]
58. Gusain, R.; Kumar, N.; Fosso-Kankeu, E.; Ray, S.S. Efficient Removal of Pb(II) and Cd(II) from Industrial Mine Water by a Hierarchical MoS2/SH-MWCNT Nanocomposite. *ACS Omega* **2019**, *4*, 13922–13935. [CrossRef] [PubMed]
59. Sepehri, S.; Kanani, E.; Abdoli, S.; Rajput, V.D.; Minkina, T.; Lajayer, B.A. Pb(II) Removal from Aqueous Solutions by Adsorption onStabilized Zero-Valent Iron Nanoparticles—A Green Approach. *Water* **2023**, *15*, 222. [CrossRef]
60. Altynbaeva, L.S.; Mashentseva, A.A.; Aimanova, N.A.; Zheltov, D.A.; Shlimas, D.I.; Nurpeisova, D.T.; Barsbay, M.; Abuova, F.U.; Zdorovets, M.V. Eco-Friendly Electroless Template Synthesis of Cu-Based Composite Track-Etched Membranes for Sorption Removal of Lead(II) Ions. *Membranes* **2023**, *13*, 495. [CrossRef] [PubMed]
61. Tariq, R.; Ali, M.; Sheikh, N.A.; Shahzad, M.W.; Xu, B.B. Deep learning artificial intelligence framework for sustainable desiccant air conditioning system: Optimization towards reduction in water footprints. *Int. Commun. Heat Mass Transf.* **2023**, *140*, 106538. [CrossRef]
62. Tariq, R.; Torres-Aguilar, C.E.; Xamán, J.; Zavala-Guillén, I.; Bassam, A.; Ricalde, L.J.; Carvente, O. Digital twin models for optimization and global projection of building-integrated solar chimney. *Build. Environ.* **2022**, *213*, 108807. [CrossRef]
63. Tariq, R.; Cetina-Quiñones, A.J.; Cardoso-Fernández, V.; Daniela-Abigail, H.L.; Soberanis, M.A.E.; Bassam, A.; De Lille, M.V. Artificial intelligence assisted technoeconomic optimization scenarios of hybrid energy systems for water management of an isolated community. *Sustain. Energy Technol. Assess.* **2021**, *48*, 101561. [CrossRef]
64. Tariq, R.; Abatal, M.; Bassam, A. Computational intelligence for empirical modeling and optimization of methylene blue adsorption phenomena using available local zeolites and clay of Morocco. *J. Clean. Prod.* **2022**, *370*, 133517. [CrossRef]

65. Vicente-Flores, M.; Güemes-Vera, N.; Chanona-Pérez, J.J.; Perea-Flores, M.d.J.; Arzate-Vázquez, I.; Quintero-Lira, A.; Sánchez-Fuentes, C.E. Study of cellular architecture and micromechanical properties of cuajilote fruits (*Parmentiera edulis* D.C.) using different microscopy techniques. *Microsc. Res. Tech.* **2021**, *84*, 12–27. [CrossRef] [PubMed]
66. Pavasant, P.; Apiratikul, R.; Sungkhum, V.; Suthiparinyanont, P.; Wattanachira, S.; Marhaba, T.F. Biosorption of Cu^{2+}, Cd^{2+}, Pb^{2+}, and Zn^{2+} using dried marine green macroalga *Caulerpa lentillifera*. *Bioresour. Technol.* **2006**, *97*, 2321–2329. [CrossRef]
67. Imran, M.; Anwar, K.; Akram, M.; Shah, G.M.; Ahmad, I.; Samad Shah, N.; Khan, Z.U.H.; Rashid, M.I.; Akhtar, M.N.; Ahmad, S.; et al. Biosorption of Pb (II) from contaminated water onto *Moringa oleifera* biomass: Kinetics and equilibrium studies. *Int. J. Phytoremediation* **2019**, *21*, 777–789. [CrossRef] [PubMed]
68. Mashentseva, A.A.; Aimanova, N.A.; Parmanbek, N.; Temirgaziyev, B.S.; Barsbay, M.; Zdorovets, M.V. *Serratula coronata* L. Mediated Synthesis of ZnO Nanoparticles and Their Application for the Removal of Alizarin Yellow R by Photocatalytic Degradation and Adsorption. *Nanomaterials* **2022**, *12*, 3293. [CrossRef] [PubMed]
69. Parmanbek, N.; Sütekin, D.S.; Barsbay, M.; Mashentseva, A.A. Hybrid PET Track-Etched Membranes Grafted by Well-Defined Poly(2-(dimethylamino)ethyl methacrylate) Brushes and Loaded with Silver Nanoparticles for the Removal of As(III). *Polymers* **2022**, *14*, 4026. [CrossRef] [PubMed]
70. Khanmohammadi, S.; Musharavati, F.; Tariq, R. A framework of data modeling and artificial intelligence for environmental-friendly energy system: Application of Kalina cycle improved with fuel cell and thermoelectric module. *Process. Saf. Environ. Prot.* **2022**, *164*, 499–516. [CrossRef]

molecules

MDPI

Article

Utilizing Black Soldier Fly Larvae to Improve Bioconversion and Reduce Pollution: A Sustainable Method for Efficient Treatment of Mixed Wastes of Wet Distiller Grains and Livestock Manure

Tao Li [1,2], Samiullah Khan [1,2], Mao Wei [1,2], Haiyin Li [1,2], Tingchi Wen [3], Jianjun Guo [1,2,*] and Daochao Jin [1,2,*]

1 Institute of Entomology, Guizhou University, Guiyang 550025, China; litaojzx@126.com (T.L.); samiullahakbar4@gmail.com (S.K.); weim111@163.com (M.W.); hyli3@gzu.edu.cn (H.L.)

2 Scientific Observing and Experimental Station of Crop Pest in Guiyang, Ministry of Agriculture and Rural Affairs, Guiyang 550025, China

3 State Key Laboratory Breeding Base of Green Pesticide and Agricultural Bioengineering, Key Laboratory of Green Pesticide and Agricultural Bioengineering, Ministry of Education, Guizhou University, Guiyang 550025, China; tingchiwen@yahoo.com

* Correspondence: author: jjguo@gzu.edu.cn (J.G.); dcjin@gzu.edu.cn (D.J.)

Abstract: Widespread environmental contamination caused by huge amounts of wastes generated by human activities has become a critical global concern that requires urgent action. The black soldier fly (BSFL) has gradually been used to treat different wastes due to high efficiency and low cost. However, little information is available regarding the treatment of mixed wastes by BSFLs. The impact of BSFLs on conversion of cow manure (COM) and pig manure (PM) via the incorporation of wet distiller grains (WDG) was assessed. Results demonstrate that the waste reduction rate was increased by 20% by incorporating 45% WDG to COM and PM. The bioconversion rate of BSFLs in COM and PM also increased from 1.20 ± 0.02% and 0.92 ± 0.02% to 10.54 ± 0.06% and 10.05 ± 0.11%, respectively. Total nitrogen content and $\delta^{15}N/^{14}N$ ratios of WDG + COM and WDG + PM were found to be significantly lower than those of COM and PM alone ($p < 0.01$). The organic matter changes during manure degradation were further analyzed by combing ultraviolet–visible spectrum (UV–vis) with excitation–emission matrix (EEM) spectroscopy techniques and fluorescence area integration (FRI) method. The UV–vis spectra results indicate that the addition of WDG to manures resulted in the decreased aromaticity and molecular weight of the waste. EEM spectra demonstrated that the accumulative $P_{i,n}$ values of regions III and V in COM, COM + WDG, PM, and PM + WDG were 58%, 49%, 52% and 63%, respectively. These results not only provide new insights into the potential of mixed wastes for BSFL treatment but also contribute to the basis for the formulation of effective management measurements that reduce and/or reuse these wastes.

Keywords: sustainable waste management; black soldier fly; livestock manure; wet distiller grains; fluorescence spectra; humic acid

1. Introduction

Globally, wastes induced by intense animal husbandry have developed to be a serious environmental issue. Increasing wastes derived from the livestock sector, in particular, continues to cause a serious burden for the environment in several developing countries [1]. For example, developing countries generated 2009 million tons of wastes in 2016, representing 64% of the world's waste; furthermore, the amounts of wastes generated appeared to have higher increased trends in developing countries than those in developed countries [2]. Approximately 587 billion tons of manure were produced by an estimation of the global livestock annually, which contains active organic matter, heavy metals, antibiotic residues, and pathogens, all of which pose a threat to the environment and public health [3–5]. It is noteworthy that animal manures contributes up to 40% of the global anthropogenic ammonia (NH_3) and nitrous oxide (N_2O) emissions [6,7]. If the hazardous substances in animal manures are not effectively treated, they will severely impact the environment and human health. To improve manure management, 59 million tons of manure was intraregionally transported between farms in Germany in 2015~2016 [8]. However, this approach is uneconomical, requiring lots of manpower and material. Subsequently, several results have demonstrated that composting and/or anaerobic fermentation could effectively treat livestock manures [9,10]. However, the development and application of these technologies have been hindered by various challenges, such as lengthy implementation periods, high costs, low profit, and the production of undesirable gases, particularly NH_3 and N_2O [11,12]. Hence, it is urgently needed to develop a more effective and ecologically acceptable method for treating animal manures.

Previous studies have demonstrated that the BSFLs are a sustainable and cost-effective option for the degradation of manures due to their short period of growth and reproduction, enormous biomass, and diverse diet [12–14]. BSFLs could decompose the carboxylic, alcohol, and aliphatic components of organic matter, and the carboxylic substances in wastes after black soldier fly treatment increased compared with those before treatment [12,15,16]. Further, comparing the organic matter before and after treatment, the organic matter after treatment obviously decomposed and converted into humic-like substances [17]. In addition, the residues obtained from the bioconversion process of manures could potentially be utilized as fertilizers in agriculture due to their superior humification levels compared with fresh manure, which could have significant benefits for soil health and fertility. Studies have shown that humic substance rates in manures treated with BSFL were significantly higher (42.45–63.74%) than that of raw manures and natural composting manures (38.78–50.31%) [12,18]. Furthermore, BSFLs are also an economic insect, whose larvae can fully mature in 2–3 weeks; compared with the price of fishmeal, each ton of dried BSFL is cheaper, approximately 5800–7800 yuan (data from The China Feed Industry Information Website) [19,20]. Previous studies have reported the low reduction rates of 56% and 12.70% for POM and COM treated by BSFL, respectively [21,22]. Furthermore, subsequent studies also reported the relatively low reduction rates of dewatered sludge and abattoir wastes after being treated by BSFL, approximately 34.93% and 46.30%, respectively [23,24]. However, these studies were limited in scope, focusing mainly on single waste, and they did not account for the potential effects of mixed wastes. Therefore, our study aimed to explore the potential of BSFL bioconversion for mixed livestock manures, such as those containing WDG, which is currently not available.

Furthermore, varied dissolved organic matter (DOM) and nitrogen losses (e.g., the emissions of NH_3 and N_2O) caused by denitrification are the most concerned during the biodegradation processes of these manures performed by BSFLs. One is the presence of DOM in the waste, which might serve as a precursor to carcinogenic disinfection induced by byproducts such as trihalomethanes and haloacetic acids [25,26]. The other, namely NH_3 emissions from animal manures during the biodegradation performed by BSFL, has been recognized as a major environmental issue, which may lead to acid rain and water eutrophication [11,27]. To enhance waste management policies, it is necessary to evaluate the variation of DOM and the rate of N depletion during waste disposal processing. DOM characteristics have been studied using integrated approaches, such as parallel factor analysis (PARAFAC) and fluorescence regional integration (FRI) technique [25,28]. The strongly correlated relationships among the aromatic degree, molecular weight, and hydrophobicity of DOM have been established using the molar absorbance coefficient of UV–vis spectra at 250–280 nm [29–31]. Furthermore, significant correlation among $SUVA_{260}$, the degree of humification aromaticity, and hydrophobicity was observed [32]; $SUVA_{280}$ reflected the aromatic compound content in organic matter [33]. In addition, the E_{250}/E_{365} ratio was inversely proportional to the degree of humification, and the molecular weight of organic matter was typically correlated with the degree of aromatic condensation and the molecular weight of DOM [30]. Moreover, a previous study further demonstrated that the initial DOM in the residues mainly consisted in the form of biological metabolites, such as soluble microbial products and tryptophan; subsequently, the humic acid (region V) and fulvic acid (region III) became predominant over time, while metabolic intermediates (region IV) were gradually decreased during these periods [34]. Additionally, two studies revealed that protein analogs and soluble microbial byproducts from livestock manures could be metabolized by BSFLs and produce humic substrates [11,35]. However, these results were obtained by integrating many studies with distinct experiment conditions, which could result in incorrect conclusions. Therefore, the transformation mechanism of DOM from animal manures performed by BSFLs during biodegradation warrants further study. Concurrently, protein-derived substances are associated with N content, and the variations of N content during the process of waste disposal by BSFLs are helpful for the understanding of composition difference of DOM. Nitrogen content has been found to be correlated with the presence of harmful gases in animal wastes, as noted in a recent study where BSFL larvae were effective in reducing the nitrogen contents in fresh pig manure [36]. Further, Chadwick et al. (2015) demonstrated that that at least 13% of NH_3-N from pig manure could be incorporated into BSFL bodies in the form of protein using the stable isotope ^{15}N in NH_3, thus reducing N_2O emissions [37]. In turn, few studies have compared the variations in N content during disposal different wastes via BSFL [11,35].

To enhance the treatment efficiency for cow and pig manures by BSFL, WDG was added in the two above-mentioned manures. The treatment efficiency of this method was further estimated, and the emissions of nitrogen-related gas were also assessed by TN content and $\varepsilon^{15}N/^{14}N$ ratio. In addition, we solely determined residual DOM from the treated WDG-added livestock manures to evaluate the difference in DOM and the degree of humification of residues using fluorescence technology and FRI analysis. These results help better understand the process of DOM transformation during BSFL treatment and provide a reference for increasing the bioconversion efficiency of livestock manure.

2. Results

2.1. BSFL Treatment Efficiency for WDG-Added Livestock Manures

In the present study, WDG were mixed into COM and PM to investigate the biodegradation treatment efficiency performed by BSFLs over a twelve-day period. As shown at Table 1, the results show that larval survival rates in the COM + WDG and PM + WDG were 89.08 ± 0.14% and 94.00 ± 0.25%, with growth rate index (GRI) values of 8.53 ± 0.13 mg/day and 7.71 ± 0.14 mg/day, respectively. In contrast, the larval survival rates in solely COM and PM were 65.42 ± 0.29% and 66.58 ± 0.14%, with GRI values of 1.40 ± 0.39 mg/day and 1.16 ± 0.07 mg/day, respectively. After adding WDG, the larval survival rate in the combined COM and PM increased by 23.66% and 27.42%, respectively, and the GRI value increased by 6.09–6.65 times. These results indicate that WDG could significantly increase the survival rates and development of BSFLs in livestock manures. Compared with cow manure treatment, PM + WDG could provide more nutrition, which is conducive to the growth of BSFLs, and hence the BSFL survival rates in the combination of WDG + PM was higher (Table S1). Similarly, after 12 days of the treatments, the reduction rates of the wastes from COM, COM + WDG, PM, and PM + WDG were 43.85 ± 0.74%, 73.41 ± 2.12%, 42.70 ± 2.43% and 67.11 ± 2.52%, with bioconversion rates of 1.20 ± 0.02%, 10.54 ± 0.06%, 0.92 ± 0.02% and 10.05 ± 0.11%, respectively. Overall, these results suggest that WDG could promote the reduction rates and the bioconversion rates of the wastes between these combinations treated by BSFLs.

Table 1. Summary of performance data of BSFLs that are reared on various feed substrates.

Treatment	Sample Rate	Total Weight (Dry Weight)	Survival Rate (%)	Growth Rate Index (mg/Days)	Substrate Reduction Rate (%)	Bioconversion Rate (%)
COM	100%COM	150 g	65.42 ± 0.29 [a]	0.27 ± 0.03 [b]	43.85 ± 0.74 [a]	1.20 ± 0.02 [b]
COM + WDG	55%COM + 45%WDG		89.08 ± 0.14 [c]	1.22 ± 0.02 [c]	73.41 ± 2.12 [c]	10.54 ± 0.06 [c]
PM	100%PM		66.58 ± 0.14 [b]	0.16 ± 0.01 [a]	42.70 ± 2.43 [a]	0.92 ± 0.02 [a]
PM + WDG	55%PM + 45%WDG		94.00 ± 0.25 [d]	1.11 ± 0.06 [d]	67.11 ± 2.52 [b]	10.05 ± 0.11 [d]

Average and standard deviation (n = 3) are displayed. Different letters indicate significant differences among different treatments ($p < 0.05$): cow manure (COM), pig manure (PM), wet distiller grains (WDG), COM blended with WDG with a ratio of 55:45 (COM + WDG), and PM blended with WDG with a ratio of 55:45 (PM + WDG).

2.2. *The Reduction in N-Relevant Harmful Gas Emission with the Addition of WDG in Livestock Manures Treated by BSFL*

The variation of TN content and $\delta^{15}N/^{14}N$ ratios in the residues of these treatments was further measured. After the wastes were digested by BSFLs, the TN content of the wastes from COM + WDG (1.89 ± 0.04%) after treatment was substantially lower than that of the solely WDG waste (3.11 ± 0.09%) (Figure S1). By contrast, there was little difference between TN content in the residual PM (2.82 ± 0.22%) and that in PM + WDG (2.59 ± 0.02%) (Figure 1A). Further, the $\delta^{15}N/^{14}N$ ratio in these signal wastes were significantly lower than in these combinations (Figure 1B).

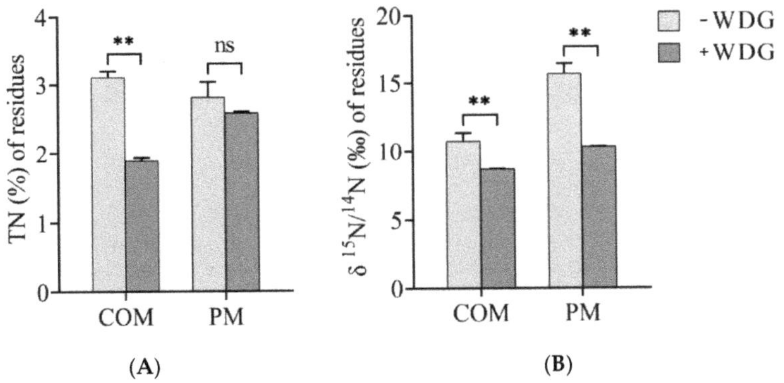

Figure 1. N-relevant harmful gas emission of BSFL bioconversion in residues of livestock manures: TN content (**A**) and $\delta^{15}N/^{14}N$ ratio (**B**) of residues. Each point represents the mean ± standard error (n = 3). ** $p < 0.01$; [ns] $p > 0.05$ (determined by one-way ANOVA followed by the LSD test). ANOVA, analysis of variance; −WDG represents manures without WDG; +WDG means feces with WDG.

2.3. *Effect of Adding WDG in Livestock Manures on the Organic Matter*

Next, we detected livestock manures with or without WDG using EEM and UV–vis spectra and then carried out FRI technical analysis to further study the effect of adding WDG to feces on organic matter.

2.3.1. EEM Spectra

Results from EEM spectra revealed that the fluorescence intensity of regions I and II in mixed samples COM + WDG and PM + WDG of the group C was higher than that of samples of solely COM and PM, indicating that WDG could increase the livestock manures' aromatic protein content (Figure 2A). After BSFL bioconversion, fluorescence intensity of regions I, II, and IV of group C became weaker than groups A and B, as well as whether or not added WDG had no impact on the overall trend of these three regions (Figure 2). At the same time, the fluorescence intensity of region V of COM + WDG was increased, indicating the generation of humic substances analogues in COM + WDG.

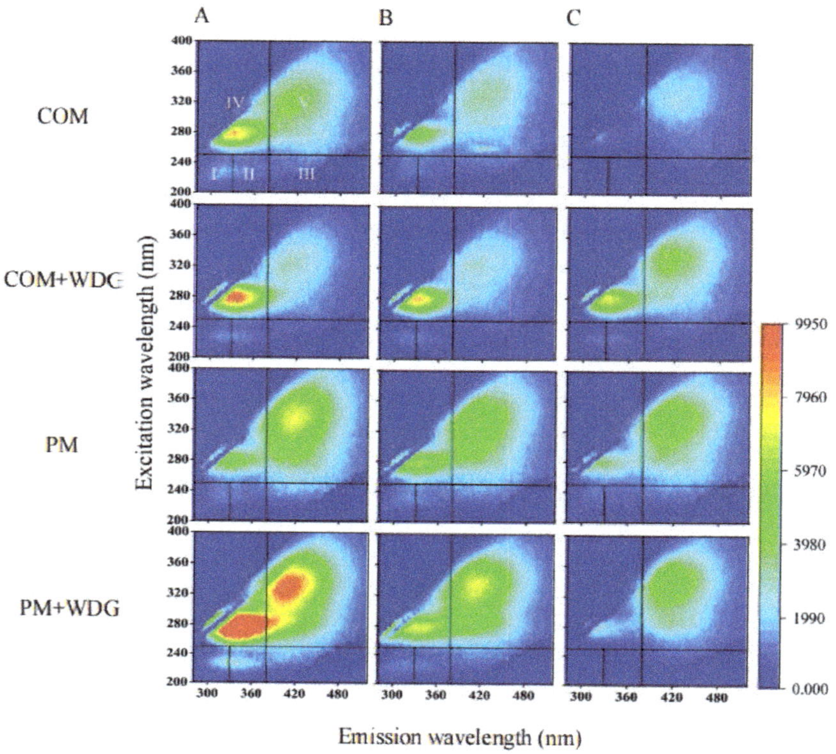

Figure 2. EEM spectra of all samples of DOM in the fresh livestock manures (**A**), control (**B**), and black soldier fly treatment manures (**C**). The excitation/emission wavelength ranges of regions (I, II, III, IV, and V) were 230–250/280–330 nm, 230–250/330–380 nm, 200–250/380–520 nm, 250–440/280–380 nm and 250–440/380–380 nm, respectively. The control group represents natural composting, and the treatment group represents using BSFL degradation waste.

The content of aromatic proteins, fulvic-acid-like substrates, soluble microbial-byproduct-like compounds, and humic-acid-like materials were visualized for each region by calculating their percentages of fluorescence response ($P_{i,n}$) (Figure 3). The accumulative $P_{i,n}$ values of regions I, II, and IV of COM, COM + WDG, and PM + WDG in group C after being treated by BSFLs were markedly lower than those of groups B and A, and the accumulative $P_{i,n}$ values of regions III and V were fundamentally higher than those of groups B and A. The results illustrate that BSFLs accelerated the consumption of protein-like substances and soluble microbial-byproduct-like compounds, as well as promoted the stabilization and humification of organic matter in livestock manures. As for group C, the accumulative $P_{i,n}$ values of regions III and V in COM, COM + WDG, PM, and PM + WDG were 58%, 49%, 52%, and 63%, respectively, suggesting that the level of livestock manures' humification of COM was decreased by adding WDG.

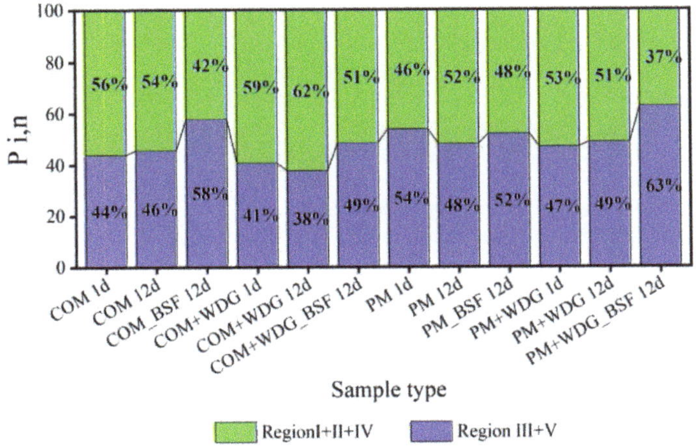

Figure 3. $P_{i,n}$ of the five regions of all samples of DOM from different treatments.

2.3.2. UV–Vis Spectra

Next, to further investigate the effect of adding WDG in manures on the varied contents of organic matter, organic matter from these restudies treated by BSFL were analyzed by using UV–vis spectra by molar absorbance coefficients and four representative parameters ($SUVA_{254}$, $SUVA_{260}$, $SUVA_{280}$, and E_{250}/E_{365}) (Figure 4). After twelve days of being treated by BSFL, the order of the molar absorbance coefficients of DOM from these residues at the range of 250–280 nm was PM + WDG > COM + WDG > COM > PM (Figure 4A), demonstrating that WDG could enhance the decomposition of livestock manures. Subsequently, the $SUVA_{254}$, $SUVA_{260}$, and $SUVA_{280}$ values of the treated groups were found to be significantly higher than the control. However, regardless of whether WDG was added or not, the E_{250}/E_{365} ratio of the treated groups was lower than the control, suggesting that the wastes' aromaticity, molecular weight, hydrophobicity, and humification level when treated by BSFL were higher than natural composting. Above all, WDG did not alter the general trend of organic matter decomposition (Figure 4B). Moreover, the $SUVA_{254}$, $SUVA_{260}$, and $SUVA_{280}$ values in livestock manures with the addition of WDG were lower than those without WDG, but the E_{250}/E_{365} ratio was higher in COM + WDG and PM + WDG than in COM and PM, and WDG decreased the aromatic polycondensation, molecular weight, and polar functional groups of DOM in livestock manures.

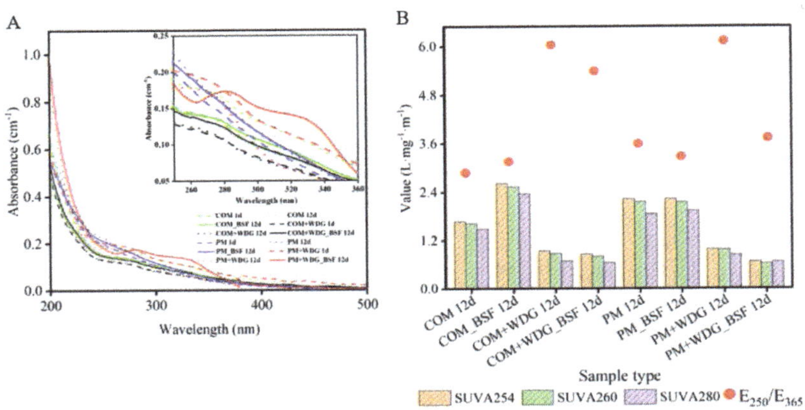

Figure 4. UV−vis spectra (**A**) of DOM in the initial groups, control, and treatment groups and UV spectral coefficient parameters (**B**) of the control and treatment groups.

2.4. Phytotoxicity Study and Micronutrient Composition of Residues

To assess the quality and phytotoxicity of the residual manures, edible rape seeds were used to evaluate residues plant toxicity (Table S2). The seed germination index (GI) of all samples was 154.76%, 123.64%, 131.97%, and 164.46%, respectively. Based on the phytotoxicity analysis, we found that all residues proved to be nontoxic and have the potential to be used as soil organic fertilizer. Further, we also tested the micronutrient composition of manure/compost produced by BSFLs, and the results show that the micronutrients included major elements (P, Ca, K, Mg, and Na) and trace elements (Zn, Cu, Fe, Mn, Mo, Ni, Co, and Cr) in residues, and most trace elements were lower than international soil safety standards (except Zn, Fe, and Mn) (Table S3). These residues might be used to improve soil nutrition due their rich nutritional content and low environmental pollution.

3. Discussion

3.1. Enhancement in BSFL Bioconversion and Livestock Manures' Reduction Rates

BSFLs have the ability to survive and grow in a variety of substrates of livestock manures [12,34]. However, the reduction and bioconversion rates of wastes from manures were essentially low when treated by BSFL, particularly for COM and PM [12,34]. These findings emphasized that it is urgently needed to optimize the treatment of livestock manures with BSFLs to enhance its efficacy as a sustainable waste management solution. In this study, we investigated the effect of adding 45% WDG into livestock manures on the reduction rates of wastes, and the results demonstrate a significant improvement in the effectiveness of BSFL treatment. The reduction rates of COM and PM without WDG were 43.85 ± 0.74% and 42.70 ± 2.43%, respectively, whereas the addition of WDG increased the reduction rates to 73.41 ± 2.12% and 67.11 ± 2.52% for COM and PM, respectively. Additionally, the BSFL bioconversion rates in COM (1.20 ± 0.02 to 10.54 ± 0.06%) and PM (0.92 ± 0.02 to 10.05 ± 0.11%) were also increased. Compared with cow manure treatment, PM + WDG could provide more nutrition, which was conducive to the growth of BSFL; hence, the addition of WDG in PM could improve BSFL survival rates (Table S1). Similar results were also observed by mixing food wastes into dewatered sludge, with an increased waste reduction rate of 34.93 to 40.75% [3–5]. Adding high-nutrient wastes to low-nutrient wastes could increase the combined nutrients and help improve the efficiency of BSFLs in wastes treatment [12]. Fruits and vegetable wastes were added into abattoir waste, which improved the reduction rate from 46.3% to 61.1% [24]. Similarly, adding soybean curd residue into cow manure, the bioconversion rate increased approximately two times, from 6.40% to 11.60% [38]. With BSFL biotreated mixed food waste, the bioconversion rate was the highest, 14.00% in the mixed wastes of coconut endosperm and soybean curd residue and 18.54% in the soybean curd residue and kitchen waste mixture [39,40]. Compared with the single-waste treatment, mixed wastes could provide more nutrition, which is conducive to the growth of BSFLs. Like other living organisms, BSFLs need nutrients to support their growth. Therefore, for higher bioconversion performance, BSFLs need to feed on organic wastes rich in digestible nutritive substances. After treating waste, the larvae could be used to feed poultry, due to BSFLs having polyunsaturated fatty acids and low heavy metal concentrations (heavy metal concentrations were even below limits of international guidelines for animal feed) [13,41]. For example, BSFL could positively increase the levels of the health-claimable omega-3 fatty acid EPA within broiler bodies, and plasma triglyceride concentrations from broiler bodies also increased in a dose-dependent manner with increasing levels of BSFL [41,42]. Additionally, after wastes were treated by BSFL, heavy metals (Cd, Co, Cr, Cu, Fe, Mn, Mo, Ni, Pb, and Zn) were not bioaccumulated in the larvae [13]. According to the study by Shi et al. (2022), these elements had a low chance to bioaccumulate/biomagnify in BSFL bodies due to multiresistance bacterial strains in the gut of BSFLs, which helped the larvae against heavy metal stress [43].

However, the waste reduction rates of solely COM and solely PM were only 12.70% and 56.00% after being treated by BSFL, respectively [21,22]. Hence, we obtained a common conclusion that different mixed wastes could enhance the reduction rate of these wastes.

At the same time, in this study, the total Kjeldahl nitrogen of the raw material of WDG (46.11 ± 0.06%) was higher than that of COM (43.17 ± 0.12%) and PM (44.48 ± 0.09%), indicating that WDG contained more protein than the livestock manures. This is consistent with the finding that protein is a necessary nutrient for larval growth and development [44]. Furthermore, the C/N ratio of raw material from WDG (26.44 ± 0.31%) was higher than that of COM (12.91 ± 0.10%) and PM (12.68 ± 0.28%). Previous studies have shown that a higher initial C/N ratio could lead to improved composting stability when mixing chicken manure with different amounts of sawdust [23]. Furthermore, we found that heavy metal levels in WDG were lower than those in COM and PM. Therefore, adding WDG to livestock manures can improve the composting conditions and reduce environmental pollution. The incorporation of WDG into livestock manures resulted in an approximately 20% improvement in waste reduction rates after BSFL treatment. These findings underscored the potential of WDG as an additive to enhance the performance of BSFL treatment in mixed livestock manure.

3.2. The Reduction in NH_3 and N_2O Emission

While NH_3 emissions from livestock manures are also one of the main environmental burdens, at the same time, NH_3 was an indirect source for N_2O [11,45]. Therefore, to further understand the emission of pollutants (e.g., NH_3 and N_2O) during the experiment of treatment by BSFL, it was necessary to emphasize the changes in N contents. In the present study, the TN content and $\delta^{15}N/^{14}N$ ratio of livestock manures before and after the experiment were analyzed. WDG was added into the livestock manure, which significantly decreased TN content and $\delta^{15}N/^{14}N$ ratio in these blended livestock manures, with TN content and $\delta^{15}N/^{14}N$ ratio from COM + WDG and COM being 1.89 ± 0.04% and 3.11 ± 0.09%, 8.68 ± 0.01‰ and 10.76 ± 0.61‰, respectively. These results illustrate that a mixture of different wastes could reduce NH_3 emission pollution more. These findings also highlighted the potential benefits of cotreatment and codigestion of livestock manures and other waste products. Mixing different waste streams might achieve more effective wastes treatment, resulting in decreased environmental pollution. Earlier studies had explained the reason why NH_3 emissions could be reduced by using BSFL bioconversion waste: approximately 25% of the N in initial fresh pig manure could be accumulated in larvae by a mass balance law [36,46]. A subsequent study further revealed that 13% of the NH_3-N derived from manures was incorporated into the larval biomass by using the stable isotope ^{15}N during manures' bioconversion with BSFL [11]. This means that BSFLs could reduce NH_3 emission by 8.97 kg when 1 ton of pig manure is treated by BSFL. And manure treatment decreased greenhouse gas (CO_2, CH_4 and N_2O) emissions from manures in EU countries by 0–17% in 2010 [47]. In our study, TN contents in the combination of COM + WDG was 1.65 times that of pure manure; we guessed that BSFLs might reduce NH3 emission by 21.39%, which means that BSFLs treating 1 ton of cow manure could reduce NH_3 emissions by 6.65 kg. However, pig manure kept similar patterns with Parodi et al. (2022). Hence, further exploration is required to quantify harmful gas emissions and better understand the underlying mechanisms of the observed reductions between TN contents and $\delta^{15}N/^{14}N$ ratio and explore the potential benefits and limitations of the combination of different wastes after treatment by BSFL.

3.3. Identification of the Composition of Organic Matter

The high humus content of livestock manures played a central role in regulating soil fertility. Hence, two spectra were used to analyze the humification degree of feces after these treatments. The EEM spectra could offer more information for the molecular configuration of organic matter from different sources. EEM spectra with the FRI method were employed to analyze all the samples, and the EEM spectra were divided into five regions using the FRI method [4,48]. Each organic matter has its corresponding region. Further, aromatic proteins, fulvic-like materials, soluble microbial-byproduct-like substances, and humic-like compounds fall in regions I and II, region III, region IV, and region V, respectively [49,50].

EEM data suggested that the fluorescence intensity of regions I, II, and IV of all samples in group C became weaker than that of group A and B after being treated by BSFL, which indicated that the proteinaceous substances and soluble microbial-byproduct-like substrates treated by BSFL could be metabolized to humic compounds, as well as whether or not added WDG had no impact on the overall trends of the decomposition of organic matter. These results are further supported by previous studies [11,51], suggesting that the simple structural organic matter obviously decomposed and converted into the humic-like substances after BSFL treatments. Moreover, the accumulative $P_{i,n}$ values of regions III and V in COM, COM + WDG, PM, and PM + WDG after BSFL treatments were 58%, 49%, 52%, and 63%, respectively, demonstrating that the level of livestock manures' humification of COM was decreased by adding WDG, and the production of humic matter in PM increased.

UV–vis spectra has been widely utilized to characterize the molecular structure of organic matter [33]. More fractions of DOM characterized by UV–vis were observed than those identified by fluorescence spectroscopy. Results from UV–vis spectra indicate that the aromaticity, molecular weight, humification, and hydrophobicity of the residues were decreased after adding WDG between these treatments. This could be attributed to the comparatively low content of humus in WDG, thus reducing the degree of humification of the feces upon addition [52]. Overall, both spectra results suggest that the addition of WDG increased the protein analogue content of the wastes' raw material and might decrease the level of humification of livestock manures after BSFL treatments.

3.4. Non-Phytotoxicity and Rich Micronutrients of Residues

Based on the phytotoxicity analysis, the GI of all residues in edible rape seeds were more than 120% and higher than the Chinese standard for organic fertilizer (GI $\geq$ 70%, NY525-2021) Hence, these residues were proved to be nontoxic and have the potential to be used as soil organic fertilizer. Further, to date, studies have shown that the seeds of eight species (white radish, cucumber, fruit radish, edible rape, round radish, hybrid cucumber, cress, and Chinese cabbage) were selected as the best choices to conduct phytotoxicity [53]. For example, Wang et al. (2011) reported that GI measured with Chinese cabbage seeds was about 107.2% for composted pig manure [54]. And the GI was about 120% for composted chicken manure [55]. Compared with their composted manure, our residues had higher GI, revealing that our residues favored the growth and reproduction of the plants.

4. Materials and Methods

4.1. Raw Materials

Pig manure (PM) and cow manure (COM) were collected from a local company (Guizhou Junong Meat Industry Co., Ltd., Guizhou, China), and WDG were provided from Huaxi district, Guiyang city, Guizhou province, China. BSFLs were obtained from a local breeding base in Bijie city, Guizhou province, China. BSFLs were reared in an artificial climate chamber under controlled conditions (temperature: 28 °C and humidity: 70%) until 10 days old as the initial larvae of the experiment. The micronutrient compositions of manures and WDG are shown in Table S1.

4.2. Experimental Design

Two livestock manures were compared in this research with and without WDG to investigate the performance of BSFLs that were reared on various feed substrates: COM, COM + WDG in a ratio of 55:45, PM, and PM + WDG in a ratio of 55:45 (PM + WDG). In order to further understand whether the addition of WDG would affect the difference in organic matter during BSFL treatments, two controlled groups were set up for spectral analysis: one control was raw materials (group A), and the second was manures from natural composting without BSFL (group B). Additionally, the bioconversion of manures of BSFLs was group C. The experimental design is shown in Supplementary Material S1.

All samples were individually placed into plastic containers (28 (length) × 20 (width) × 17 (height) cm), and then 150 g (dry matter) of livestock manures was put into each container separately (Table 1). Approximately 0.2 g of BSFLs (10 days old) was added into each container and mixed with each treatment at a larval density of 1 larva/cm^3. Moreover, each individual was weighed (approximately 6.80 ± 0.87 mg). Furthermore, the feeding rate was equivalent to 24 mg dry matter/larva/day and then incubated in an indoor manual climatic box (each treatment was replicated three times) at the Institute of Entomology, Guizhou University. Meanwhile, all samples were turned over daily and weighed for insect and residual manures [36]. After 12 days, 25% of the larval pupae in the mixtures were observed, and then the experiment was completed. Pupae and larvae were separated from the mixtures and weighed, before storing them at −80 °C [56].

4.3. *Survival and Growth of BSFLs on Livestock Manures*

At the beginning of the experiment, we weighed and counted the selected initial 10-day-old larvae. After the experiment, the final number and weight of the surviving larvae were recorded. The final weight of the residues were calculated by the difference between the total weight of the rearing container, the weight of the empty container, and the weight of mature larvae [13]. The efficiency of bioconversion performed by BSFL, the survival rate and growth rate index (GRI) of larvae, the substrate reduction rate (SR), and the bioconversion rate (BCR) of total organic wastes were calculated according to Equations (1)–(4) [48,57].

$$\text{GRI (mg/days)} = (L_{end} - L_0)/t \tag{1}$$

where t is days of rearing; L_0 is initial weight of ten-day- old larvae; and L_{end} is total weight of mature larvae.

$$\text{Survival rate (\%)} = 100 \times nL_{end}/nL_0 \tag{2}$$

nL_0 is the number of larvae at the beginning, and nL_{end} is the number of larvae at the end

$$\text{SR (\%)} = ((S - R)/S) \times 100 \tag{3}$$

S is the weight of total substrate; R is the weight of total residues

$$\text{BCR (\%)} = (L_{DM}/S_{DM}) \times 100 \tag{4}$$

L_{DM} is the total dry weight of mature larvae; S_{DM} is the dry weight

4.4. *DOM Extraction and Spectral Analysis*

To obtain the content of DOM, samples of groups A, B, and C and Milli-Q water were mixed at a ratio of 1:10 (*w*/*w*, dry weight) [58,59] and then shaken at 25 °C for 24 h, followed by centrifuging at 4390× *g* for 15 min. After centrifugation, the supernatant was filtered using a 0.45 µm filter. The organic carbon (DOC) concentration of the filtrate was analyzed by an automated TOC analyzer (elementar vario TOC select, Alimenta, Germany) with three measurements per sample and an analytical precision < 2% [60]. Finally, the filtrate was diluted to 10 mg/L with Milli-Q water according to the measured DOC concentration, and then the samples were stored at 4 °C. Subsequently, these storage samples were further analyzed by ultraviolet–visible (UV–vis) spectra (by UV–vis spectrophotometer (BlueStar A, Beijing Labtech Instruments Co., Ltd. Beijing, China)), and the excitation–emission matrix (EEM) of fluorescence spectra was measured (by a fluorescence spectrophotometer (F-4500, Japan)) [12].

The UV–vis spectra were detected using a UV–vis spectrophotometer with a quartz cuvette with an optical range length of 1 cm in the range of 200–600 nm, while the samples were put into the quartz cuvette, with Milli-Q water used as a blank. Meanwhile, the absorbance ratios at 254, 260, and 280 nm were calculated according to the methods of Song et al. (2015) and Zhang et al. (2022), noted as $SUVA_{254}$, $SUVA_{260}$, and $SUVA_{280}$, respectively [32,33]. The UV absorbance ratios at 250 and 365 nm were noted as E_{250}/E_{365}.

A fluorescence spectrophotometer (F-4500) was carried out to measure the excitation–emission matrix (EEM) of the fluorescence spectra. The slit widths of the emission and excitation monochromators were both adjusted to 10 nm; subsequently, the scanning speed was set to 1200 nm/min, and the emission and excitation wavelengths were detected in the range of 280–520 nm and 200–400 nm, respectively. Additionally, the increments of excitation and emission wavelengths were adjusted to 5 nm and 2 nm, respectively [61]. Lastly, the FRI method was used to analyze the EEM spectra, and the EEM spectrum of DOM was divided into five contiguous regions (I, II, III, IV, and V), while the excitation/emission wavelength ranges were 230–250/280–330 nm, 230–250/330–380 nm, 200–250/380–520 nm, 250–440/280–380 nm, and 250–440/380–380 nm, respectively. The volume (Φ_i) beneath region "i" of the EEM was calculated by the following equation [28,59]:

$$\Phi_i = \int_{ex}\int_{em} I_{\lambda ex,\, \lambda em}\, d_{\lambda ex} d_{\lambda em} \tag{5}$$

For discrete data, the volume (Φi) was expressed by

$$\Phi_i = \sum\nolimits_{ex}\sum\nolimits_{em} I_{\lambda ex,\, \lambda em}\Delta\lambda_{ex}\Delta\lambda_{em} \tag{6}$$

in which $\Delta\lambda_{ex}$ is the excitation wavelength interval, and $\Delta\lambda_{em}$ is the emission wavelength interval. $I_{\lambda ex,\, \lambda em}$ was the fluorescence intensity at each excitation–emission wavelength pair.

However, since the divided regions were not equal, the volume integral of each region would have some effect on the interpretation of the whole data because of the different regions, and we thus normalized their area of the regions separately to eliminate this effect. The normalized excitation–emission area volumes ($\Phi_{i,n}$, $\Phi_{T,n}$) and percent fluorescence response ($P_{i,n}$) were calculated as follows:

$$\Phi_{i,n} = MF_i\Phi_i \tag{7}$$

$$\Phi_{T,n} = \sum\nolimits^{5} \Phi_{i,n} \tag{8}$$

$$P_{i,n} = \Phi_{i,n}/\Phi_{T,n} \times 100\% \tag{9}$$

in which MF_i is called multiplication factor.

4.5. Analysis of TP, TN, and $\delta^{15}N/^{14}N$

All samples were freeze-dried at −80 °C for several days. The dried samples were ground and sieved through a 200 mesh. These sieved samples were stored at −20 °C for TP, TN, and $\delta^{15}N/^{14}N$ analysis. The content of TP was analyzed using the combustion method. TP and TN content was analyzed using an organic elemental analyzer (vario MACRO, Elimontal, Germany). The $\delta^{15}N/^{14}N$ ratio of these samples was analyzed using a gas isotope ratio mass spectrometer (MAT253, Thermo Fisher, Waltham, MA, USA).

4.6. Phytotoxicity Anlysis

We selected edible rape seeds (seed germination rate greater than 98%, Henan Lvyuan Agricultural Technology Co., Ltd., Zhengzhou, China) to assess residue phytotoxicity. First, 10 g of residue (from wastes after being treated by BSFL) and 100 mL deionized water were shaken at 170 rpm for 30 min and then filtered through 0.45 μm filter paper, and the filtrate was collected. Then, seed germination index was determined [53]. Every treatment had three duplicates.

$$GI = (A_1/A_2) \times (B_1/B_2) \times 100 \tag{10}$$

A_1 represents the number of germinated seeds in the extract; A_2 represents the number of germinated seeds in control; B_1 represents root length in extract; and B_2 represents root length in control.

4.7. Micronutrient Composition Anlysis

The heavy metal content of raw materials and residues was determined. All samples needed to be dried in a freeze-dryer (FDU-2110, EYELA, Tokyo, Japan) and ground through a 200-mesh screen before analysis. The heavy metal content was measured by the method proposed by the Chinese Ministry of Ecology and Environment [62]. The metal concentrations were detected by inductively coupled plasma mass spectrometry (ICP-MS, Perkin Elmer NexION 300X, Waltham, MA, USA).

5. Data Analysis

Statistical calculations and data analysis of all samples were performed using SPSS 22.0 (the significance coefficient was set at $p < 0.05$ or $p < 0.01$). The physical and chemical parameters of all samples were also analyzed using Origin 2021, and EEM spectral data were collected and modified with MATLAB 2020a and FRI analysis.

6. Conclusions

The major findings were that adding 45% of WDG into livestock manures treated by BSFL could significantly improve waste reduction rates by approximately 20%. The mixture of WDG and manures treated by BSFL might result in decreasing NH_3 emissions 1.65 times compared with cow manure alone. Our results suggest that the utilization of mixed wastes could also enhance the bioconversion rate of BSFLs and help reduce levels of harmful matter pollution. These findings contribute to a better understanding of the potential benefits of utilizing mixed wastes by the incorporation of WDG wastes with relatively high nutritional value, in order to minimize environmental pollution by accelerating a reduction in low-nutrient-value wastes as a beneficial strategy for sustainable waste management practices. Most importantly, this study provides an economical and efficient method for the treatment of combined wastes.

Supplementary Materials: The following supporting information can be downloaded at: https://www.mdpi.com/article/10.3390/molecules28155735/s1, Figure S1: N-relevant harmful gas emission after manure treated by BSFL; Table S1: Selected characteristics of raw materials in this study; Table S2: The phytotoxicity study of waste residues in rape; Table S3: Micro nutrient composition of manure/compost produced by BSFL.

Author Contributions: T.L., data curation, formal analysis, validation, writing—original draft, methodology, and writing—review and editing; S.K., validation, writing—original draft, and writing—review and editing; M.W., formal analysis and validation; H.L., validation, writing—review and editing; T.W., formal analysis and validation; J.G. and D.J.: funding acquisition, supervision, methodology, and writing—review and editing. All authors have read and agreed to the published version of the manuscript.

Funding: This project was funded by three funds, namely Guizhou Provincial Science and Technology Projects (Qiankehe Zhicheng [2021]180), Guizhou Provincial Science and Technology Projects (Qiankehe Pingtai Rencai-GCC [2022]029-1) and Lancang-Mekong Cooperation Special Fund Projects (JiaoWaiSiYa [2020]619).

Institutional Review Board Statement: Exclude this statement.

Informed Consent Statement: Informed consent was obtained from all subjects involved in the study.

Data Availability Statement: Not sharing research data due to privacy reasons.

Conflicts of Interest: The authors declare no conflict of interest.

Availability of Data and Materials: The datasets analyzed in the current study are available from the corresponding author upon reasonable request.

Sample Availability: Not applicable.

References

1. Qian, Y.; Song, K.; Hu, T.; Ying, T. Environmental status of livestock and poultry sectors in China under current transformation stage. *Sci. Total Environ.* **2018**, *622*, 702–709. [CrossRef]
2. Kaza, S.; Yao, L.; Bhada-Tata, P.; Van Woerden, F. *What A Waste 2.0: A Global Snapshot of Solid Waste Management to 2050*; World Bank Publications: Washington, DC, USA, 2018.
3. Afazeli, H.; Jafari, A.; Rafiee, S.; Nosrati, M. An investigation of biogas production potential from livestock and slaughterhouse wastes. *Renew. Sustain. Energy Rev.* **2014**, *34*, 380–386. [CrossRef]
4. Chen, X.; Zhao, Y.; Zeng, C.; Li, Y.; Zhu, L.; Wu, J.; Chen, J.; Wei, Z. Assessment contributions of physicochemical properties and bacterial community to mitigate the bioavailability of heavy metals during composting based on structural equation models. *Bioresour. Technol.* **2019**, *289*, 121657. [CrossRef]
5. Gao, X.; Tan, W.; Zhao, Y.; Wu, J.; Sun, Q.; Qi, H.; Xie, X.; Wei, Z. Diversity in the mechanisms of humin formation during composting with different materials. *Environ. Sci. Technol.* **2019**, *53*, 3653–3662. [CrossRef] [PubMed]
6. Galloway, J.N.; Dentener, F.J.; Capone, D.G.; Boyer, E.W.; Howarth, R.W.; Seitzinger, S.P.; Asner, G.P.; Cleveland, C.C.; Green, P.; Holland, E.A. Nitrogen cycles: Past, present, and future. *Biogeochemistry* **2004**, *70*, 153–226. [CrossRef]
7. Oenema, O.; Wrage, N.; Velthof, G.L.; van Groenigen, J.W.; Dolfing, J.; Kuikman, P.J. Trends in global nitrous oxide emissions from animal production systems. *Nutr. Cycl. Agroecosyst.* **2005**, *72*, 51–65. [CrossRef]
8. Klages, S.; Aue, C.; Reiter, K.; Heidecke, C.; Osterburg, B. Catch crops in lower saxony-more than 30 years of action against water pollution with nitrates: All in vain? *Agriculture* **2022**, *12*, 447. [CrossRef]
9. Xiao, X.; Mazza, L.; Yu, Y.; Cai, M.; Zheng, L.; Tomberlin, J.K.; Yu, J.; van Huis, A.; Yu, Z.; Fasulo, S. Efficient co-conversion process of chicken manure into protein feed and organic fertilizer by *Hermetia illucens* L. (diptera: Stratiomyidae) larvae and functional bacteria. *J. Environ. Manag.* **2018**, *217*, 668–676. [CrossRef]
10. Kim, C.-H.; Ryu, J.; Lee, J.; Ko, K.; Lee, J.-Y.; Park, K.Y.; Chung, H. Use of black soldier fly larvae for food waste treatment and energy production in asian countries: A review. *Processes* **2021**, *9*, 161. [CrossRef]
11. Parodi, A.; Yao, Q.; Gerrits, W.J.; Mishyna, M.; Lakemond, C.M.; Oonincx, D.G.; Van Loon, J.J. Upgrading ammonia-nitrogen from manure into body proteins in black soldier fly larvae. *Resour. Conserv. Recycl.* **2022**, *182*, 106343. [CrossRef]
12. Wang, Q.; Ren, X.; Sun, Y.; Zhao, J.; Awasthi, M.K.; Liu, T.; Li, R.; Zhang, Z. Improvement of the composition and humification of different animal manures by black soldier fly bioconversion. *J. Clean. Prod.* **2021**, *278*, 123397. [CrossRef]
13. Bohm, K.; Hatley, G.A.; Robinson, B.H.; Gutiérrez-Ginés, M.J. Black soldier fly-based bioconversion of biosolids creates high-value products with low heavy metal concentrations. *Resour. Conserv. Recycl.* **2022**, *180*, 106149. [CrossRef]
14. Liu, T.; Awasthi, M.K.; Wang, X.; Awasthi, S.K.; Jiao, M.; Shivpal, V.; Zhou, Y.; Liu, H.; Zhang, Z. Effects of further composting black soldier fly larvae manure on toxic metals and resistant bacteria communities by cornstalk amendment. *Sci. Total Environ.* **2022**, *806*, 150699. [CrossRef]
15. Li, Q.; Zheng, L.; Qiu, N.; Cai, H.; Tomberlin, J.K.; Yu, Z. Bioconversion of dairy manure by black soldier fly (Diptera: Stratiomyidae) for biodiesel and sugar production. *Waste Manag.* **2011**, *31*, 1316–1320. [CrossRef] [PubMed]
16. He, X.S.; Xi, B.; Li, X.; Pan, H.W.; An, D.; Bai, S.G.; Li, D.; Cui, D.Y. Fluorescence excitation-emission matrix spectra coupled with parallel factor and regional integration analysis to characterize organic matter humification. *Chemosphere* **2013**, *93*, 2208–2215. [CrossRef] [PubMed]
17. Xi, B.D.; He, X.S.; Wei, Z.M.; Jiang, Y.H.; Li, M.X.; Li, D.; Li, Y.; Dang, Q.L. Effect of inoculation methods on the composting efficiency of municipal solid wastes. *Chemosphere* **2012**, *88*, 744–750. [CrossRef]
18. Elhag, O.; Zhou, D.; Song, Q.; Soomro, A.A.; Cai, M.; Zheng, L.; Yu, Z.; Zhang, J. Screening, expression, purification and functional characterization of novel antimicrobial peptide genes from *Hermetia illucens* (L.). *PLoS ONE* **2017**, *12*, e0169582. [CrossRef]
19. Mwaniki, Z.; Shoveller, A.K.; Huber, L.A.; Kiarie, E.G. Complete replacement of soybean meal with defatted black soldier fly larvae meal in Shaver White hens feeding program (28–43 wks of age): Impact on egg production, egg quality, organ weight, and apparent retention of components. *Poult. Sci.* **2020**, *99*, 959–965. [CrossRef]
20. China Feed Industry Information Network. *Prices of Imported and Domestically Produced Fish Meal in China in February*; China Feed Industry Information Network: Beijing, China, 2023.
21. Wang, C.; Guo, X.; Deng, H.; Dong, D.; Tu, Q.; Wu, W. New insights into the structure and dynamics of actinomycetal community during manure composting. *Appl. Microbiol. Biotechnol.* **2014**, *98*, 3327–3337. [CrossRef] [PubMed]
22. Gold, M.; Cassar, C.M.; Zurbrügg, C.; Kreuzer, M.; Boulos, S.; Diener, S.; Mathys, A. Biowaste treatment with black soldier fly larvae: Increasing performance through the formulation of biowastes based on protein and carbohydrates. *Waste Manag.* **2020**, *102*, 319–329. [CrossRef]
23. Gao, M.; Liang, F.; Yu, A.; Li, B.; Yang, L. Evaluation of stability and maturity during forced-aeration composting of chicken manure and sawdust at different c/n ratios. *Chemosphere* **2010**, *78*, 614–619. [CrossRef] [PubMed]
24. Lalander, C.; Diener, S.; Zurbrügg, C.; Vinnerås, B. Effects of feedstock on larval development and process efficiency in waste treatment with black soldier fly (*Hermetia illucens*). *J. Clean. Prod.* **2019**, *208*, 211–219. [CrossRef]
25. Imai, A.; Fukushima, T.; Matsushige, K.; Kim, Y.H. Fractionation and characterization of dissolved organic matter in a shallow eutrophic lake, its inflowing rivers, and other organic matter sources. *Water Res.* **2001**, *35*, 4019–4028. [CrossRef]

26. Zhang, F.; Wang, Y.; Chu, Y.; Gao, B.; Yue, Q.; Yang, Z.; Li, Q. Reduction of organic matter and trihalomethane formation potential in reclaimed water from treated municipal wastewater by coagulation and adsorption. *Chem. Eng. J.* **2013**, *223*, 696–703. [CrossRef]
27. Strokal, M.; Kroeze, C.; Wang, M.; Bai, Z.; Ma, L. The marina model (model to assess river inputs of nutrients to seas): Model description and results for China. *Sci. Total Environ.* **2016**, *562*, 869–888. [CrossRef] [PubMed]
28. Chen, W.; Westerhoff, P.; Leenheer, J.A.; Booksh, K. Fluorescence excitation-emission matrix regional integration to quantify spectra for dissolved organic matter. *Environ. Sci. Technol.* **2003**, *37*, 5701–5710. [CrossRef] [PubMed]
29. Chin, Y.-P.; Aiken, G.R.; Danielsen, K.M. Binding of pyrene to aquatic and commercial humic substances: The role of molecular weight and aromaticity. *Environ. Sci. Technol.* **1997**, *31*, 1630–1635. [CrossRef]
30. Peuravuori, J.; Pihlaja, K. Molecular size distribution and spectroscopic properties of aquatic humic substances. *Anal. Chim. Acta* **1997**, *337*, 133–149. [CrossRef]
31. Nishijima, W.; Speitel, G.E., Jr. Fate of biodegradable dissolved organic carbon produced by ozonation on biological activated carbon. *Chemosphere* **2004**, *56*, 113–119. [CrossRef]
32. Zhang, X.; Li, Y.; Ye, J.; Chen, Z.; Ren, D.; Zhang, S. The spectral characteristics and cadmium complexation of soil dissolved organic matter in a wide range of forest lands. *Environ. Pollut.* **2022**, *299*, 118834. [CrossRef]
33. Song, C.; Li, M.; Xi, B.; Wei, Z.; Zhao, Y.; Jia, X.; Qi, H.; Zhu, C. Characterisation of dissolved organic matter extracted from the bio-oxidative phase of co-composting of biogas residues and livestock manure using spectroscopic techniques. *Int. Biodeterior. Biodegrad.* **2015**, *103*, 38–50. [CrossRef]
34. Ma, J.J.; Jiang, C.L.; Tao, X.H.; Sheng, J.L.; Sun, X.Z.; Zhang, T.Z.; Zhang, Z.J. Insights on dissolved organic matter and bacterial community succession during secondary composting in residue after black soldier fly larvae (*Hermetia illucens* L.) bioconversion for food waste treatment. *Waste Manag.* **2022**, *142*, 55–64. [CrossRef]
35. Murphy, K.R.; Hambly, A.; Singh, S.; Henderson, R.K.; Baker, A.; Stuetz, R.; Khan, S.J. Organic matter fluorescence in municipal water recycling schemes: Toward a unified parafac model. *Environ. Sci. Technol.* **2011**, *45*, 2909–2916. [CrossRef] [PubMed]
36. Parodi, A.; Gerrits, W.J.; Van Loon, J.J.; De Boer, I.J.; Aarnink, A.J.; Van Zanten, H.H. Black soldier fly reared on pig manure: Bioconversion efficiencies, nutrients in the residual material, greenhouse gas and ammonia emissions. *Waste Manag.* **2021**, *126*, 674–683. [CrossRef] [PubMed]
37. Chadwick, D.; Wei, J.; Yan'an, T.; Guanghui, Y.; Qirong, S.; Qing, C. Improving manure nutrient management towards sustainable agricultural intensification in china. *Agric. Ecosyst. Environ.* **2015**, *209*, 34–46. [CrossRef]
38. Rehman, K.U.; Hollah, C.; Wiesotzki, K.; Rehman, R.U.; Rehman, A.U.; Zhang, J.; Zheng, L.; Nienaber, T.; Heinz, V.; Aganovic, K. Black soldier fly, *Hermetia illucens* as a potential innovative and environmentally friendly tool for organic waste management: A mini-review. *Waste Manag. Res.* **2023**, *41*, 81–97. [CrossRef]
39. Lim, J.W.; Mohd-Noor, S.N.; Wong, C.Y.; Lam, M.K.; Goh, P.S.; Beniers, J.; Oh, W.D.; Jumbri, K.; Ghani, N.A. Palatability of black soldier fly larvae in valorizing mixed waste coconut endosperm and soybean curd residue into larval lipid and protein sources. *J. Environ. Manag.* **2019**, *231*, 129–136. [CrossRef]
40. Hao, J.; Liu, S.; Luo, A.; Zhao, J.; Shi, S.; Zhang, Y.; Li, C. Assessing nursery-finishing pig manures on growth of black soldier fly larvae. *Animals* **2023**, *13*, 452. [CrossRef]
41. De Souza Vilela, J.; Alvarenga, T.I.R.C.; Andrew, N.R. Technological quality, amino acid and fatty acid profile of broiler meat enhanced by dietary inclusion of black soldier fly larvae. *Foods* **2021**, *10*, 297. [CrossRef]
42. Seyedalmoosavi, M.M.; Dannenberger, D.; Pfuhl, R. Lipid metabolism, fatty acid composition and meat quality in broilers supplemented with increasing levels of defrosted black soldier fly larvae. *J. Insects Food Feed* **2023**, *9*, 583–598. [CrossRef]
43. Shi, Z.H.; Zhang, J.; Jiang, Y.J.; Wen, Y.T.; Gao, Z.H.; Deng, W.H.; Yin, Y.M.; Zhu, F. Two low-toxic Klebsiella pneumoniae strains from gut of black soldier fly *Hermetia illucens* are multi-resistance to sulfonamides and cadmium. *Environ. Pollut.* **2022**, *312*, 120062. [CrossRef] [PubMed]
44. Deng, B.; Zhu, J.; Wang, G.; Xu, C.; Zhang, X.; Wang, P.; Yuan, Q. Effects of three major nutrient contents, compost thickness and treatment time on larval weight, process performance and residue component in black soldier fly larvae (*Hermetia illucens*) composting. *J. Environ. Manag.* **2022**, *307*, 114610. [CrossRef] [PubMed]
45. Gerber, P.J.; Steinfeld, H.; Henderson, B.; Mottet, A.; Opio, C.; Dijkman, J.; Falcucci, A.; Tempio, G. *Tackling Climate Change through Livestock: A Global Assessment of Emissions and Mitigation Opportunities*; Food and Agriculture Organization: Rome, Italy, 2013.
46. Parodi, A.; De Boer, I.J.; Gerrits, W.J.; Van Loon, J.J.; Heetkamp, M.J.; Van Schelt, J.; Bolhuis, J.E.; Van Zanten, H.H. Bioconversion efficiencies, greenhouse gas and ammonia emissions during black soldier fly rearing—A mass balance approach. *J. Clean. Prod.* **2020**, *271*, 122488. [CrossRef]
47. Hou, Y.; Velthof, G.L.; Lesschen, J.P.; Staritsky, I.G.; Oenema, O. Nutrient recovery and emissions of ammonia, nitrous oxide, and methane from animal manure in Europe: Effects of manure treatment technologies. *Environ. Sci. Technol.* **2017**, *51*, 375–383. [CrossRef] [PubMed]
48. Baker, A. Fluorescence excitation-emission matrix characterization of some sewage-impacted rivers. *Environ. Sci. Technol.* **2001**, *35*, 948–953. [CrossRef] [PubMed]
49. Ahmad, S.; Reynolds, D. Monitoring of water quality using fluorescence technique: Prospect of on-line process control. *Water Res.* **1999**, *33*, 2069–2074. [CrossRef]

50. Coble, P.G.; Green, S.A.; Blough, N.V.; Gagosian, R.B. Characterization of dissolved organic matter in the black sea by fluorescence spectroscopy. *Nature* **1990**, *348*, 432–435. [CrossRef]
51. He, X.S.; Xi, B.D.; Jiang, Y.H.; Li, M.X.; Yu, H.B.; An, D.; Yang, Y.; Liu, H.L. Elemental and spectroscopic methods with chemometric analysis for characterizing composition and transformation of dissolved organic matter during chicken manure composting. *Environ. Technol.* **2012**, *33*, 2033–2039. [CrossRef]
52. Bhadra, R.; Muthukumarappan, K.; Rosentrater, K. Physical and chemical characterization of fuel ethanol coproducts relevant to value-added uses. *Cereal Chem.* **2010**, *87*, 439–447. [CrossRef]
53. Yang, Y.; Wang, G.Y.; Li, G.X.; Ma, R.N.; Kong, Y.L.; Yuan, J. Selection of sensitive seeds for evaluation of compost maturity with the seed germination index. *Waste Manag.* **2021**, *136*, 238–243. [CrossRef]
54. Wang, D.M.; Wu, D.; Li, J.; Zhang, L.L. Effects of water extraction ratio of pig manure and its compost on seed germinative indicators. *J. Agro Environ. Sci.* **2011**, *30*, 579–584.
55. Wang, G.; Kong, Y.; Liu, Y.; Li, D.; Zhang, X.; Yuan, J.; Li, G. Evolution of phytotoxicity during the active phase of co-composting of chicken manure, tobacco powder and mushroom substrate. *Waste Manag.* **2020**, *114*, 25–32. [CrossRef] [PubMed]
56. Liu, Y.; Feng, J.; Wang, Y.; Lv, J.; Li, J.; Guo, L.; Min, Y. Fermented corn-soybean meal mixed feed modulates intestinal morphology, barrier functions and cecal microbiota in laying hens. *Animals* **2021**, *11*, 3059. [CrossRef] [PubMed]
57. Diener, S.; Zurbrügg, C.; Tockner, K. Conversion of organic material by black soldier fly larvae: Establishing optimal feeding rates. *Waste Manag. Res.* **2009**, *27*, 603–610. [CrossRef] [PubMed]
58. Gigliotti, G.; Kaiser, K.; Guggenberger, G.; Haumaier, L. Differences in the chemical composition of dissolved organic matter from waste material of different sources. *Biol. Fertil. Soils* **2002**, *36*, 321–329. [CrossRef]
59. He, X.S.; Xi, B.D.; Jiang, Y.H.; He, L.S.; Li, D.; Pan, H.W.; Bai, S.G. Structural transformation study of water-extractable organic matter during the industrial composting of cattle manure. *Microchem. J.* **2013**, *106*, 160–166. [CrossRef]
60. Komatsu, K.; Onodera, T.; Kohzu, A.; Syutsubo, K.; Imai, A. Characterization of dissolved organic matter in wastewater during aerobic, anaerobic, and anoxic treatment processes by molecular size and fluorescence analyses. *Water Res.* **2020**, *171*, 115459. [CrossRef]
61. Wang, Q.; Awasthi, M.K.; Zhao, J.; Ren, X.; Wang, M.; Li, R.; Wang, Z.; Zhang, Z. Utilization of medical stone to improve the composition and quality of dissolved organic matter in composted pig manure. *J. Clean. Prod.* **2018**, *197*, 472–478. [CrossRef]
62. Ministry of Ecology and Environment of the People's Republic of China. *Soil and Sediment-Determination of Aqua Regia Extracts of 12 Metal Elements-Inductively Coupled Plasma Mass Spectrometry*; China Environmental Press: Beijing, China, 2016.

molecules

MDPI

Article

Synthesis of Iron on Carbon Foam for Use in the Removal of Phenol from Aqueous Solutions

Siphesihle Praise-God Khumalo *, David Lokhat , Chante Jasmine-Tre Anwar and Huvin Reddy

School of Engineering, Discipline of Chemical Engineering, University of KwaZulu-Natal, Durban 4000, South Africa

* Correspondence: spgkhumalo@gmail.com

Abstract: The potential use of magnetic nanopowder for phenol adsorption mobilised on natural grain carbon foam from an aqueous solution was studied. Phenolic compounds are priority pollutants with high toxicity even at low concentrations. A magnetic nanopowder was synthesised by dissolving an iron sponge in nitric acid to produce iron nitrate, which was added to a natural grain mixture with flour as the main ingredient. The synthesised carbon foam was investigated for the effects of initial concentration, time, and TEM (transmission electron microscopy) characterisation. The phenol adsorption increased as the iron content of the carbon foam and the initial concentration increased. A kinetic study showed that the phenol adsorption data adequately covered all the carbon foam samples tested using an equation corresponding to a pseudo-first order chemical reaction. The Freundlich, Langmuir, and Temkin equations were tested for modelling the adsorption isotherms at equilibrium, and it was concluded that the Temkin model fit the experimental data adequately. Due to its exceptional physical and chemical properties, carbon magnetic nanopowder is regarded as an outstanding pollutant absorber in environmental investigations. R^2 values derived from the pseudo-first-order model exceed 0.99. $R^2 > 0.94$ indicates that the Freundlich isotherm provides the best fit to the equilibrium data.

Keywords: adsorption; phenol; carbon foam

1. Introduction

In recent years, the removal of phenol has become recognised as an important area of research within the discipline of chemical engineering [1]. Phenolic compounds are high on the priority pollutants list because they can pose a significant risk even in small amounts [2]. Phenol that is contained in wastewater can form complicated compounds by combining with the metal ions that are discharged from other industries [3]. Phenols are used to manufacture phenolic resins, epoxy resins, adhesives, and polyamide for a variety of commercial applications. Coal conversion at high temperatures, petroleum refining, resins, and plastics all produce wastewater containing phenolic pollutants. Aromatic hydroxy compounds, which are toxic at high concentrations and suspected carcinogens, are a top pollutant. Therefore, phenol must be removed from industrial effluents prior to their discharge into waterways [4]. Even at low concentrations, phenolic chemicals are harmful. If these compounds are released without first being handled, it is probable that significant health problems may be posed to both human and animal populations as well as to aquatic ecosystems [5]. The establishment of rigorous discharge regulations for phenols has been mandated by international regulatory organisations to make certain that the environment will remain habitable [6].

To properly extract phenolic compounds from wastewater before it is disposed of in water resources, a variety of technologies are used [7]. Typical removal methods include steam distillation, liquid–liquid extraction, adsorption, solid-phase extraction, wet air oxidation, catalytic wet air oxidation, and biodegradation [8]. Electrochemical oxidation, photo-oxidation, ozonation, UV/H_2O_2, the Fenton reaction, membrane processes, and enzymatic treatment are among the most advanced methods for removing phenols [9]. Physical, chemical, and biological treatment are the three conventional categories for the elimination of phenolic contaminants from aqueous solutions [10].

Adsorption is the most efficient method for removing organic and inorganic contaminants from wastewater [11] because it is simple to set up, inexpensive, does not require much time, and the adsorbent used in the process is not harmful to the environment and can be recovered and reused without a reduction in efficiency [12]. Therefore, the hunt for inexpensive and readily accessible adsorbents has prompted numerous researchers to seek more cost-effective and effective strategies for employing natural and synthetic materials as adsorbents [13]. The most extensively employed adsorbent is activated carbon. It has a high capacity for the adsorption of organic molecules [14].

Carbon foam's properties, such as its specific surface area, surface functional groups, pore size distribution, and other surface characteristics, may affect its ability to absorb phenol and how it does so.

The primary advantages of carbon as an adsorbent over other techniques are its ability to remove both organic and inorganic chemicals by batch or column methods and its regeneration after repeated usage. Recent research, however, has proven that nanoparticles have highly specific surface areas [15].

Agriculture wastes such as coir pith, waste wood, orange peel, bagasse, coffee husk, pinecone, sunflower seed hull, coconut tree, hazelnut husks, rice hulls, oil palm shell, coconut husk and pine-fruit shell were effectively converted into activated carbon. Carbonisation is the process of removing non-carbon components, such as oxygen and hydrogen, in gaseous form by pyrolysis. However, the adsorption capacity of the resulting carbonised product is still limited, necessitating the immobilisation of nanoparticles to increase its adsorption [7]. Even though various research has been published on the use of metal nanoparticles as catalysts, this article focuses on the application as a sorbent, which is less demanding and may be robustly made using low-cost materials.

Magnetic nanopowder is thought to be one of the most effective and cost-efficient adsorbents for removing organic and inorganic pollutants from aquatic environments [16]. It has a high surface area and a high porosity, and it is cheap, abundant, made from renewable sources, stable at high temperatures, and resistant to all chemicals [1]. Hence, in this study, the use of magnetic nanopowder for phenol adsorption on natural grain carbon foam to reduce the concentration of phenol in diverse solutions was investigated. In addition to altering the phenol content in the solution while maintaining the nanoparticle mass, stirring time, and stirring speed constant, additional studies were conducted to determine the impact of nanoparticle contact time with the phenol solution.

In this study, the carbon precursor and magnetic nanopowder precursor were mixed and carbonised in a single vessel. When the natural grain is baked and carbonised at the same time, a porous structure is formed that can support the iron oxide particles well.

2. Results and Discussion

2.1. Characterisation of Carbon Foam

The presence of iron accelerates the action of yeast, namely, the consumption of sugars in the grain combination and the creation of carbon dioxide, which is responsible for the formation of pores in the carbonised material. The presence of iron nanoparticles enhances the adsorption capabilities of an adsorbent. Due to their ultra-fine diameters and increased surface areas, these iron oxide nanoparticles have demonstrated a remarkable capacity for the adsorption of pollutants with high reactivity, high removal efficiencies, and quick operation [17]. Phenol adsorption to negatively charged iron nanoparticles is predominantly promoted by electrostatic interactions and is very sensitive to pH and ionic strength changes. Anionic (negatively charged) iron oxide nanoparticles dispersed in deionised water and cationic phenol, which adsorbs onto the nanoparticles' surface and initiates their aggregation. In the presence of a magnetic field, dipolar interactions between primordial agglomerates induce agglomeration [18]. The existence of cavities in the carbon structure is advantageous for the adsorption process because they allow phenol molecules to penetrate the adsorbent. Figure 1 shows carbon foam samples after carbonisation from the furnace. Figure 2a–e depicts the TEM micrographs of the iron immobilised in natural grain carbon foam. It is evident that the addition of iron nitrate particles to carbon foam produced a structure that is more amorphous and porous, see Figure 2. The surface of the material featured geometrical properties, such as an irregular form, large agglomerates, and a rough surface, which would provide extra sites for adsorbing heavy metal ions [19]. Figure 2a,b,d illustrate the findings of a TEM investigation at 200 nm magnification, which demonstrated that iron oxide was successfully distributed across the surface of the carbon foam. The presence of iron oxide is indicated by the number of dark crystallites in the image. If the iron concentration in the sample is too high, as depicted in Figure 2c,e, agglomeration may occur, which enlarges the iron particles and reduces the sorbent's surface area and adsorptive properties [20]. These characteristics have been demonstrated to be helpful for synthetic carbon foam.

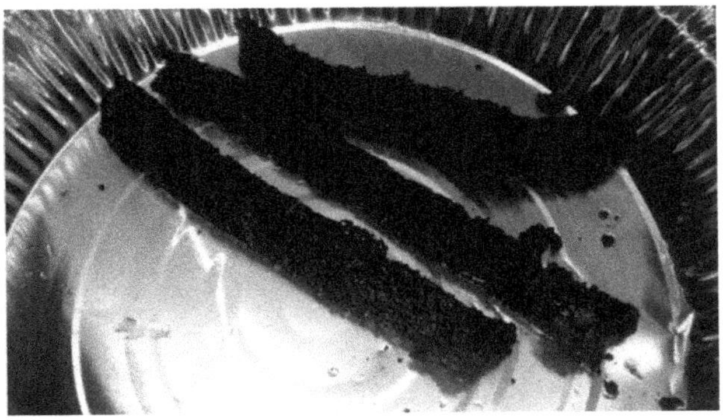

Figure 1. Photograph of carbon foam from the furnace.

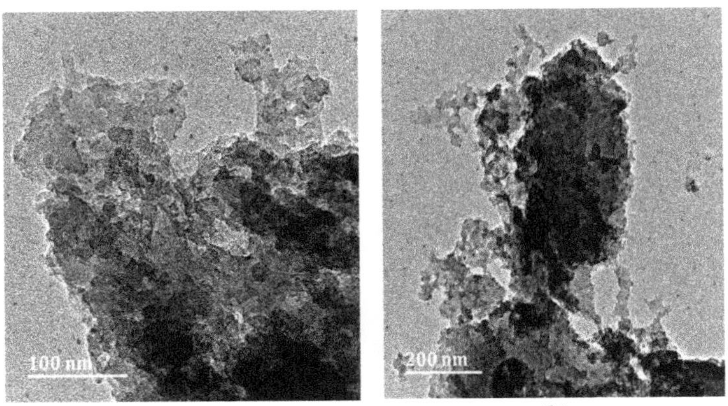

(**a**) (**b**)

Figure 2. *Cont.*

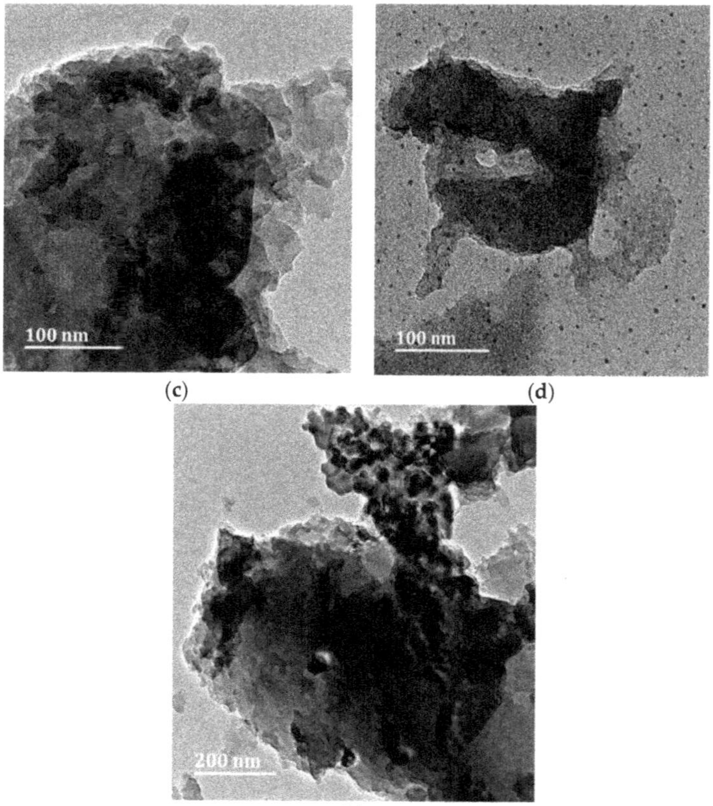

(c) (d)

(e)

Figure 2. TEM image conducted on a Jeol 2100 HRTEM operating at 200 KV: (**a**) Sample 1, (**b**) Sample 2, (**c**) Sample 3, (**d**) Sample 4, and (**e**) Sample 5.

2.2. Effect of Initial Concentration

Shaking the adsorbent–adsorbate solution until equilibrium is reached for increments of the initial adsorbate concentration at a similar adsorbent dosage and time can be used to study the effect of the initial concentration [21]. Adsorption depends on the initial concentration of an adsorbate, since lowering the initial concentration of phenol molecules lowers the surface area. As the initial concentration increases, the driving force of mass transfer increases, enhancing adsorption [22]. Figure 3 shows the effect of the initial concentration of phenol, where initial concentrations of 25, 50, 75, and 100 mg/L were adsorbed by iron immobilised on natural grain carbon foam in four different samples. There was an increase in adsorption with an increase in the initial concentration at room temperature with a contact time of 10 min and 25 mg/L of solution. The amount of available surface area on adsorbents goes down as the initial concentration increases. This means that more adsorbate is taken up and less phenol is removed.

The following equation was used to calculate the amount of phenol adsorbed [19]:

$$q_t = \frac{(C_0 - C) \times v}{w} \quad (1)$$

where q_t (mg/g) is the adsorption density, C_0 (mg/L) is the initial concentration of aqueous solution, C (mg/L) denotes the final aqueous solution concentration (mg/L), v (L) the volume of solution, and w (g) the mass of adsorbent.

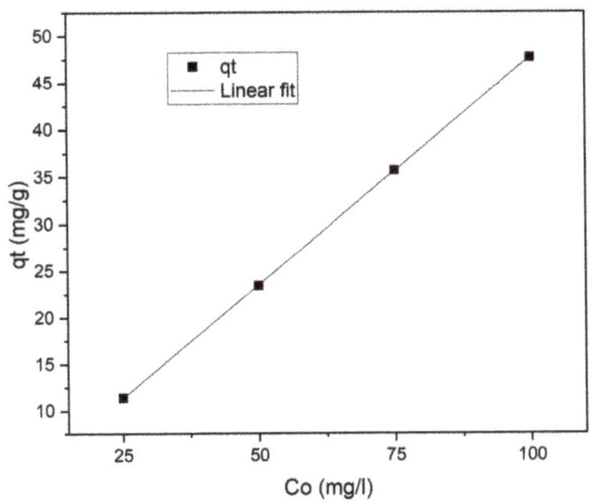

Figure 3. The effect of initial concentration on adsorption.

Four phenol solutions were prepared with concentrations ranging from 25, 50, 75, and 100 mg/L. A plot of the percentage of phenol adsorbed versus the initial concentration was made for each sample of nanopowder in each solution concentration. Figure 4 indicates that the percentage of phenol adsorbed increased for each sample as the initial concentration increased. This demonstrates that more light is being absorbed by a solution of greater concentration. This makes sense because when the concentration of a solution goes up, there are more particles for the light to hit, so there are more chances for absorption [23].

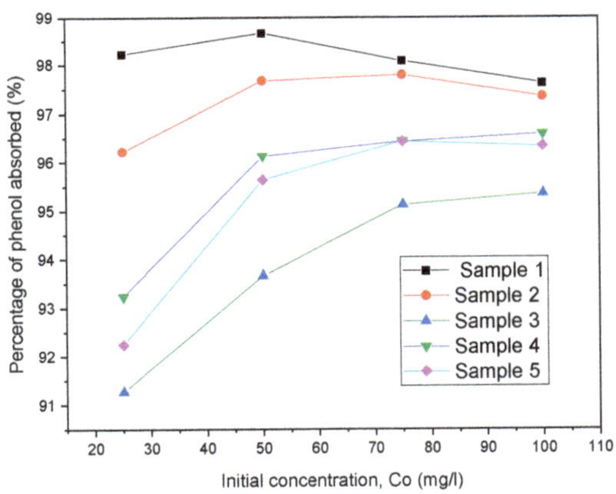

Figure 4. The effect of initial concentration on percentage of phenol adsorbed (50 mL of phenol solution and 0.1 magnetic nanopowder).

2.3. *The Effect of Time on Adsorption*

Adsorbate rate rises with longer contact times because it deposits on the accessible adsorption sites on the adsorbent material; nevertheless, longer contact times will not result in an increase in uptake [24]. The amount of adsorbate adsorbed onto the adsorbent is in equilibrium with the amount of adsorbate desorbing from the adsorbent. The effectiveness of adsorbate removal is substantially impacted by the length of time that the two components are in contact [3].

Figure 5 displays the results obtained, and for each sample, phenol adsorption increased with contact time. The magnetite solution displayed the least phenol adsorption. This can be attributed to the uneven distribution of the solid particles in the initial organic mixture. The sample of undissolved iron sponges yielded similar results to the previous sample for the same reasons. The three samples that made use of liquid iron nitrate yielded better results, as there was more even distribution in the organic mixture, resulting in a sample of constant composition. An increase in the amount of iron sponge added to the mixture increased phenol adsorption. A clarification for a decline in the adsorption rate after 15 min is that the solutions are approaching equilibrium. This can be attributed to the large number of vacant sites on the magnetic nanoparticle surface, which are gradually occupied over time as adsorption takes place.

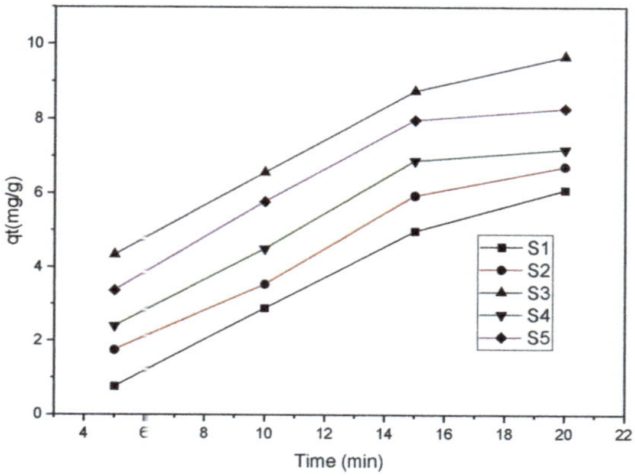

Figure 5. The effect of contact time on the adsorption of phenol at concentration of 75 mg/L and 100 mg.

2.4. Mechanism

Using kinetic models, the rate of the adsorption process and the rate-controlling step are explored to fully comprehend the phenomena involved in phenol adsorption on magnetic nanopowder. In addition, it is crucial to consider the fact that the adsorption mechanism is dependent on the physical and/or chemical properties of the adsorbent as well as the mass transport process. To investigate the mechanism of phenol adsorption onto magnetic nanopowder, pseudo-first-order, pseudo-second order, and intraparticle diffusion models are considered to fit the experimental data obtained from batch studies, and the correlation coefficient is regarded as a measure of the relationship between the experimental data and these proposed models.

Since intraparticle diffusion was not the only rate-limiting step in the adsorption of phenol onto iron supported by carbon foam, the adsorption rate was controlled by a multistep elementary reaction mechanism. Other processes may also affect the adsorption rate, and they may all be active concurrently. Consequently, the Temkin isotherm best characterises the phenol adsorption process, which suggests multilayer adsorption over a heterogeneous surface with varied energy distribution. On the adsorption isotherm, the Temkin model reflects the features of indirect adsorbate–adsorbent interactions. Due to adsorbate–adsorbent interactions, it is believed that the heat of adsorption of all molecules in the layer reduces linearly with coverage. Due to adsorbate–adsorbent interactions, the adsorption of all molecules in the layer reduces linearly with increasing layer coverage. Phenol adsorption on negatively charged iron nanoparticles is primarily facilitated by electrostatic interactions and is extremely sensitive to changes in pH and ionic strength. Nanoparticles of anionic (negatively charged) iron oxide are distributed in deionised water with cationic phenol, which adsorbs on the nanoparticles' surface and triggers their aggregation. In the presence of a magnetic field, primordial agglomerates form due to dipolar interactions.

The mechanism for phenol adsorption on carbon foam as an adsorbent is based on their mainly sp2-dominated electronic structures. As a result, one could speculate that the adsorption mechanism for this system involves interactions between delocalised electrons on the carbon surface structure and electrons from the aromatic ring [25].

2.5. Kinetic Modelling

Lagergren pseudo-first order and pseudo-second order models were used for the investigation of modelling of adsorption kinetics. The nonlinear form of pseudo-first order is given by the following equation:

$$q_t = q_e(1 - e^{-K_1 t}) \quad (2)$$

The nonlinear form of pseudo-second order is given by the following equation:

$$q_t = \frac{q_e{}^2 K_2 t}{1 + K_2 q_e t} \quad (3)$$

where q_t (mg/g) is the amount absorbed at time t, q_e (mg/g) is the amount remaining after equilibrium of adsorption, and K_1 and K_2 are the pseudo-first and pseudo-second order model rate constants, expressed in min^{-1} and g/mg/min, respectively. Tables 1 and 2 illustrates the calculated values of q_e, K_1, and K_2, and the regression coefficient R^2 values. Figure 6 shows the plot of the pseudo-first order and pseudo-send order. The validity of the model is judged by evaluating the correlation coefficients R^2 and the comparability of the experimental and calculated values of q_e. Considering these findings, it is possible to conclude that the pseudo-first-order kinetic model gives a stronger correlation for phenol adsorption onto magnetic nanopowder immobilised on carbon foam than the other models. The results indicate that R^2 is closer to 1 for pseudo-first order compared to the pseudo-second order model for all five samples.

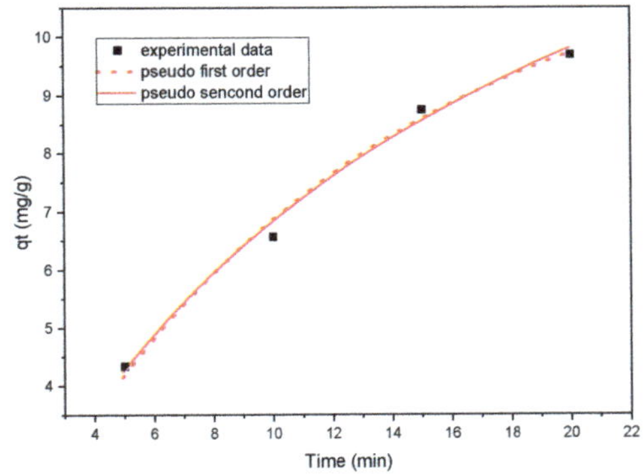

Figure 6. Nonlinear fit of pseudo-first order and pseudo-second order for phenol adsorption of sample 3 at a 75 mg/L of concentration and 100 mg sorbent.

Table 1. Pseudo-first order parameters.

	Exp. qe	Calc. qe	K_1	R^2
Sample 1		39.99×10^2	7.64×10^{-8}	0.953
Sample 2	7.3	26.21	0.015	0.974
Sample 3	6.5	11.76	0.872	0.992
Sample 4	9.8	11.84	0.050	0.963
Sample 5	8.5	10.74	0.080	0.976

Table 2. Pseudo-second order parameters.

	Exp. qe	Calc. qe	K_2	R^2
Sample 1		46.54×10^2	1.42×10^{-8}	0.95
Sample 2	7.3	50.49	1.58×10^{-4}	0.97
Sample 3	6.5	17.22	3.79×10^{-4}	0.99
Sample 4	9.8	19.88	1.54×10^{-3}	0.96
Sample 5	8.5	16.34	3.41×10^{-3}	0.97

2.6. Intraparticle Diffusion

The intraparticle diffusion resistance was evaluated utilising the intraparticle particle diffusion model by Weber and Morris [26], given by the following equation:

$$q_t = k_{id}t^{1/2} + c \tag{4}$$

where q_t (mg/g) is the adsorption amount at time t (min), k_{id} (mg/g/min$^{1/2}$) is the adsorption rate constant of the intraparticle diffusion model, and c is a constant related to the thickness of the boundary layer. The plot illustrated in Figure 7 is linear, and the sorption process is solely regulated by intra-particle diffusion since the plot of q_t versus $t^{1/2}$ yields a straight line. The difference in mass transfer rates between the start and final stages of adsorption may be the cause of the plot's departure from the origin. If intraparticle diffusion is involved in the sorption process, a graph of adsorbate uptake vs. the square root of time would have a linear connection, and intraparticle diffusion would be the rate-regulating step if this line passes through the origin [27].

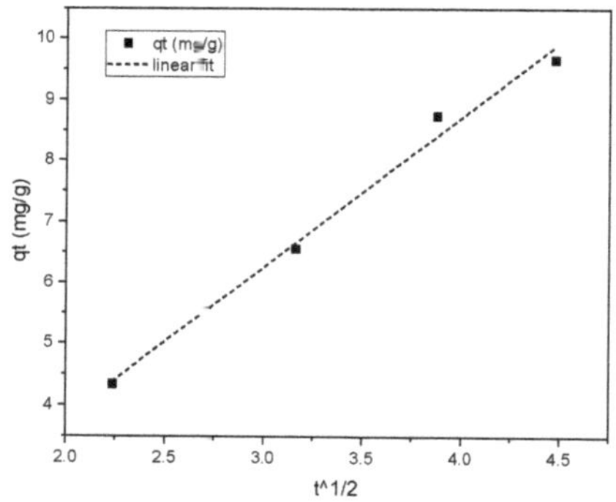

Figure 7. Plot of the intraparticle diffusion model for phenol onto iron supported by carbon foam (75 mg/L of concentration, and 100 mg sorbent).

Furthermore, such a departure from the origin of the straight line suggests that pore diffusion is not the only factor affecting rate [28].

2.7. Equilibrium Adsorption

Adsorption equilibrium occurs between the adsorbed molecules and the adsorbent surface when an adsorbate is in contact with the adsorbent. The adsorption isotherm is the equilibrium relationship between the amount of adsorbed adsorbate (q_e) and the residual adsorbate concentration (C_e) at a constant temperature. In general, adsorption isotherms provide information on the affinity and the binding energy between the adsorbate and the adsorbent, on the adsorption capacity, and on the surface phase, which may be considered a monolayer or multilayer. The modelling of the adsorption isotherms consists of describing the exponential data using theoretical or empirical mathematical equations and allowing the determination of isotherm parameters to compare the efficiency of different adsorbents.

Three adsorption equilibrium isotherm models were tested in the present research, namely, the Freundlich, Langmuir, and Temkin isotherm models. The empirical Freundlich model, which is known to be satisfactory for low concentrations and is based on sorption on the heterogeneous surface, is expressed by the following equation:

$$q_e = K_F C_e^{1/n} \tag{5}$$

where C_e (mg/L) is the equilibrium concentration of adsorbate, q_e (mg/g) is the amount of adsorption at the equilibrium. K_F and n are Freundlich constants related to the adsorption capacity and adsorption intensity, respectively. All the parameters of the isotherm models were calculated from the nonlinear fitting of q_e versus C_e on Origin Lab software. The Temkin model reflects the properties of indirect adsorbate–adsorbent interactions on the adsorption isotherm. It is assumed that the heat of adsorption of all molecules in the layer decreases linearly with coverage due to adsorbate–adsorbent interactions. The Temkin model is expressed by the following equation:

$$q_e = \frac{RT}{b} ln(AC_e) \tag{6}$$

where q_e (mg/g) is the amount of adsorption at the equilibrium, C_e (mg/L) is the equilibrium concentration of adsorbate, T (K) is the temperature in Kelvin, R (J/mol/K) is the universal gas constant, b (J/mol) is related to the heat of adsorption and K_T is the equilibrium binding constant corresponding to the maximum binding energy.

The Langmuir isotherm equation is expressed as follows:

$$q_e = \frac{q_m K_L C_e}{1 \pm K_L C_e} \tag{7}$$

where q_e (mg/m) is the monolayer adsorption capacity, C_e (mg/L) is the equilibrium concentration of adsorbate, q_m (mg/g) is the monolayer adsorption capacity, and K_L is the Langmuir constants.

Figure 8 illustrates the curves of the Freundlich, Temkin and Langmuir models using Equations (5)–(7), respectively. Table 3 depicts the calculated parameters of the Freundlich and Temkin isotherms, the R^2 values obtained by the nonlinear fitting method. Based on the R^2 value comparisons, the Freundlich model represents a better fit of the experimental data at equilibrium compared to both the Langmuir and Temkin models. Thus, the Temkin isotherm best describes the phenol adsorption process, which indicates multilayer adsorption on a heterogeneous surface with different energy distribution. The Freundlich constant, n, is a measure of adsorption intensity. A value of 1/n was found to be between 0 and 1, indicating that Freundlich was also favourable for the adsorption of phenol [29].

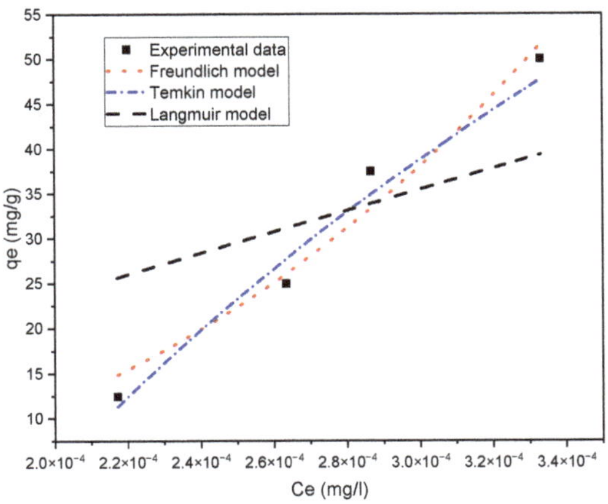

Figure 8. Fitting of adsorption isotherms into experimental data for sample 3.

Table 3. Temkin, Freundlich, and Langmuir model parameters.

Sample	Temkin			Freundlich			Langmuir		
	b	K_T	R^2	n	K_F	R^2	q_m	K_L	R^2
1	38.76	7920.14	0.98	1.97	16,564.30	0.91	82750.9	1.96	0.78
2	30.89	5480.35	0.94	2.35	60,792.30	0.87	88,020.60	1.19	0.66
3	29.16	5260.43	0.98	2.91	401,434.60	0.97	105,489.00	1.12	0.61
4	24.53	4494.55	0.97	2.99	577,825.00	0.93	159,254.50	0.67	0.52
5	23.1	4075.37	0.94	2.92	252,312.60	0.87	83,231.50	1.27	0.55

3. Materials and Methods

3.1. Materials and Chemicals

The following materials were used: deionised water, phenol, iron oxide nanoparticles, iron sponge, nitric acid, magnetite, flour, and yeast.

3.2. Preparation of Carbon Foam

Phenol removal is a topic of particular interest in the field of preventing and getting rid of pollution because it is toxic to water quality and can hurt people even in small amounts. Adsorption is still one of the best ways to get rid of something. It became possible when magnetic nanopowder was made in a carbon foam suspension. The 5 grammes of dry yeast was dissolved in 115 mL of water with continuous stirring. Once the yeast and water were completely dissolved, the mixture was added to 300 g of flour. A varying amount of iron nitrate was added into the flour–yeast mixture. This was performed by mixing iron sponge with nitric acid to make iron nitrate, which was then added to a mixture of natural grains to bake. The mixture was then mixed with water, and the mechanical mixer was used to knead the mixture until it formed a paste. The porous structure was made by letting the paste ferment in an oven at 35 °C for an hour. The bread was then allowed to bake in an oven set at 180 °C for 40 min. Thereafter, the bread was allowed to dry in an oven at 80 °C for 18 h. A U-tube furnace under argon gas conditions was set up. The dry bread was placed into the furnace and allowed to carbonise at 10 °C/min. The oven was set to heat up at a rate of 10 °C per minute and reach a maximum temperature of 600 °C. The holding time for the maximum temperature was set at 120 min. The sample was allowed to cool in the argon atmosphere until it reached room temperature. When different amounts of liquid iron nitrate were added to the flour, yeast, and water mix, four different organic samples were made. To make a second sample, magnetite was added to the dry ingredients, and then water was added. Table 4 depicts the iron contents for each sample.

Table 4. Iron and magnetite added into carbon form for each sample.

Sample 1	Sample 2	Sample 3	Sample 4	Sample 5
Magnetite	3 g pure iron	3 g iron sponge dissolved in nitric acid	1 g iron sponge dissolved in nitric acid	2 g of iron sponge dissolved in nitric acid

3.3. Adsorption of Phenol

The beaker-based adsorption studies were conducted in batch model, where 0.1 g of iron oxide nanoparticles was added to 50 mL of the 25 mg/L solution, and the resulting mixture was agitated with an overhead stirrer at 500 rpm for 10 min. Filtration was then used to remove the nanoparticles from the solution, and the clear filtrate was placed into the photocell for testing. This digital absorbance and transmittance data were collected so that the solution's ultimate concentration could be determined using the calibration curve. The stages were performed for each concentration of phenol and water solution at room temperature, roughly 25 °C.

4. Conclusions

The amount of phenol adsorbed for these tests all increased with the increase in initial concentration. Adsorption also increased with the increase in iron content in the carbon form, and the magnetite sample had lower adsorption capabilities compared to other samples. The adsorption values increased rapidly with time on the q_t (mg/g) versus time (min) graph, and then they started to level out as the solutions got closer to equilibrium. This was attributed to the fact that there were a lot of open sites on the magnetic nanoparticle surface, which over time, as adsorption occurred, gradually filled up. This study demonstrates that magnetic nanopowder immobilised on carbon foam significantly reduces phenol concentrations. The correlation between the adsorption data and the pseudo-first-order equation is the highest. The Temkin adsorption model is more suitable for describing the phenol adsorption equilibrium data.

Author Contributions: Conceptualization, S.P.-G.K., D.L., C.J.-T.A. and H.R.; methodology, S.P.-G.K., D.L.; investigation, S.P.-G.K., D.L., C.J.-T.A. and H.R.; resources, C.J.-T.A. and H.R.; writing—review and editing, S.P.-G.K., D.L., C.J.-T.A. and H.R.; visualization, S.P.-G.K.; supervision, D.L.; funding acquisition, D.L.; S.P.-G.K. wrote the main manuscript and prepared all figures. All authors have read and agreed to the published version of the manuscript.

Funding: This research was funded by National Research of South Africa. Grant number 11626 and 120828.

Institutional Review Board Statement: Not applicable.

Informed Consent Statement: Not applicable.

Data Availability Statement: Not applicable.

Conflicts of Interest: The authors declare no conflict of interest.

Sample Availability: Samples of the compounds are available from the authors.

References

1. Fathy Mubarak, M.; Maher Ahmed, A.; Saad Gabr, S. Nanoporous carbon materials toward phenolic compounds adsorption. In *Nanopores*; IntechOpen: London, UK, 2021.
2. Bagheri, B.; Hosseini, S.A.; Mehrizadeh, H. A Comparison of the Catalytic Activity of Cu-X_2 (X = Mn, Co) Nano Mixed Oxides toward Phenol Remediation from Wastewater by Catalytic Wet Peroxide Oxidation. *J. Water Environ. Nanotechnol.* **2020**, *5*, 139–146.
3. Fierro, V.; Torné-Fernández, V.; Montané, D.; Celzard, A. Adsorption of phenol onto activated carbons having different textural and surface properties. *Microporous Mesoporous Mater.* **2007**, *111*, 276–284. [CrossRef]
4. Kandisa, R.V.; Saibaba, K.N.; Shaik, K.B.; Gopinath, R. Dye Removal by Adsorption: A Review. *J. Bioremediat. Biodegrad.* **2016**, *7*, 371. [CrossRef]
5. Adeleye, A.S.; Conway, J.R.; Garner, K.; Huang, Y.; Su, Y.; Keller, A.A. Engineered nanomaterials for water treatment and remediation: Costs, benefits, and applicability. *J. Chem. Eng.* **2016**, *286*, 640–662.
6. Villegas LG, C.; Mashhadi, N.; Chen, M.; Mukherjee, D.; Taylor, K.E.; Biswas, N. A Short Review of Techniques for Phenol Removal from Wastewater. *Curr. Pollut. Rep.* **2016**, *2*, 157–167. [CrossRef]
7. Asencios, Y.J.O.; Lourenço, V.S.; Carvalho, W.A. Removal of phenol in seawater by heterogeneous photocatalysis using activated carbon materials modified with TiO_2. *Catalysis Today* **2022**, *388–389*, 247–258. [CrossRef]
8. Pirillo, S.; Pedroni, V.; Rueda, E.; Ferreira, M.L. Elimination of dyes from aqueous solutions using iron oxides and chitosan as adsorbents: A comparative study. *Química Nova* **2009**, *35*, 1239–1244. [CrossRef]
9. Mohammadi, S.; Kargari, A.; Sanaeepur, H.; Abbassian, K.; Najafi, A.; Mofarrah, E. Phenol removal from industrial wastewaters: A short review. *Desalination Water Treat.* **2015**, *53*, 2215–2234. [CrossRef]
10. Oyelude, E.O.; Owusu, U.R. Adsorption of Methylene Blue from Aqueous Solution Using Acid Modified Calotropis Procera Leaf Powder. *J. Appl. Sci. Environ. Sanit.* **2011**, *6*, 477–484.
11. Shahid, M.; Farooqi, Z.H.; Begum, R.; Arif, M.; Irfan, A.; Azam, M. Extraction of cobalt ions from aqueous solution by microgels for in-situ fabrication of cobalt nanoparticles to degrade toxic dyes: A two fold-environmental application. *Chem. Phys. Lett.* **2020**, *754*, 137645. [CrossRef]
12. Turco, A.; Monteduro, A.G.; Mazzotta, E.; Maruccio, G.; Malitesta, C. An Innovative Porous Nanocomposite Material for the Removal of Phenolic Compounds from Aqueous Solutions. *Nanomaterials* **2018**, *8*, 334. [CrossRef] [PubMed]

13. Gupta, S.; Babu, B.V. Experimental, kinetic, equilibrium and regeneration studies for adsorption of Cr(VI) from aqueous solutions using low cost adsorbent (activated flyash). *Desalination Water Treat.* **2010**, *20*, 168–178. [CrossRef]
14. Ho, Y.S.; McKay, G. Kinetic models for the sorption of dye from aqueous solution by wood. *Process. Saf. Environ. Prot.* **1998**, *76*, 183–191. [CrossRef]
15. Arif, M. Complete life of cobalt nanoparticles loaded into cross-linked organic polymers: A review. *R. Soc. Chem.* **2022**, *12*, 15447. [CrossRef] [PubMed]
16. Negrescu, A.M.; Killian, M.S.; Raghu, S.N.; Schmuki, P.; Mazare, A.; Cimpean, A. Metal Oxide Nano-particles as an Adsorbent for Removal of Heavy Metals. *J. Adv. Chem. Eng.* **2015**, *5*, 274.
17. Arif, M.; Munawar, K.; Ali, S.; Raza, H.; Ashfaq, M. Synthesis and Optical Study of Sensitive and Selective Detection of Fe(III) ions Based on Solvents. *Egytian J. Chem.* **2020**, *63*, 4029–4038. [CrossRef]
18. Talbot, D.; Queiros Campos, J.; Checa-Fernandez, B.L.; Marins, J.A.; Lomenech, C.; Hurel, C.; Godeau, G.D.; Raboisson-Michel, M.; Verger-Dubois, G.; Obeid, L.; et al. Adsorption of Organic Dyes on Magnetic Iron Oxide Nanoparticles. Part I: Mechanisms and Adsorption-Induced Nanoparticle Agglomeration. *ACS Omega* **2021**, *6*, 19086–19098. [CrossRef] [PubMed]
19. Zygmunt Marczenko, M B. *Separation, Preconcentration and Spectrophotometry in Inorganic Analysis*; Elsevier: Amsterdam, The Netherlands, 2000; Volume 10.
20. Taromi, A.A.; Kaliaguine, S. Hydrodeoxygenation of triglycerides over reduced mesostructured Ni/γ- alumina catalysts prepared via one-pot sol-gel route for green diesel production. *Appl. Catal.* **2018**, *558*, 140–149. [CrossRef]
21. Rendo, D. Adsorption of Methylene Blue Dye using Fe_3O_4 Magnetized Natural Zeolite Adsorbent. *J. Kim. Sains Apl.* **2021**, *24*, 51–57. [CrossRef]
22. Selvi, P.P. Mass Transfer Enhancement for CO_2 Absorption in Structured Packed Absorption Column. *J. Chem. Soc. Pak.* **2019**, *41*, 820–824.
23. Asokan, K. *Mass Transfer Concepts*; University Press: New Delhi, India, 2011.
24. El-Sayed, Y.; Bandosz, T.J. A Study of Acetaldehyde Adsorption on Activated Carbons. *J. Colloid Interface Sci.* **2001**, *242*, 44–51. [CrossRef]
25. Luz-Asunción, M.D.; Sánchez-Mendieta, V.; Martínez-Hernández, A.L.; Castaño, V.M.; Velasco-Santos, C. Adsorption of phenol from aqueous solutions by carbon nanomaterials of one and two dimensions: Kinetic and equilibrium studies. *J. Nanomater.* **2015**, *16*, 422. [CrossRef]
26. Weber, W.J.; Morris, J.C. Kinetics of Adsorption on Carbon from Solution. *J. Sanit. Eng. Div.* **1963**, *89*, 31–59. [CrossRef]
27. Banat, F.; Al-Bashir, B.; Al-Asheh, S.; Hayajneh, O. Adsorption of Phenol by Bentonite. *Environ. Pollut.* **2000**, *107*, 391–398. [CrossRef] [PubMed]
28. El Maguana, Y.; Elhadiri, N.; Bouchdoug, M.; Benchanaa, M.; Jaouad, A. Activated Carbon from Prickly Pear Seed Cake: Optimization of Preparation Conditions Using Experimental Design and Its Application in Dye Removal. *Int. J. Chem. Eng.* **2019**, *2019*, 8621951. [CrossRef]
29. Medjor, W.; Wepuaka, C.; Godwill, S. Spectrophotometric Determination of Phenol in Natural Waters by Trichloromethane Extraction Method after Steam Distillation. *Int. Res. J. Pure Appl. Chem.* **2015**, *7*, 150–156. [CrossRef]

molecules

MDPI

Article

Application of Synthesized Vanadium–Titanium Oxide Nanocomposite to Eliminate Rhodamine-B Dye from Aqueous Medium

Mohamed R. Elamin [1,*], Babiker Y. Abdulkhair [1,2], Nuha Y. Elamin [1,2], Khalid H. Ibnaouf [3], Hajo Idriss [3], Rafia Bakheit [4] and Abueliz Modwi [5,*]

1 Chemistry Department, College of Science, Imam Mohammad Ibn Saud Islamic University (IMSIU), Riyadh 13318, Saudi Arabia
2 Chemistry Department, Faculty of Science, Sudan University of Science and Technology (SUST), Khartoum P.O. Box 407, Sudan
3 Physics Department, College of Science, Imam Mohammad Ibn Saud Islamic University (IMSIU), Riyadh 13318, Saudi Arabia
4 Department of Physics, College of Sciences and Arts, Qassim University, Unaizah 51911, Saudi Arabia
5 Department of Chemistry, College of Science and Arts, Qassim University, Ar Rass 52571, Saudi Arabia
* Correspondence: mrabuzaid@imamu.edu.sa (M.R.E.); abuelizkh81@gmail.com (A.M.)

Abstract: In this study, a V@TiO_2 nanocomposite is examined for its ability to eliminate carcinogenic Rhodamine (Rh-B) dye from an aqueous medium. A simple ultrasonic method was used to produce the nanosorbent. In addition, V@TiO_2 was characterized using various techniques, including XRD, HRTEM, XPS, and FTIR. Batch mode studies were used to study the removal of Rh-B dye. In the presence of pH 9, the V@TiO_2 nanocomposite was able to remove Rh-B dye to its maximum extent. A correlation regression of 0.95 indicated that the Langmuir model was a better fit for dye adsorption. Moreover, the maximum adsorption capacity of the V@TiO_2 nanocomposite was determined to be 158.8 mg/g. According to the thermodynamic parameters, dye adsorption followed a pseudo-first-order model. Based on the results of the study, a V@TiO_2 nanocomposite can be reused for dye removal using ethanol.

Keywords: V@TiO_2 nanocomposite; Rh-B dye removal; removal mechanism; regeneration

1. Introduction

Industrial wastewater contains artificial pigments that constitute an environmental risk; thus, toxic dyes should permanently be removed from the aquatic system [1]. Dumping colored wastewater into receiving waterways has become a severe environmental issue worldwide [2]. The release of wool, rug, pulp, sheet, and textile effluents frequently tint the collecting waters for kilometers in the direction of the source [3]. The color is visually unappealing and inhibits the passage of light into the water, lowering the effectiveness of photosynthesis in aquatic plants and negatively affecting their development [4]. The extensive use of dyes in the fabric, publishing, latex, beauty product, plastic, and leather sectors produces a massive volume of colored wastewater [5–7]. Rhodamine B has a comprehensive application, encompassing textiles, food, laser technology, biomarkers, molecular probes, sensitizers, electrochemical-luminescence, and solar panels [8]. rhodamine B (Rh-B) dye is considered a skin-irritating agent, neurotoxic, chronically toxic to aquatic organisms, and carcinogenic to humans. [9]. Thus, purifying water bodies from Rhodamine B is an urgent task to protect the environment and human health. One of the most popular cationic water-soluble organic dyes is Rh-B [10]; it is toxic and lethal to aquatic environments [11]. The numerous procedures for decolorization are categorized as chemical, physical, and microbial techniques [12–14]. Physical treatments involve sorption, ozonation, and membrane processes [15]. While chemical procedures comprise oxidation, precipitation/co-precipitation, and photochemical processes, biological techniques include aerobic decomposition, bacterial destruction, and biosorption [16]. Even though these approaches are successful, they have several drawbacks, including increased chemical consumption and sludge formation, and they are expensive [17].

Among all the dye treatment procedures, adsorption is reported to be efficient and economical [18]. Activated carbon is an efficient adsorbent for removing dyes from industrial sewage effluents; nevertheless, its cost limits its usage [19]. Alongside traditional adsorbents, various economical nonconventional adsorbents have been demonstrated to eliminate dyes efficiently [20]. For the removal of dyes, investigations, including the analysis of practical and inexpensive adsorbents produced from available resources, are increasingly relevant [21]. Employing nanotechnology to clean contaminated environments has shown to be advantageous, saving a lot and lowering pollution levels to tolerable levels [22–24]. Metallic nanoparticles and nanocomposites are regarded as potential adsorbents with high dye removal capability due to their sensitivity, permeability, and recyclability within each known sorbent [25,26]. Lately, metal-based nanocomposites such as MgO/TeSe, $(Y_2O_3)n$–ZnO, GO-TiO_2, Fly-Ash@Fe_3O_4, and Fe_2O_3–TiO_2–graphene have been used to eliminate various dyes from water systems [27–33]. Several proposals for nanoparticle production have involved mechanical, physiochemical, and biological procedures [34–36]. Furthermore, nanostructured and composite materials can be employed to capture or destroy dye contamination in an aqueous environment [37–39]. V-TiO_2 was utilized as the base material to synthesize triple composites and used mostly as a photocatalyst [40–42].

This research aimed to establish a straightforward method for producing V-TiO_2 nanomaterial. The synthesized V-TiO_2 will be analyzed by physical means and introduced as a practical, environmentally friendly, and inexpensive composite to get rid of the coloring dyes from water.

2. Experimental Section

2.1. Fabrication of V@TiO_2 Sorbent

For the V-TiO_2 sorbent, 0.0338 moles of TiO_2 nanoparticles (Sigma Aldrich) were dispersed in 0.12 L isopropanol solvent using an ultrasonic bath for 20.0 min. Then, 0.00165 moles of V_2O_5 were added to the milky TiO_2 solution, and the mixture was sonicated for a further 40.0 min with vigorous stirring (550 rpm). After the mixtures were blended, they were heated for 20 h at 90 °C in an electric dry oven, and the resulting nanoparticles were ground before being calcined. The greenish-white powders were annealed for 2 h at 145 °C.

2.2. Sorbent Characteristics

The structural properties of the as-synthesized sample were investigated by X-ray diffraction (XRD) utilizing a Rigaku diffractometer operated with Cu K radiation (λ = 0.15406 nm) at 40 kV and 40 mA. Fourier transform infrared (FT-IR) spectra were acquired with the KBr pellet (400 to 4000 cm^{-1}) on a Thermo Scientific Nicolet 380 Fourier transform spectrometer. Using a JEM-2100 high-resolution transmission electron microscope (TEM), the microscopic characteristics of the materials were studied. X-ray photoelectron spectroscopy (XPS) was utilized to investigate surface chemistry using an RBD upgraded PHI-5000C ESCA system (Perkin Elmer) with Al K radiation ($h\upsilon = 1486.6$ eV).

2.3. Rh-B Sorption Studies

The batch test protocol was employed to investigate the Rh-B dye adsorption onto the V@TiO_2 nanocomposite. In 0.025 L small bottles, 5 to 100 ppm concentrations of Rh-B dye in distilled water were mixed with 10 mg of the nanocomposite. The mixture solution was constantly agitated for 20 h. Following the establishment of equilibrium with the aqueous phase, the nanocomposite was isolated by filtering, and Rh-B dye concentrations were measured using a UV-vis spectrophotometer. The amount of adsorbed Rh-B dye at any given time (in minutes) and adsorption equilibrium magnitudes Q_t and Q_e (mg/g) were calculated using the same equation (Equation (1)):

$$Q_t = V(C_0 - C_t)/m \tag{1}$$

where V is the volume of the solution in liters; C_0, C_e, and C_t are the starting concentration, equilibrium concentration, and concentration of the Rh-B dye in solution, respectively, in milligrams per liter; and m is the mass of the adsorbent (g). The contact times data were used to investigate the Rh-B sorption via the pseudo-first-order, and pseudo-second-order models (PSFOM, PSSOM) as expressed in equations 2 and 3, respectively. Also, the sorption control mechanism was studied utilizing the liquid film and the intraparticle diffusion models (IDM and LDM) presented by Equations (4) and (5) [43]:

$$q_t = q_e\left(1 - exp^{-K_1 \cdot t}\right) \tag{2}$$

$$q_t = \frac{k_2 \cdot q_e^2 \cdot t}{1 + k_2 \cdot q_e \cdot t} \tag{3}$$

$$\ln(1 - \mathrm{F}) = -\mathrm{K_{LF}} * \mathrm{t} \tag{4}$$

$$\mathrm{q_t} = \mathrm{K_{IP}} * \mathrm{t}^{\frac{1}{2}} + \mathrm{C_i} \tag{5}$$

where q_e (mg g^{-1}) represents q_t at equilibrium and k_1 (min^{-1}) and k_2 (g mg^{-1} min^{-1}) are the rate adsorption constants for the PSFO and PSSO models, which have been calculated from the slope and intercept values, respectively. The LFDM and IPDM constants were represented as K_{IP} (mg g^{-1} min$^{-0.5}$) and K_{LF} (min^{-1}), and both were computed from their slope values. C_i (mg g^{-1}) is a boundary-layer-thickness factor [44].

2.4. pH Point of Zero Charges Experiment

In each flask, with a pH ranging from 1 to 12, ten milligrams of V@TiO_2 nanocomposite and ten milliliters of a 0.1 mole/L sodium chloride (NaCl) solution were added. A trace amount of hydrochloric acid or sodium hydroxide was incorporated into the solution in order to bring about a change in the pH. In order to achieve equilibrium, these bottles were left on a multi-stirrer at room temperature for exactly one hour; then, the pH levels of the solutions were measured. The point zero charges were found by comparing the initial and final pH values against the original pH graph.

3. Results and Discussions

3.1. XRD Analysis of V@TiO_2 Nanocomposite

X-ray diffraction (XRD) examination was conducted to verify the produced V@TiO_2 nanocomposite. Figure 1 depicts XRD data from generated samples pertaining to the structural properties of the V@TiO_2 nanocomposite. Peaks of phase obtained at 2*theta* = 27.51, 36.19, 44.19, and 54.42 are consistent with rutile phase (110), (101), (111), and (211), respectively. On the other hand, the appearance of diffraction peaks at 41.42° and 56.66° corresponds to orthorhombic phase planes (020) and (012) of V_2O_5, which is consistent with JCPDS Card No. 01-0359 [45]. By means of the Debye–Scherer equation [46], the average crystal size of the V@TiO_2 nanocomposite was determined to be 33.83 nm, with a d-spacing of 3.236 Å.

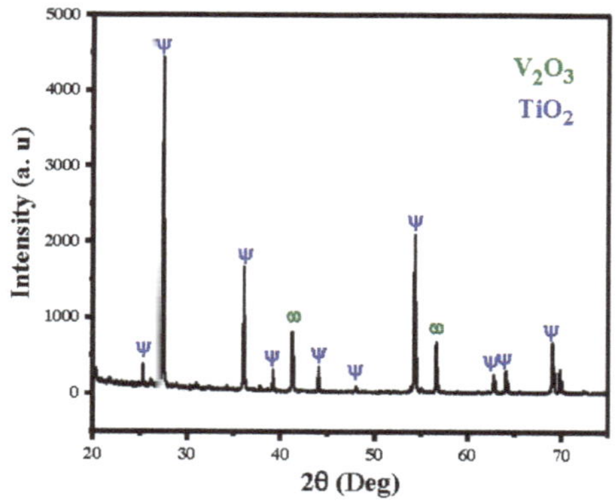

Figure 1. XRD of the fabricated V@TiO_2 nanocomposite.

3.1.1. Morphological Observations

The TEM images of the V@TiO_2 sorbent at different magnifications confirmed the morphology and indicated the elemental distribution, as shown in Figure 2a–c. The semi-spherical form of the TiO_2 nanoparticles is seen clinging to the V_2O_5 nanomaterials, as shown in the TEM pictures. The elemental distribution of V, Ti, and O in the synthesized nanomaterials is characterized by means of EDX spectra (Figure 2d). The EDX analysis sheds light on the stoichiometry of oxygen, vanadium, and titanium, as well as the proportions of these elements and the actual integration of V_2O_5 into the TiO_2 lattice. The transmission electron microscope (TEM) gave an estimate of 30–100 nm for the typical particle size of the sorbent. The value, however, was comparable to the crystallite size determined by XRD.

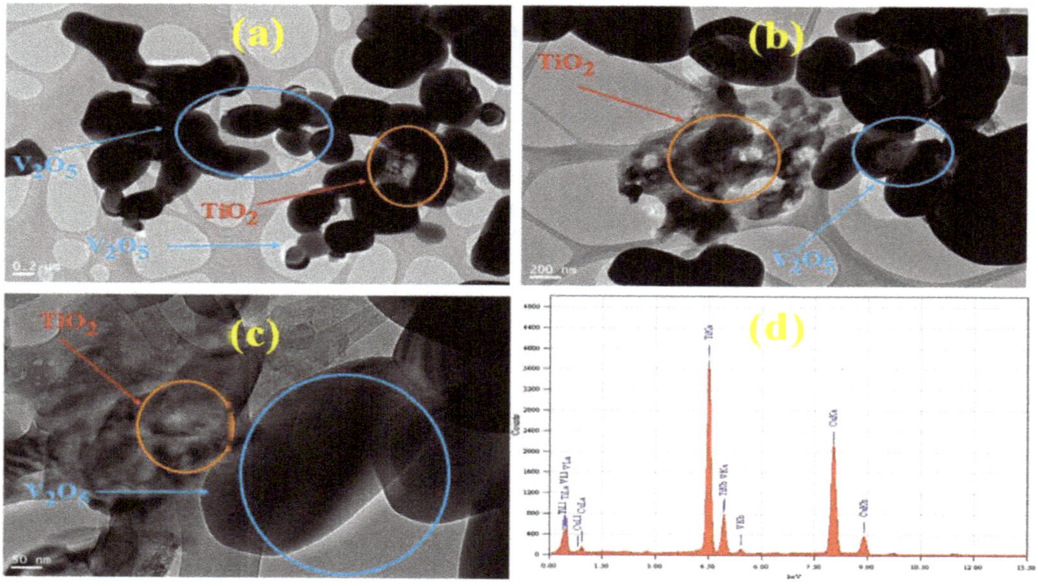

Figure 2. (**a**–**c**) TEM images at different magnifications and (**d**) EDX of the V@TiO_2 nanocomposite.

3.1.2. XPS of V@TiO_2 Nanocomposite

Figure 3a–c depict O1s, Ti 2p, and V2p core-level spectra of the V_2O_5–TiO_2 samples. The O1s peak fit into two Gaussian peaks at 529.8 eV, which arose from the lattice O

(Ti–O–Ti) and surface-absorbed hydroxyl groups of Ti–OH (Figure 3a) [47]. Due to the self-orbital coupling effect, the Ti 2p peak was subdivided into Ti 2p3/2 (458.3 eV) and Ti 2p1/2 (464.1 eV), as shown in Figure 3c. They were symmetrical with a normal Gaussian peak shape, indicating that Ti^{4+} was predominantly present in the undoped sample. Due to the influence of the O1s satellite peak, only the V 2p3/2 peak was displayed in the current experiment. The V doped in the TiO_2 nanoparticles (Figure 3d) consisted of two chemical states, V^{5+} and V^{4+}, with energies of 514.9 and 527.33 eV, respectively. The literature [48] indicates that V dopants can exist in surficial VO^{2+}, surficial V_2O_5 islands, interstitial, and replacement V ions. In light of the XRD and TEM analyses, as well as the fact that V doping can result in the O1s peak shifting, it is plausible to assume that the V ions existed in a substituted form and not as the V_2O_5 isolated phase, as suggested by the literature [49].

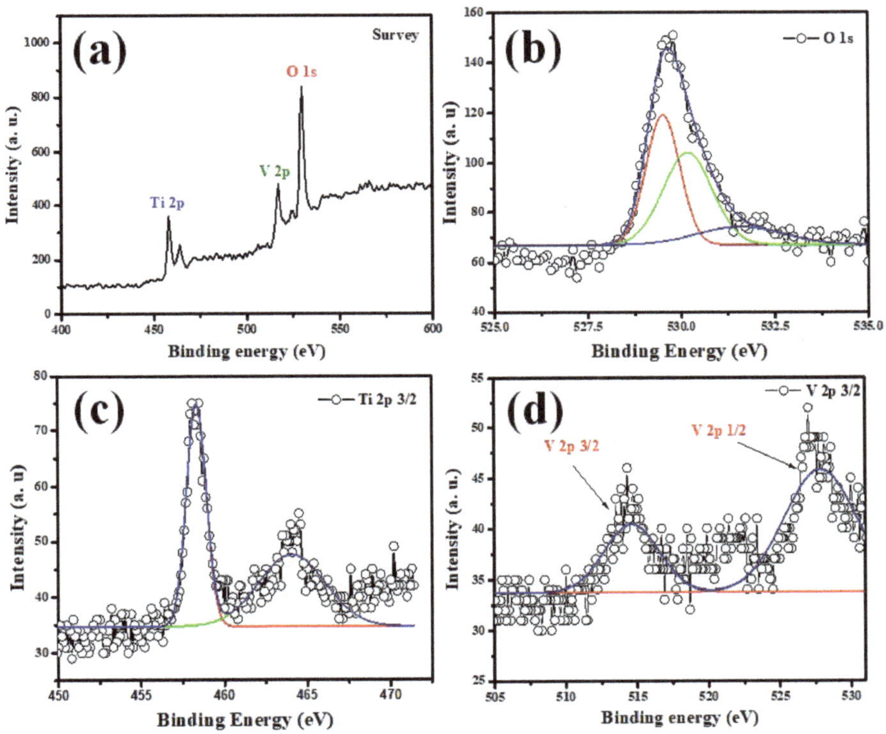

Figure 3. Extraordinary resolution XPS spectra of (**a**) Survey (**b**) O1s, (**c**) Ti 2p 3/2, and (**d**) V 2p 3/2 of V_2O_5-TiO_2 nanohybrid.

Usually, due to the similar radii of V^{4+} and Ti^{4+} species, V^{4+} ions can typically only integrate into the TiO_2 lattice by substituting Ti^{4+} ions. Therefore, the Ti-O-V bond, which results in oxygen vacancy and a high electron production capability, is formed by sharing the oxygen atoms of the V^{4+} ions in the TiO_2 lattice. Furthermore, the presence of more V^{4+} ions leads to the creation of more O_2 radicals, the most powerful oxidizing species in photocatalysis [50,51]. As a result, the existence of V^{4+} contributes significantly to the improvement of photocatalytic activity [52].

3.1.3. pHzc of V@TiO_2 Nanocomposite

The pH of the point of zero charge (pHzc) of the V@TiO_2 nanocomposite was evaluated using the pH drift method to completely comprehend the pH effect on the adsorptive removal of dye pollution from an aqueous medium. According to the obtained results, the pHZPC value of the V@TiO_2 nanocomposite was 4.2. At this pH (pHzc = 4.2), the surface charge on the V@TiO_2 nanocomposite is zero, whereas the surface charge is positive at pH values less than 4.2 and negative at pH values greater than the pH_{ZPC} (Figure 4). The influence of pH on Rh-B dye removal is a significant factor for the adsorption technique, as it modifies not only the active sites on the V@TiO_2 surface of the adsorbent that are capable of Rh-B dye bonding, but also the solubility of Rh-B dye in the aqueous solution.

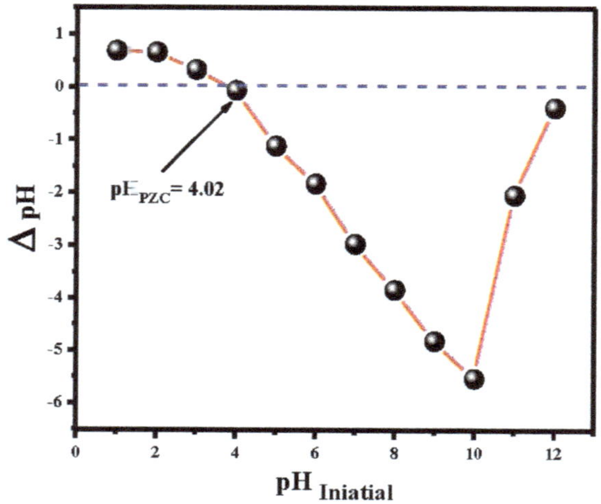

Figure 4. pH zero charge of the V@TiO_2 nanocomposite.

3.1.4. Surface Characteristics of the V@TiO_2 Nanocomposite

From the surface property investigation, the isotherm and pore size distribution of the manufactured V@TiO_2 nanocomposite are depicted in Figure 5a. The resulting isotherm is of class IV with H1 hysteresis loops, which seem to be characteristic of nanostructured and mesoporous materials [53,54]. Figure 5b shows a BET surface area of 16 m^2/g and a pore volume of 0.013 cc/g, along with a pore size distribution with a median of roughly 27.6 nm.

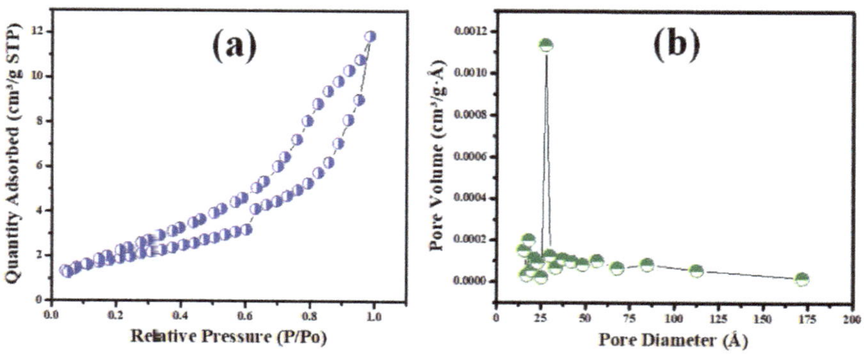

Figure 5. (**a**) The nitrogen adsorption–desorption isotherm and (**b**) the pore distribution of the V@TiO_2 nanocomposite.

3.2. Adsorption of Rd-B by V@TiO_2 Sorbent

Figure 6a depicts the effect of contact duration on Rh-B dye sorption on V@TiO_2, with experimental q_t values of 138 mg g^{-1} and adsorption equilibrium points of 4.0 h. It is worth mentioning that nearly 95% of these uptakes occurred during the first 60 min of interaction. In addition, the impact of the concentrations of Rh-B on sorption by V@TiO_2 was tested. Figure 6b illustrates that the q_t value increased proportionally up to 75 mg L^{-1}, after which the inflation occurred, indicating the suitability of a 1:4 sorbent solution ratio up to 75 mg L^{-1}. Typically, q_t values of 133.4 mg g^{-1} from 100 mg L^{-1} Rh-B solutions imply the applicability of the V@TiO_2 for treating industrial effluents with high pollutant concentrations. Additionally, the semi-complete Rh-B dye removal by V@TiO_2 from the 10 mg L^{-1} solutions demonstrates its effectiveness in treating contaminated water resources. Furthermore, the decrease of qt values from the same concentrations as the temperature increased showed the exothermic nature of removing Rh-B dye by a V@TiO_2 sorbent.

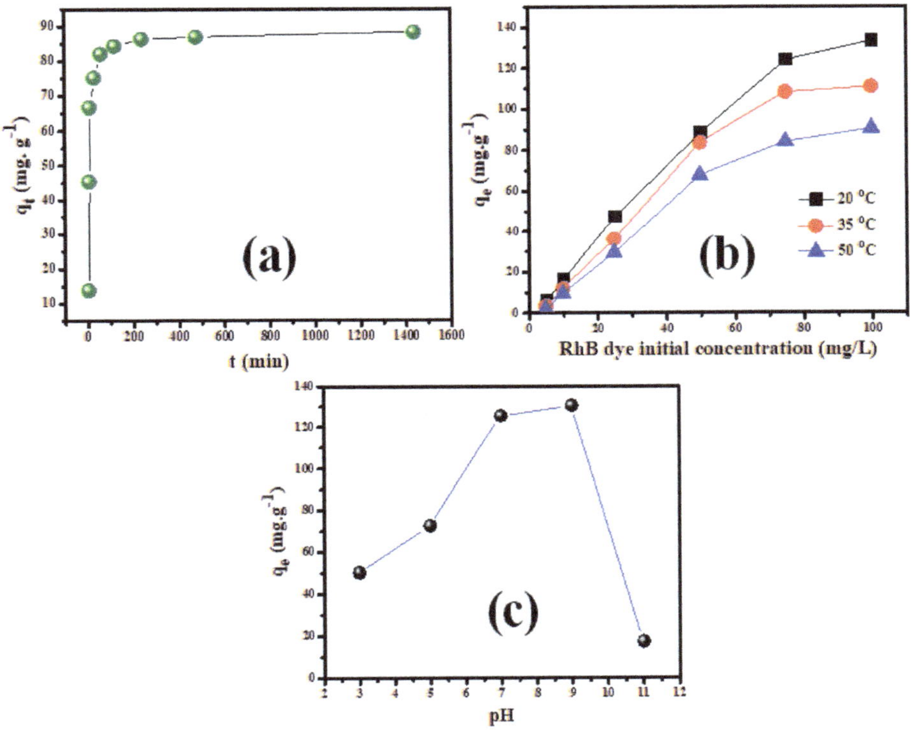

Figure 6. (**a**) The contact time study for the adsorption of Rh-B dye on the synthesized V@TiO_2, (**b**) the influence of a solution's pH on Rh-B dye sorption, (**b**) the impact of feeding concentration on the Rh-B sorption on V@TiO_2 at 20 °C, 35 °C, and 50 °C for (**a**) Rh-B dye, and (**c**) The zero-charge investigation of V@TiO_2.

Furthermore, the pH impact on Rh-B adsorption on V@TiO_2 was evaluated (Figure 6c). The obtained results show the suitability of mild alkaline media for removing Rh-B dye. The decrease in sorption capacity for V@TiO_2 in strongly acidic media is probably due to the protonation of oxygen atoms in the nanocomposite and/or turning part of the oxides into soluble salts. On the other hand, the hydroxyl groups in strong alkaline media may compete with Rh-B dye for sorbent sites and/or repulse it away from the V@TiO_2 surface [55,56].

3.2.1. Adsorption Kinetics

The linear regression plots of the PSFOM, PSSOM, IDM, and LDM for Rh-B dye sorption on V@TiO_2 are illustrated in Figure 7. The k_1, k_2, K_{IDM}, and K_{LDM} values gathered in Table 1 were computed utilizing the extracted regression parameters (slope and intercept) [43,57]. The obtained results revealed that Rh-B sorption on V@TiO_2 fit the PSFOM model, which may explain their relatively short-time equilibrium. Further, the investigation of the rate-control step showed that the intraparticle diffusion step controlled the adsorption of Rh-B onto the V@TiO_2 surface. These findings indicate that Rh-B dye has a higher affinity toward the V@TiO_2 surface and imply fast pore-diffusion during the Rh-B dye removal by V@TiO_2 [58].

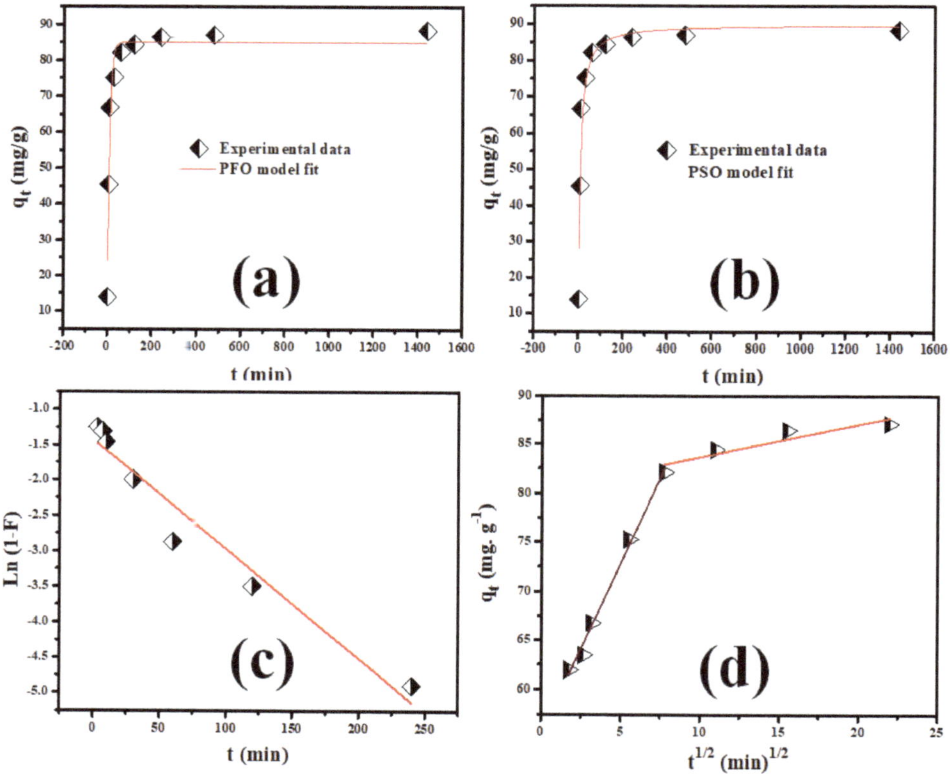

Figure 7. (**a**,**b**) The PSFO and PSSO sorption rate order investigations for Rh-B sorption on the V@TiO_2 nanocomposite; (**c**,**d**) LFD and IPD sorption mechanism investigations for Rh-B sorption on the V@TiO_2 nanocomposite.

Table 1. The kinetic parameters for Rh-B adsorption on the V@TiO_2 nanocomposite from 50 mg L^{-1} pollutant solution.

Adsorption Kinetic						
Adsorption Rate Order						
qe exp. (mg g^{-1})	**PSFO**			**PSSO**		
	qe cal. (mg g^{-1})	**R^2**	**k_1**	**qe cal. (mg g^{-1})**	**R^2**	**k_2**
88.367	88.7	0.947	0.114	90.000	0.927	0.002
Adsorption mechanism						
	IPDM			**LFDM**		
	K_{IP} (mg g^{-1} $min^{0.5}$)	**C (mg g^{-1})**	**R^2**	**K_{LF} (min^{-1})**		**R^2**
Stage 1	3.4950	55.38	0.9959	0.015		0.956
Stage 2	0.3369	80.256	0.9306			

3.2.2. Adsorption Isotherms

The Langmuir and Freundlich models (LIM and FIM) were the most used isotherm models for describing adsorption processes. Both linearized models (Equations (6) and (7)) were utilized to analyze Rh-B sorption on synthesized V@TiO_2.

$$q_e = \left(\frac{K_l q_m C_e}{1 + q_m C_e}\right) \quad (6)$$

$$q_e = K_F \cdot C_e^{\frac{1}{n}} \quad (7)$$

In these equations, Ce (mg L^{-1}) is the equilibrium solution concentration, q_m is the computed maximum sorption capacity, $1/n$ is the Freundlich adsorption intensity, and KL and KF are the LIM and FIM constants, respectively [59]. The linear fits of LIM and FIM studies of Rh-B sorption on synthesized V@TiO_2 are shown in Figure 8. Better fitting to LIM was observed in the data presented in Table 2. The Rh-B sorption showed preferential sorption, as indicated by the 1/n value below unity, which was also consistent with its PFOM agreement [60–62].

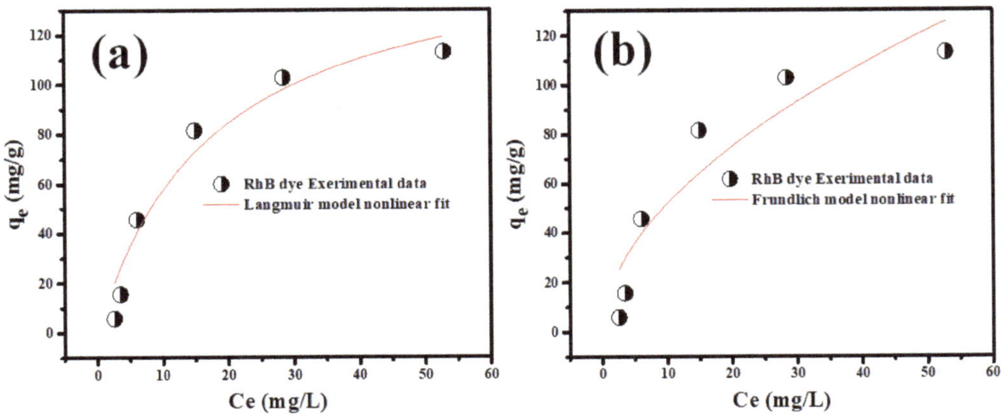

Figure 8. The (**a**) LIM and (**b**) FIM investigations for the adsorption of Rh-B on the synthesized V_2O_5@TiO_2 at 20 °C.

Table 2. Removal parameters of the V@TiO_2 sorbent compared with the diverse **Rh-B** dye adsorbent nanomaterials.

Nanomaterials	Adsorption Capacity (mg g^{-1})	References
Halloysite HU	8.37	[63]
alpha alumina (α-Al_2O_3)	52.0	[64]
L-Asp capped Fe_3O_4 NPs	7.7	[65]
Magnetic $ZnFe_2O_4$	12.1	[66]
MWCNT-COOH	42.68	[67]
Humic acid functionalized MNPs	161.8	[68]
Sodium montmorillonite	42.19	[69]
Lignocellulose	82.34	[70]
NiO nanoparticles	111	[71]
NiZnAl-LDH nano-sheets	97.09	[72]
V@TiO_2 nanocompostie	158.8	This paper

As shown in Table 2, the performance of the V@TiO_2 nanocomposite in adsorbing Rh-B dye has been further evaluated and compared to that of other adsorbents that have been reported in the literature. First, the equilibrium time for the V@TiO_2 nanocomposite is shorter. These obtained results show that the Rh-B dye leaves the aqueous solution quickly. Further, the V@TiO_2 nanocomposite has a more significant adsorption capacity than the other nanostructures, i.e., 158.8 mg/g compared to 7–161 mg/g, as documented in Table 2.

3.2.3. Adsorption Thermodynamics

The thermodynamics of Rh-B removal by the V@TiO_2 were inspected. The slope and intercept extracted from the plot of Equation (8) (Figure 9) were utilized in computing the enthalpy and entropy (ΔS^o and ΔH^o). The Gibbs free energy (ΔG^o) was obtained by applying the ΔS^o and ΔH^o values in Equation (9). The ideal gas constant (R) was used as 0.0081345 kJ mol^{-1} for calculating these parameters, and the findings are summarized in Table 3.

$$\ln K_c = \frac{\Delta H^{\,o}}{RT} + \frac{\Delta S^{\,o}}{R} \tag{8}$$

$$\Delta\, G^{\,o} = \Delta\, H^{\,o} - T\,\Delta\, S^{\,o} \tag{9}$$

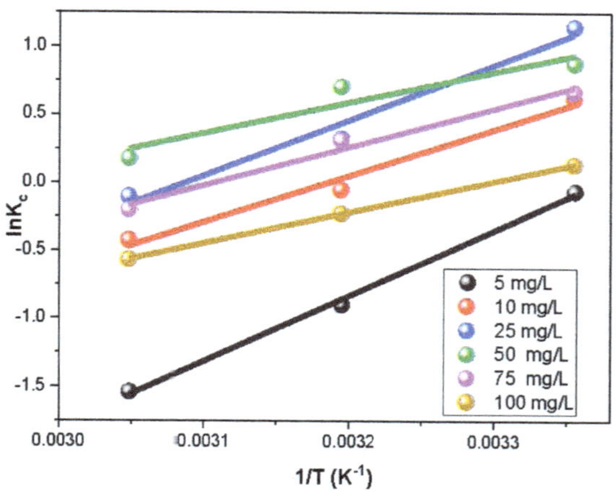

Figure 9. Thermodynamic study of the sorption of Rh-B on the synthesized V@TiO_2 at 298, 313, and 328 °K.

Table 3. The isotherms and thermodynamic findings for the Rh-B sorption on the V@TiO_2 nanocomposite.

Adsorption Isotherms					
Langmuir			Freundlich		
R^2	K_L (L mg^{-1})	q_m (mg g^{-1})	R^2	K_f (L mg^{-1})	n^{-1} (a.u.)
0.952	0.056	158.8	0.880	0.536	0.065
Thermodynamic parameters					
Fed conc. (mg L^{-1})	ΔH^o (kJmol^{-1})	ΔS^o (kJmol^{-1})	ΔG^o (kJmol^{-1}) 298 K	ΔG^o (kJmol^{-1}) 313 K	ΔG^o (kJmol^{-1}) 328 K
10	−28.450	−0.091	−1.439	−0.079	1.280
25	−34.026	−0.105	−2.714	−1.138	chem1.280
50	−18.856	−0.055	−2.335	−1.503	−0.672
75	−23.573	−0.073	−1.739	−0.640	0.460
100	−18.979	−0.063	−0.339	0.599	1.537

Table 3 shows that Rh-B sorption possessed negative ΔG^o values, indicating the sorption's spontaneity and exothermic nature. Also, the negative ΔH^o values corroborate that V@TiO_2 removed the Rh-B dye via a physisorption process [73–77].

3.3. Rh-B Dye Adsorption Mechanism

Several variables, such as the functional group of adsorbents, the pH of a solution, the surface charge on the particles, the porosity, and the nature of the adsorbate, can affect the adsorption process. Rh-B dye was adsorbed on the V@TiO_2 nanocomposite surface in this study. Point zero charges (surface charge) and FTIR (functional groups) that existed on the V@TiO_2 sorbent could explain the likely mechanism of the adsorption process. Hydrogen bonding and electrostatic interactions might have contributed to the adsorption of the Rh-B dye (Figure 10a). In Figure 10a of the FTIR spectrum, the absorption peak at 2309 cm^{-1} due to C–N stretching vibrations that appeared after the adsorption of Rh-B dye demonstrates its role in the adsorption of dye molecules. Due to the presence of C–H and C=C functional groups in the spectra of adsorbed Rh-B dye, the absorption band between 1539.58 and 1448 cm^{-1} was observed. The narrow peak between 1000 and 900 cm^{-1} is attributable to the C–H bond, which indicates their participation in the elimination of Rh-B dye. A prominent broad peak at 521 cm^{-1}, which shifted lower to 472 cm^{-1}, correlates to V@TiO_2 nanocomposite stretching; FTIR of the dye-loaded V@TiO_2 nanocomposite revealed a considerable drop in the peak, indicating its significance in the elimination of Rh-B dye from an aqueous solution.

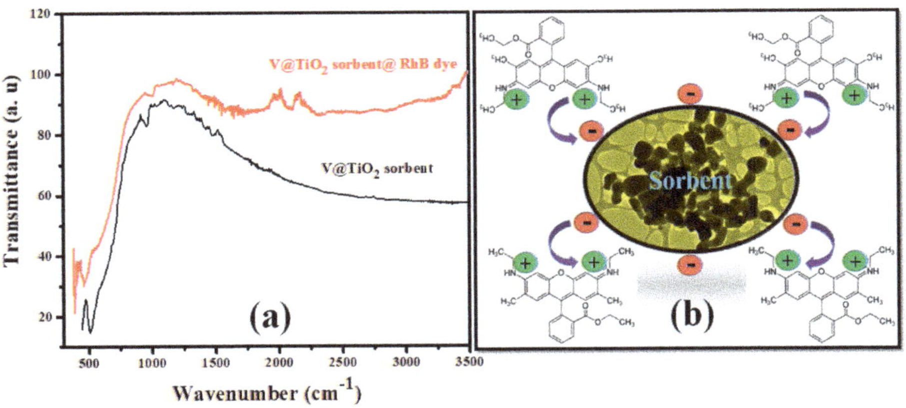

Figure 10. (**a**) FTIR spectra of the V@TiO_2 nanocomposite and Rh-B @ V@TiO_2 nanocomposite; (**b**) the suggested removal mechanism of Rh-B dye by the V@TiO_2 nanocomposite.

The pHpzc study determined that the pHpzc exists at a pH of 4.2, above which the surface of V@TiO_2 is negatively charged, which may increase the sorption by withdrawing the positive N^+–$(CH_3)_2$ group on Rh-B (Figure 10a); this hypothesis indicates high participation of electrostatic attractions in Rh-B removal by V@TiO_2. Furthermore, the adsorption of Rh-B dye on the surface of the V@TiO_2 sorbent is also the result of interactions involving H-H bonding and dipole-dipole interactions. In addition, the isothermal studies demonstrate that the multilayer adsorption of the Rh-B molecules is physical sorption; this result is in line with the ΔH^o values less than 80 $kJmol^{-1}$. Figure 10b depicts the subsequent adsorption of Rh-B dye molecules onto the V@TiO_2 sorbent. Rh-B dye is attracted to the nanocomposites via hydrogen bonds and electrostatic interactions.

3.4. V@TiO_2 Nanomaterials Regeneration

The regeneration of V@TiO_2 nanomaterials was studied by removing the Rh-B dye using absolute ethanol. During the regeneration process, ethanol was used as the desorption agent (Figure 11). The rapid discoloration of absolute ethanol filtrate confirmed Rh-B dye desorption from the V@TiO2 sorbent. The sorbent was re-immersed in a freshly produced Rh-B dye solution with the same concentration and volume (5 mg/L, 75 mL). A total of five repetitions of this step were performed. Figure 11 illustrates the effectiveness of Rh-B dye desorption by ethanol from the V@TiO_2 sorbent. The efficacy of the V@TiO_2 nanocomposite to remove dyes decreased with each desorption cycle.

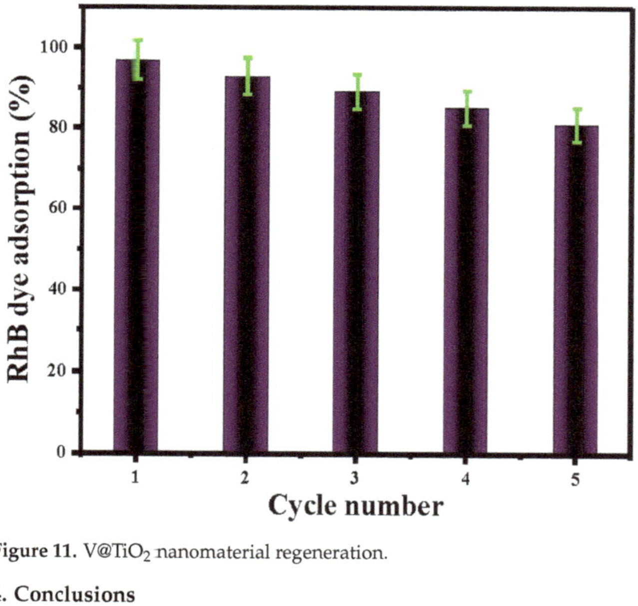

Figure 11. V@TiO_2 nanomaterial regeneration.

4. Conclusions

Nanostructured V@TiO2 is synthesized from metal oxide using ultrasonic synthesis techniques. A study of the adsorption properties of Rh-B dye on V@TiO_2 was conducted. According to XRD, the V@TiO_2 nanomaterials had a crystal size of 33.83 nm. In aqueous media, the nanosorbent of V@TiO_2 was demonstrated to be capable of eliminating Rh-B dye molecules with a maximum adsorption capacity of 158.8 mg/g. Furthermore, the thermodynamic analysis showed that such reactions were highly spontaneous and endothermic. The Langmuir technique emphasized the adsorption monolayer. As a result, the adsorption kinetic could be fitted with a pseudo-first-order reaction model. Additionally, 100% ethanol was chosen as the best selective desorption eluent for reusing the adsorbent. Hence, the regeneration of V@TiO_2 sorbent is a cost-effective method for treating effluents from the textile industry.

Author Contributions: A.M.: conceptualization, writing—original draft, preparation; M.R.E.: supervision, writing, review of the final version; B.Y.A.: writing, review of the final version; N.Y.E.: validation, writing, review; K.H.I.: writing, review; R.B. and H.I.: review. All authors have read and agreed to the published version of the manuscript.

Funding: The Deanship of Scientific Research at Imam Mohammad Ibn Saud Islamic University funded this work through Research Group no. RG-21-09-74.

Institutional Review Board Statement: Not applicable.

Acknowledgments: The authors extend their appreciation to the Deanship of Scientific Research at Imam Mohammad Ibn Saud Islamic University for funding this work.

Conflicts of Interest: The authors declare that they have no known competing financial interests or personal relationships that could have appeared to influence the work reported in this paper.

References

1. Tkaczyk, A.; Mitrowska, K.; Posyniak, A. Synthetic organic dyes as contaminants of the aquatic environment and their implications for ecosystems: A review. *Sci. Total Environ.* **2020**, *717*, 137222. [CrossRef] [PubMed]
2. Ehalt Macedo, H.; Lehner, B.; Nicell, J.; Grill, G.; Li, J.; Limtong, A.; Shakya, R. Distribution and characteristics of wastewater treatment plants within the global river network. *Earth Syst. Sci. Data* **2022**, *14*, 559–577. [CrossRef]
3. Rápó, E.; Tonk, S. Factors affecting synthetic dye adsorption; desorption studies: A review of results from the last five years (2017–2021). *Molecules* **2021**, *26*, 5419. [CrossRef] [PubMed]
4. Peng, X.; Lin, Q.; Liu, B.; Huang, S.; Yan, W.; Zhang, L.; Ge, F.; Zhang, Y.; Wu, Z. Effect of submerged plant coverage on phytoplankton community dynamics and photosynthetic activity in situ. *J. Environ. Manag.* **2022**, *301*, 113822. [CrossRef] [PubMed]
5. Mugdha, A.; Usha, M. Enzymatic treatment of wastewater containing dyestuffs using different delivery systems. *Sci. Rev. Chem. Commun.* **2012**, *2*, 31–40.
6. Pang, Y.L.; Abdullah, A.Z. Current status of textile industry wastewater management and research progress in Malaysia: A review. *Clean–Soil Air Water* **2013**, *41*, 751–764. [CrossRef]
7. Gürses, A.; Açıkyıldız, M.; Güneş, K.; Gürses, M.S. Dyes and pigments: Their structure and properties. In *Dyes and Pigments*; Springer: Berlin/Heidelberg, Germany, 2016; pp. 13–29.
8. Butman, M.F.; Gushchin, A.A.; Ovchinnikov, N.L.; Gusev, G.I.; Zinenko, N.V.; Karamysheva, S.P.; Krämer, K.W. Synergistic effect of dielectric barrier discharge plasma and TiO_2-pillared montmorillonite on the degradation of rhodamine B in an aqueous solution. *Catalysts* **2020**, *10*, 359. [CrossRef]
9. Dire, D.J.; Wilkinson, J.A. Acute exposure to rhodamine B. *J. Toxicol. Clin. Toxicol.* **1987**, *25*, 603–607. [CrossRef]
10. Ali Khan, M.; Siddiqui, M.R.; Otero, M.; Alshareef, S.A.; Rafatullah, M. Removal of rhodamine b from water using a solvent impregnated polymeric dowex 5wx8 resin: Statistical optimization and batch adsorption studies. *Polymers* **2020**, *12*, 500. [CrossRef]
11. Wabaidur, S.M.; Khan, M.A.; Siddiqui, M.R.; Alothman, Z.A.; Al-Ghamdi, M.; Al-Sohami, H. Dodecyl sulfate chain anchored bio-char to sequester triaryl methane dyes: Equilibrium, kinetics, and adsorption mechanism. *Desal. Water Treat* **2017**, *67*, 357–370. [CrossRef]
12. Bhatia, D.; Sharma, N.R.; Singh, J.; Kanwar, R.S. Biological methods for textile dye removal from wastewater: A review. *Crit. Rev. Environ. Sci. Technol.* **2017**, *47*, 1836–1876. [CrossRef]
13. Hao, O.J.; Kim, H.; Chiang, P.-C. Decolorization of wastewater. *Crit. Rev. Environ. Sci. Technol.* **2000**, *30*, 449–505. [CrossRef]
14. Zouari, N. Decolorization of olive oil mill effluent by physical and chemical treatment prior to anaerobic digestion. *J. Chem. Technol. Biotechnol. Int. Res. Process Environ. AND Clean Technol.* **1998**, *73*, 297–303. [CrossRef]
15. Muhamad, M.S.; Salim, M.R.; Lau, W.J.; Yusop, Z. A review on bisphenol A occurrences, health effects and treatment process via membrane technology for drinking water. *Environ. Sci. Pollut. Res.* **2016**, *23*, 11549–11567. [CrossRef] [PubMed]
16. Akcil, A.; Erust, C.; Ozdemiroglu, S.; Fonti, V.; Beolchini, F. A review of approaches and techniques used in aquatic contaminated sediments: Metal removal and stabilization by chemical and biotechnological processes. *J. Clean. Prod.* **2015**, *86*, 24–36. [CrossRef]
17. Rajasulochana, P.; Preethy, V. Comparison on efficiency of various techniques in treatment of waste and sewage water–A comprehensive review. *Resour. Effic. Technol.* **2016**, *2*, 175–184. [CrossRef]
18. Jain, R.; Mathur, M.; Sikarwar, S.; Mittal, A. Removal of the hazardous dye rhodamine B through photocatalytic and adsorption treatments. *J. Environ. Manag.* **2007**, *85*, 956–964. [CrossRef]
19. Bakar, N.A.; Othman, N.; Yunus, Z.M.; Altowayti, W.A.H.; Tahir, M.; Fitriani, N.; Mohd-Salleh, S.N.A. An insight review of lignocellulosic materials as activated carbon precursor for textile wastewater treatment. *Environ. Technol. Innov.* **2021**, *22*, 101445. [CrossRef]
20. Crini, G.; Lichtfouse, E.; Wilson, L.D.; Morin-Crini, N. Conventional and non-conventional adsorbents for wastewater treatment. *Environ. Chem. Lett.* **2019**, *17*, 195–213. [CrossRef]
21. Adegoke, K.A.; Bello, O.S. Dye sequestration using agricultural wastes as adsorbents. *Water Resour. Ind.* **2015**, *12*, 8–24. [CrossRef]

22. Modwi, A.; Khezami, L.; Ghoniem, M.; Nguyen-Tri, P.; Baaloudj, O.; Guesmi, A.; AlGethami, F.; Amer, M.; Assadi, A. Superior removal of dyes by mesoporous MgO/g-C3N4 fabricated through ultrasound method: Adsorption mechanism and process modeling. *Environ. Res.* **2022**, *205*, 112543. [CrossRef] [PubMed]
23. Aissa, B.; Khezami, L.; Taha, K.; Elamin, N.; Mustafa, B.; Al-Ayed, A.; Modwi, A. Yttrium oxide-doped ZnO for effective adsorption of basic fuchsin dye: Equilibrium, kinetics, and mechanism studies. *Int. J. Environ. Sci. Technol.* **2022**, *9*, 9901–9914. [CrossRef]
24. Kerrami, A.; Khezami, L.; Bououdina, M.; Mahtout, L.; Modwi, A.; Rabhi, S.; Bensouici, F.; Belkacemi, H. Efficient photodegradation of azucryl red by copper-doped TiO2 nanoparticles—Experimental and modeling studies. *Environ. Sci. Pollut. Res.* **2021**, *28*, 57543–57556. [CrossRef] [PubMed]
25. Toghan, A.; Abd El-Lateef, H.M.; Taha, K.K.; Modwi, A. Mesoporous TiO_2@ g-C3N4 composite: Construction, characterization, and boosting indigo carmine dye destruction. *Diam. Relat. Mater.* **2021**, *118*, 108491. [CrossRef]
26. Idriss, H. Decolorization of malachite green in aqueous solution via MgO nanopowder. *J. Optoelectron. Biomed. Mater. Vol.* **2021**, *13*, 183–192.
27. Idriss, H.; Alakhras, A.; El Khair, H. Malachite green removal from aqueous solutions by MgO (86%). Se (7%). Te (7%) nanocomposites. *Chalcogenide Lett.* **2021**, *18*, 629–638.
28. Modwi, A.; Aissa, B.; Taha, K.K.; Khezami, L.; El Ghoul, J.; Al-Ayed, A.S.; Bououdina, M. Fabrication of (Y2O3) n–ZnO nanocomposites by high-energy milling as potential photocatalysts. *J. Mater. Sci. Mater. Electron.* **2021**, *32*, 3415–3430. [CrossRef]
29. Ahmed, A.S.; Ahamad, T. Ahmad, N.; Khan, M.Z. Removal enhancement of acid navy blue dye by GO-TiO_2 nanocomposites synthesized using sonication method. *Mater. Chem. Phys.* **2019**, *238*, 121906. [CrossRef]
30. Zhang, J.-J.; Qi, P.; Li, J.; Zheng, X.-C.; Liu, P.; Guan, X.-X.; Zheng, G.-P. Three-dimensional Fe_2O_3–TiO_2–graphene aerogel nanocomposites with enhanced adsorption and visible light-driven photocatalytic performance in the removal of RhB dyes. *J. Ind. Eng. Chem.* **2018**, *61*, 407–415. [CrossRef]
31. Harja, M.; Lupu, N.; Chiriac, H.; Herea, D.-D.; Buema, G. Studies on the Removal of Congo Red Dye by an Adsorbent Based on Fly-Ash@ Fe3O4 Mixture. *Magnetochemistry* **2022**, *8*, 125. [CrossRef]
32. Le, T.T.T.; Tran, T.D. Photocatalytic degradation of rhodamine B by C and N codoped TiO_2 nanoparticles under visible-light irradiation. *J. Chem.* **2020**. [CrossRef]
33. Sriram, G.; Bendre, A.; Altalhi, T.; Jung, H.-Y.; Hegde, G.; Kurkuri, M. Surface engineering of silica based materials with Ni–Fe layered double hydroxide for the efficient removal of methyl orange: Isotherms, kinetics, mechanism and high selectivity studies. *Chemosphere* **2022**, *287*, 131976. [CrossRef] [PubMed]
34. Modwi, A.; Al-Ayed, A.S.; Mustafa, D.E.; Bagabas, A.A.; Elamin, M.; Algethami, F.K.; Arasheed, R.; Alfaifi, M.Q.; Alqarni, A.; Alotaibi, F. Ultrasound-assisted green biosynthesis of ZnO nanoparticles and their photocatalytic application. *Z. Für Nat. A* **2021**, *76*, 535–547. [CrossRef]
35. Yang, F.; Song, Y.; Hui, A.; Wang, Q.; Wang, A. Facile biosynthesis of ZnO/attapulgite nanocomposite for enhanced antimicrobial performance. *Mater. Lett.* **2022**, *323*, 132549. [CrossRef]
36. Jimoh, A.; Akpeji, B.; Azeez, S.; Ayipo, Y.; Abdulsalam, Z.; Adebayo, Z.; Ajao, A.; Zakariyah, A.; Elemike, E. Biosynthesis of Ag and TiO_2 nanoparticles and the evaluation of their antibacterial activities. *Inorg. Chem. Commun.* **2022**, *141*, 109503. [CrossRef]
37. Wafi, M.A.E.; Ahmed, M.; Abdel-Samad, H.S.; Medien, H. Exceptional removal of methylene blue and p-aminophenol dye over novel TiO_2/RGO nanocomposites by tandem adsorption-photocatalytic processes. *Mater. Sci. Energy Technol.* **2022**, *5*, 217–231. [CrossRef]
38. Alsalme, A.; Galal, A.; El-Sherbeny, E.F.; Soltan, A.; Abdel-Messih, M.; Ahmed, M. Fabrication of S-scheme TiO_2/g-C_3N_4 nanocomposites for generation of hydrogen gas and removal of fluorescein dye. *Diam. Relat. Mater.* **2022**, *122*, 108819. [CrossRef]
39. Al-Mamun, M.R.; Karim, M.N.; Nitun, N.A.; Kader, S.; Islam, M.S.; Khan, M.Z.H. Photocatalytic performance assessment of GO and Ag co-synthesized TiO_2 nanocomposite for the removal of methyl orange dye under solar irradiation. *Environ. Technol. Innov.* **2021**, *22*, 101537. [CrossRef]
40. Zhang, W.; Lan, Y.; Ma, M.; Chai, S.; Zuo, Q.; Kim, K.-H.; Gao, Y. A novel chitosan–vanadium-titanium-magnetite composite as a superior adsorbent for organic dyes in wastewater. *Environ. Int.* **2020**, *142*, 105798. [CrossRef]
41. Liang, X.; Zhong, Y.; Zhu, S.; Zhu, J.; Yuan, P.; He, H.; Zhang, J. The decolorization of Acid Orange II in non-homogeneous Fenton reaction catalyzed by natural vanadium–titanium magnetite. *J. Hazard. Mater.* **2010**, *181*, 112–120. [CrossRef]
42. Sanchez Mendez, M.; Lemarchand, A.; Traore, M.; Perruchot, C.; Sassoye, C.; Selmane, M.; Nikravech, M.; Ben Amar, M.; Kanaev, A. Photocatalytic Activity of Nanocoatings Based on Mixed Oxide V-TiO_2 Nanoparticles with Controlled Composition and Size. *Catalysts* **2021**, *11*, 1457. [CrossRef]
43. Almufarij, R.S.; Abdulkhair, B.Y.; Salih, M.; Alhamdan, N.M.J.N. Sweep-Out of Tigecycline, Chlortetracycline, Oxytetracycline, and Doxycycline from Water by Carbon Nanoparticles Derived from Tissue Waste. *Nanomaterials* **2022**, *12*, 3617. [CrossRef] [PubMed]
44. Elamin, M.R.; Ibnaouf, K.H.; Elamin, N.Y.; Adam, F.A.; Alolayan, A.H.; Abdulkhair, B.Y.J.I. Spontaneous Adsorption and Efficient Photodegradation of Indigo Carmine under Visible Light by Bismuth Oxyiodide Nanoparticles Fabricated Entirely at Room Temperature. *Inorganics* **2022**, *10*, 65. [CrossRef]
45. Navyashree, G.; Hareesh, K.; Sunitha, D.; Nagabhushana, H.; Nagaraju, G. Photocatalytic degradation performance of Nd3+ doped V2O5 nanostructures. *Mater. Res. Express* **2018**, *5*, 095007. [CrossRef]

46. Modwi, A.; Abbo, M.; Hassan, E.; Houas, A. Effect of annealing on physicochemical and photocatalytic activity of Cu 5% loading on ZnO synthesized by sol–gel method. *J. Mater. Sci. Mater. Electron.* **2016**, *27*, 12974–12984.
47. Yao, T.; Wang, H. Metal-organic framework derived vanadium-doped TiO_2@ carbon nanotablets for high-performance sodium storage. *J. Colloid Interface Sci.* **2021**, *604*, 188–197. [CrossRef] [PubMed]
48. Liu, T.; Yao, T.; Li, L.; Zhu, L.; Wang, J.; Li, F.; Wang, H. Embedding amorphous lithium vanadate into carbon nanofibers by electrospinning as a high-performance anode material for lithium-ion batteries. *J. Colloid Interface Sci.* **2020**, *580*, 21–29. [CrossRef]
49. Yin, J.; Yang, H.; Kong, W.; Man, J.; Zhou, Z.; Feng, W.; Sun, J.; Wen, Z. Highly compacted TiO_2/C micospheres via in-situ surface-confined intergrowth with ultra-long life for reversible Na-ion storage. *J. Colloid Interface Sci.* **2021**, *582*, 526–534. [CrossRef]
50. Cha, W.; Chin, S.; Park, E.; Yun, S.-T.; Jurng, J. Photocatalytic performance of V_2O_5/TiO_2 materials prepared by chemical vapor condensation and impregnation method under visible-light. *Powder Technol.* **2014**, *258*, 352–357. [CrossRef]
51. Chen, C.; Ma, W.; Zhao, J. Semiconductor-mediated photodegradation of pollutants under visible-light irradiation. *Chem. Soc. Rev.* **2010**, *39*, 4206–4219. [CrossRef]
52. Putluru, S.S.R.; Schill, L.; Godiksen, A.; Poreddy, R.; Mossin, S.; Jensen, A.D.; Fehrmann, R. Promoted V_2O_5/TiO_2 catalysts for selective catalytic reduction of NO with NH3 at low temperatures. *Appl. Catal. B Environ.* **2016**, *183*, 282–290. [CrossRef]
53. Li, L.-X.; Xu, D.; Li, X.-Q.; Liu, W.-C.; Jia, Y. Excellent fluoride removal properties of porous hollow MgO microspheres. *New J. Chem.* **2014**, *38*, 5445–5452. [CrossRef]
54. Kruk, M.; Jaroniec, M. Gas adsorption characterization of ordered organic– inorganic nanocomposite materials. *Chem. Mater.* **2001**, *13*, 3169–3183. [CrossRef]
55. Yang, W.; Lu, Y.; Zheng, F.; Xue, X.; Li, N.; Liu, D. Adsorption behavior and mechanisms of norfloxacin onto porous resins and carbon nanotube. *Hemical Eng. J.* **2012**, *179*, 112–118. [CrossRef]
56. Lin, P.; Zhang, Y.; Zhang, X.; Chen, C.; Xie, Y.; Suffet, I.H. The influence of chlorinated aromatics' structure on their adsorption characteristics on activated carbon to tackle chemical spills in drinking water source. *Front. Environ. Sci. Eng.* **2015**, *9*, 138–146. [CrossRef]
57. Naghizadeh, A.; Karimi, A.; Derakhshani, E.; Esform, A. Single-walled carbon nanotubes (SWCNTs) as an efficient adsorbent for removal of reactive dyes from water solution: Equilibrium, kinetic, and thermodynamic. *Environ. Qual. Manag.* **2022**, *31*, 133–140. [CrossRef]
58. Sallal, H.A.; Abdul-Hameed, A.A.; Othman, F. Preparation of Al_2O_3/MgO Nano-Composite Particles for Bio-Applications. *Eng. Technol. J.* **2020**, *38*, 586–593. [CrossRef]
59. Elamin, M.R.; Abdulkhair, B.Y.; Elzupir, A.O. Removal of ciprofloxacin and indigo carmine from water by carbon nanotubes fabricated from a low-cost precursor: Solution parameters and recyclability. *Ain Shams Eng. J.* **2023**, *14*, 101844. [CrossRef]
60. Acharya, J.; Sahu, J.; Mohanty, C.; Meikap, B.C. Removal of lead (II) from wastewater by activated carbon developed from Tamarind wood by zinc chloride activation. *Chem. Eng. J.* **2009**, *149*, 249–262. [CrossRef]
61. Salam, M.A.; Al-Zhrani, G.; Kosa, S.A. Chemistry, E. Removal of heavy metal ions from aqueous solution by multi-walled carbon nanotubes modified with 8-hydroxyquinoline: Kinetic study. *J. Ind. Eng. Chem.* **2014**, *20*, 572–580. [CrossRef]
62. Largitte, L.; Pasquier, R. A review of the kinetics adsorption models and their application to the adsorption of lead by an activated carbon. *Chem. Eng. Res. Des.* **2016**, *109*, 495–504. [CrossRef]
63. Wierzbicka, E.; Kusmierek, K.; Swi, A.; Legocka, I. Efficient rhodamine b dye removal from water by acid-and organo-modified halloysites. *Minerals* **2022**, *12*, 350. [CrossRef]
64. Yen Doan, T.H.; Minh Chu, T.P.; Dinh, T.D.; Nguyen, T.H.; Tu Vo, T.C.; Nguyen, N.M.; Nguyen, B.H.; Pham, T.D. Adsorptive removal of rhodamine B using novel adsorbent-based surfactant-modified alpha alumina nanoparticles. *J. Anal. Methods Chem.* **2020**. [CrossRef] [PubMed]
65. Belachew, N.; Tadesse, A.; Kahsay, M.H.; Meshesha, D.S.; Basavaiah, K. Synthesis of amino acid functionalized Fe3O4 nanoparticles for adsorptive removal of Rhodamine B. *Appl. Water Sci.* **2021**, *11*, 1–9. [CrossRef]
66. Liu, Y.; Liu, Y.-J. Biosorption isotherms, kinetics and thermodynamics. *Sep. Purif. Technol.* **2008**, *61*, 229–242. [CrossRef]
67. Oyetade, O.A.; Nyamori, V.O.; Martincigh, B.S.; Jonnalagadda, S.B. Effectiveness of carbon nanotube–cobalt ferrite nanocomposites for the adsorption of rhodamine B from aqueous solutions. *RSC Adv.* **2015**, *5*, 22724–22739. [CrossRef]
68. Ojemaye, M.O.; Okoh, A.I. Multiple nitrogen functionalized magnetic nanoparticles as an efficient adsorbent: Synthesis, kinetics, isotherm and thermodynamic studies for the removal of rhodamine B from aqueous solution. *Sci. Rep.* **2019**, *9*, 1–13. [CrossRef]
69. Peng, L.; Qin, P.; Lei, M.; Zeng, Q.; Song, H.; Yang, J.; Shao, J.; Liao, B.; Gu, J. Modifying Fe_3O_4 nanoparticles with humic acid for removal of Rhodamine B in water. *J. Hazard. Mater.* **2012**, *209*, 193–198. [CrossRef]
70. Konicki, W.; Siber, D.; Narkiewicz, U. Removal of Rhodamine B from aqueous solution by $ZnFe_2O_4$ nanocomposite with magnetic separation performance. *Pol. J. Chem. Technol.* **2017**, *19*, 65–74. [CrossRef]
71. Motahari, F.; Mozdianfard, M.R.; Salavati-Niasari, M.J.P.S.; Protection, E. Synthesis and adsorption studies of NiO nanoparticles in the presence of H2acacen ligand, for removing Rhodamine B in wastewater treatment. *Process Saf. Environ. Prot.* **2015**, *93*, 282–292. [CrossRef]
72. Nazir, M.A.; Najam, T.; Jabeen, S.; Wattoo, M.A.; Bashir, M.S.; Shah, S.S.A.; ur Rehman, A. Facile synthesis of Tri-metallic layered double hydroxides (NiZnAl-LDHs): Adsorption of Rhodamine-B and methyl orange from water. *Inorg. Chem. Commun.* **2022**, *145*, 110008. [CrossRef]

73. Letshwenyo, M.W.; Mokgosi, S. Investigation of water treatment sludge from drinking water treated with Zetafloc 553I coagulant for phosphorus removal from wastewater. *J. Environ. Manag.* **2021**, *282*, 111909. [CrossRef] [PubMed]
74. Demirbas, O.; Calimli, M.H.; Kuyuldar, E.; Alma, M.H.; Nas, M.S.; Sen, F.J.B. Equilibrium, kinetics, and thermodynamic of adsorption of enzymes on diatomite clay materials. *BioNanoScience* **2019**, *9*, 474–482. [CrossRef]
75. Aarab, N.; Hsini, A.; Essekri, A.; Laabd, M.; Lakhmiri, R.; Albourine, A. Removal of an emerging pharmaceutical pollutant (metronidazole) using PPY-PANi copolymer: Kinetics, equilibrium and DFT identification of adsorption mechanism. *Groundw. Sustain. Dev.* **2020**, *11*, 100416. [CrossRef]
76. Vasudevan, S.; Lakshmi, J.J.W.S. Studies relating to an electrochemically assisted coagulation for the removal of chromium from water using zinc anode. *Water Sci. Technol. Water Supply* **2011**, *11*, 142–150. [CrossRef]
77. Inglezakis, V.J.; Zorpas, A.A. Heat of adsorption, adsorption energy and activation energy in adsorption and ion exchange systems. *Desalination Water Treat.* **2012**, *39*, 149–157. [CrossRef]

molecules

MDPI

Review

Unlocking the Potential of Gallic Acid-Based Metal Phenolic Networks for Innovative Adsorbent Design

Shella Permatasari Santoso [1,2,3,†], Artik Elisa Angkawijaya [4,†], Kuan-Chen Cheng [5,6,7,8,†], Shin-Ping Lin [9,10,11], Hsien-Yi Hsu [12,13], Chang-Wei Hsieh [14], Astrid Rahmawati [15], Osamu Shimomura [15] and Suryadi Ismadji [1,2,3,*]

1 Chemical Engineering Department, Faculty of Engineering, Universitas Katolik Widya Mandala Surabaya, Jl. Kalijudan 37, Surabaya 60114, East Java, Indonesia; shella@ukwms.ac.id or shella_p5@yahoo.com
2 Chemical Engineering Master Program, Widya Mandala Surabaya Catholic University, Kalijudan 37, Surabaya 60114, East Java, Indonesia
3 Collaborative Research Center for Zero Waste and Sustainability, Jl. Kalijudan 37, Surabaya 60114, East Java, Indonesia
4 RIKEN Center for Sustainable Resource Science, Yokohama 230-0045, Japan; artikelisa.angkawijaya@riken.jp
5 Institute of Biotechnology, National Taiwan University, #1 Roosevelt Rd., Sec. 4, Taipei 10617, Taiwan; kccheng@ntu.edu.tw
6 Department of Optometry, Asia University, 500, Lioufeng Rd., Wufeng, Taichung 41354, Taiwan
7 Graduate Institute of Food Science and Technology, National Taiwan University, 1 Roosevelt Rd., Sec. 4, Taipei 10617, Taiwan
8 Department of Medical Research, China Medical University Hospital, China Medical University, 91 Hsueh-Shih Rd., Taichung 40402, Taiwan
9 School of Food Safety, Taipei Medical University, 250 Wu-Hsing Street, Taipei 11031, Taiwan; splin0330@tmu.edu.tw
10 TMU Research Center for Digestive Medicine, Taipei Medical University, 250 Wu-Hsing Street, Taipei 11031, Taiwan
11 Research Center of Biomedical Device, Taipei Medical University, 250 Wu-Hsing Street, Taipei 11031, Taiwan
12 School of Energy and Environment, Department of Materials Science and Engineering, Centre for Functional Photonics (CFP), City University of Hong Kong, Kowloon Tong, Hong Kong, China; sam.hyhsu@cityu.edu.hk
13 Shenzhen Research Institute of City University of Hong Kong, Shenzhen 518057, China
14 Department of Food Science and Biotechnology, National Chung Hsing University, 145 Xingda Rd., South Dist., Taichung 40227, Taiwan; welson@nchu.edu.tw
15 Department of Applied Chemistry, Osaka Institute of Technology, 5-16-1 Omiya, Ashahi-ku, Osaka 535-8585, Japan; veronica.astrid26@gmail.com (A.R.); osamu.shimomura@oit.ac.jp (O.S.)
* Correspondence: suryadiismadji@yahoo.com
† These authors contributed equally to this work.

Abstract: Metal phenolic networks (MPNs) have attracted significant attention due to their environmentally benign nature, broad compatibility, and universal adhesive properties, making them highly effective for modifying adsorbent surfaces. These supramolecular complexes are formed through the coordination of metal ions with natural phenolic ligands, resulting in stable structures while retaining the active adsorption sites of the ligands, thereby enhancing the adsorption performance of unmodified substrates. Among various MPNs, metal ion gallic acid (GA) networks are particularly well-known for their exceptional stability, biological activity, and superior adsorption ability. This review offers a comprehensive examination of GA-based MPN adsorbents, focusing on their formation chemistry, characterization techniques, and applications. The coordination chemistry underlying the stability of GA–metal complexes is analyzed through equilibrium studies, which are critical for understanding the robustness of MPNs. The main analytical methods for assessing metal ligand interactions are discussed, along with additional characterization techniques for evaluating adsorbent properties. This review also explores various synthesis and performance enhancement strategies for GA-based MPN adsorbents, including stand-alone MPNs, MPN-mediated mesoporous materials, MPN-MOF composites, and MPN-coated substrates. By consolidating current advancements in MPN-based adsorbents and offering fundamental insights into their chemistry and characterization, this

https://doi.org/10.3390/molecules30061218

review serves as a valuable resource for researchers seeking to develop stable, functional metal-organic materials. It aims to drive innovation in sustainable and efficient adsorbent technologies for diverse environmental and industrial applications.

Keywords: surface modification; adsorbent composite; chelate complex; metal phenolic coating; metal complex

1. Introduction

Surface modification is a widely adopted strategy to introduce specific functional groups into adsorbents, enhancing their adsorption capacity [1,2], selectivity [3], reactivity [1,4], and stability [5]. This approach offers a practical and economical alternative to commercial activated carbon, which is often expensive. By modifying low-cost and locally available materials such as clays, zeolites, or cellulosic biomasses, adsorbents with increased active adsorption sites and improved capacity can be developed. Among the various modification techniques, surface coating stands out as a simple yet effective method. In particular, metal-phenolic network (MPN) coatings have emerged as a promising and sustainable approach [6–10]. In adsorption-based applications, MPN coatings introduce additional active sites through their metal and organic components [6,11], while their supramolecular structure ensures robust interconnections between the adsorbent, metal, and organic modifier. This not only enhances adsorption performance but also prevents leaching and secondary contamination, making MPN coatings a viable solution for sustainable adsorbent development.

MPNs are formed through the coordination of phenolic compounds with metal ions, creating stable metal-organic complexes. Due to their universal adherence properties, MPNs offer remarkable versatility in surface modification across various materials, including adsorbents [12,13]. Tannic acid has become the most extensively studied phenolic compound for MPN preparation, particularly following the pioneering work of Ejima et al. (2013) [14]. However, the predominant focus on tannic acid-based MPNs has overshadowed the potential of other phenolic compounds, such as gallic acid (GA), which remains underexplored despite its promising structural and functional attributes.

Gallic acid (GA), chemically known as 3,4,5-trihydroxybenzoic acid, is a naturally occurring polyphenol with a planar molecular structure. It exhibits an aromatic ring and galloyl group, which consists of one carboxylic group and three hydroxyl phenol groups. This galloyl group enables GA to function as a strong organic linker, chelating metal ions through a coordination self-assembly process [15–17]. This interaction leads to the formation of metal complexes with adhesive properties, capable of forming thin films on various substrate surfaces upon coordination with metal ions [18].

While previous reviews have extensively covered MPNs in general [13,19], studies specifically dedicated to GA-based MPNs remain scarce. This review provides the first comprehensive evaluation of GA-based MPNs for adsorbent preparation. The review begins by elucidating the coordination chemistry of GA-based metal complexes, offering critical insights into metal ligand interactions and the selection criteria for organic linkers. Next, the review examines the stability of GA-based metal ligand complexes, a fundamental yet often overlooked aspect of MPN chemistry that directly influences adsorbent performance. Additionally, key characterization techniques used to confirm metal-complex formation are discussed.

Furthermore, this review explores the potential of GA-based MPNs in adsorbent preparation, summarizing various MPN-based adsorbents derived from GA, including standalone MPN adsorbents, MPN-mediated mesoporous materials, MPN-MOF composites, and MPN-coated substrates. By integrating fundamental principles with practical insights, this review addresses a critical knowledge gap and offers a structured framework for researchers developing sustainable and high-performance MPN-based adsorbents. Through an in-depth discussion of the coordination chemistry and stability of GA-based MPNs, it establishes a foundation for future advancements in MPN-modified adsorption materials.

2. Evolution of Gallic Acid-Based Metal Complexes

The study of gallic acid-based metal complexes has evolved from fundamental coordination chemistry to advanced applications in material science, therapeutics, and environmental technology. Figure 1 illustrates a timeline highlighting the evolution of gallic acid-based metal complexes and their applications in multifunctional fields.

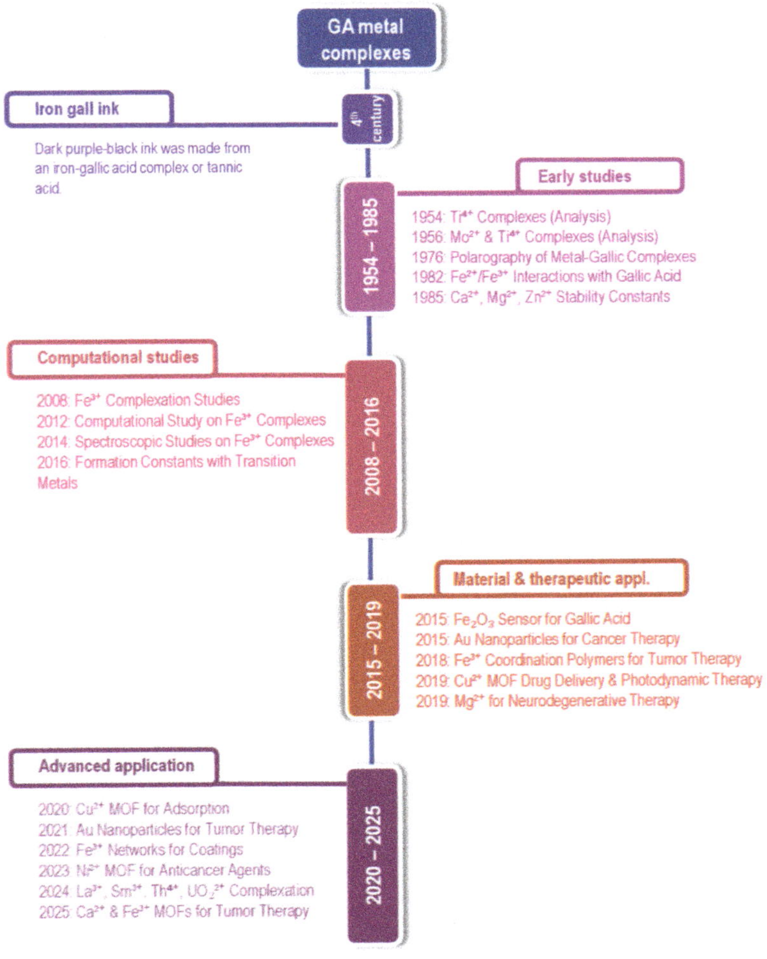

Figure 1. Timeline illustrating the development of gallic acid-based metal complexes, compiled from the publication list in Supplementary Information Table S1.

Early research (1954–1985) focused on elucidating the coordination chemistry of gallic acid-metal interactions, particularly with Ti^{4+} [20,21], Mo^{2+} [22], Fe^{2+}, Fe^{3+} [23], Ca^{2+}, Mg^{2+}, and Zn^{2+} [24,25]. With advancements in computational modeling and spectroscopy between 2008 and 2016, researchers refined their understanding of transition metal coordination [17], protonation equilibria [26], and electron distribution in GA-based metal complexes [27]. These studies extended beyond analytical chemistry to applications in cultural heritage preservation [28] and antioxidant research [29]. Due to its high metal-binding affinity, GA has also been explored for metal detection in solutions and soil, playing a crucial role in phytoremediation, where GA-rich plants enhance metal sequestration from contaminated environments [30].

Between 2015 and 2020, research expanded into biomedical and material science applications. GA-metal complexes were employed in electrochemical sensors [29], cancer therapy [31], and neurodegenerative treatments [32]. The emergence of metal-organic frameworks (MOFs) and metal phenolic networks (MPNs) revolutionized drug delivery and photodynamic therapy [33]. MPNs, in particular, gained attention for their biocompatibility, self-assembly properties, and redox-responsive behavior, making them highly effective in biomedical coatings and controlled drug release [34]. During this period, adsorption applications also gained prominence, particularly in environmental remediation. GA-based MOFs, such as Zr- and Cu-based frameworks [35–37], demonstrated excellent performance in enhanced mass spectrometry and dye removal from wastewater, leveraging GA's strong affinity for metal ions and organic pollutants.

Recent studies have focused on MOFs for wastewater treatment, targeted drug delivery, and sustainable materials. The development of gallic acid-calcium grafts [38] and bismuth-gallic acid MOFs [39] has enhanced tumor therapy and tissue regeneration. Meanwhile, MPNs have evolved into tunable coatings, mimicking natural protective barriers and expanding their applications in biomedicine [40] and advanced materials. Adsorption technologies have continued to advance, with AI-optimized green synthesis of GA-based MOFs enhancing efficiency in heavy metal and dye removal [41]. Additionally, GA's role in metal sequestration and phytoremediation continues to grow, reinforcing its importance in environmental sustainability.

The evolution of GA-based metal complexes reflects their versatility, from historical ink formulations and metal detection to cutting-edge biomedical, adsorption, and environmental applications. Among these developments, MPNs and MOFs stand out as transformative innovations, unlocking new possibilities in biomedicine, coatings, adsorption technologies, and sustainable materials.

3. Gallic Acid-Based Metal Complexes, an MPN or MOF?

MOFs and MPNs are supramolecular materials composed of metal ions or clusters coordinated with multidirectional organic ligands. Their shared precursors often lead to the misconception that they are interchangeable, yet their structural and synthetic distinctions are significant. MOFs are highly crystalline, exhibiting well-organized frameworks that provide exceptional porosity and internal surface area [42]. Their formation typically requires heat and pressure [43–45], though some MOFs can be synthesized under milder conditions, often resulting in varied particle morphologies [46,47]. In contrast, MPNs are amorphous coordination compounds that form spontaneously under ambient conditions [19]. They exist as molecular species in solution but can aggregate into solid structures at high precursor concentrations or prolonged reaction times. Unlike MOFs, which often adopt well-defined geometric shapes (e.g., cubic, octahedral, or tetrahedral), MPNs generally lack distinct morphological features due to their cluster-based assembly.

For GA-based metal complexes, their behavior aligns more closely with MPNs. While GA-metal systems can remain in solution, they can also form solid frameworks resembling MOFs. For instance Yilmaz (2024) reported the formation of two metal complexes with nanoflower-like structures, synthesized by reacting GA and copper at 4 °C for 3 days, and the other by reacting GA and zinc at 25 °C for 24 h [48]. These materials exhibited broad-spectrum antimicrobial activity and anticancer properties. The solid formation of these materials likely resulted from the uninterrupted clusterization of metal-GA networks, which aggregated into supramolecular structures (Figure 2a,b). Similarly, Santoso et al. (2021) synthesized spherical MPN particles from Cu^{2+} and GA, with N-functionalization achieved through glycine addition (Figure 2c) [37]. Meanwhile, Sharma et al. (2019) produced Cu-gallate solids with nanorod morphology by reacting copper acetate and GA in DMF with a micellar surfactant at 70–80 °C for 12 h [33]. The resulting material exhibited polycrystalline XRD patterns akin to MIL-53 and aggregated rod-like morphology (Figure 2d,e).

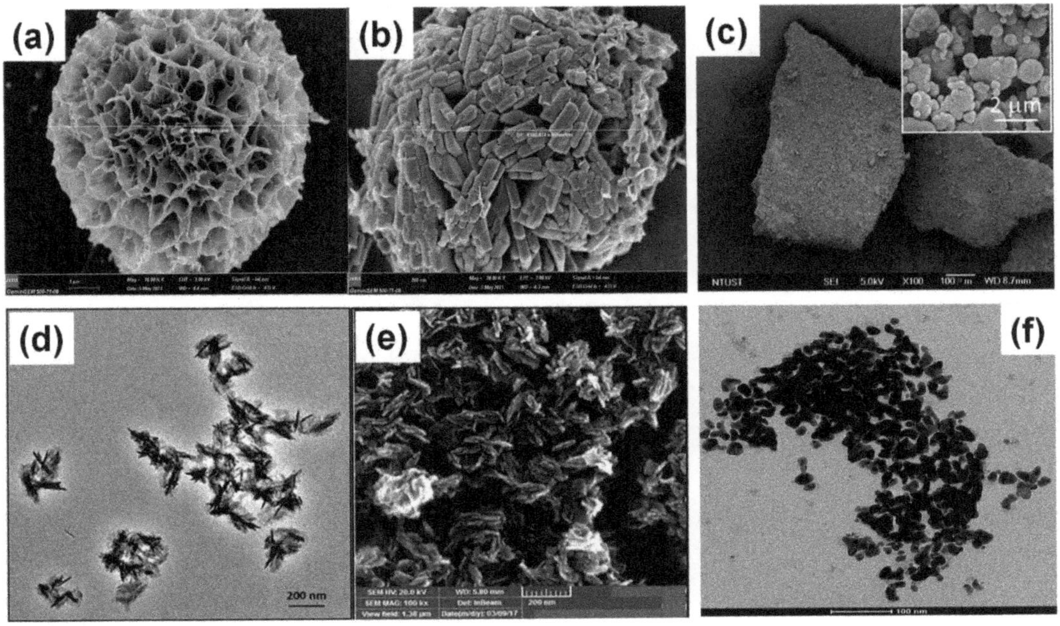

Figure 2. (**a**,**b**) Morphological characterization of gallic acid (GA)-based metal complexes. SEM images of supramolecular materials formed through coordination between GA and copper (**a**) and GA and zinc (**b**). Images reproduced from Yilmaz (2024) [48], under a Creative Commons 4.0 license. (**c**) SEM image of N-grafted GA-copper MPN. Image adapted from Santoso et al. (2021) [37], licensed under CC-BY. (**d**,**e**) TEM and SEM images of copper-gallic acid metal-organic framework. Image reprinted from Sharma et al. (2019) [33], with permission from the American Chemical Society. (**f**) TEM image of iron gallate particles. Image reproduced from Yadav et al. (2019) [49], with permission from Elsevier B.V. All rights reserved.

MOFs are renowned for their exceptionally high internal surface areas, which contribute to their superior porosity and gas adsorption properties. Notable examples include Zr-MOF NU-1103 (6550 m^2/g) [50], Cu-MOF HKUST-1 (1850 m^2/g) [51], and Cu-tbo-MOF-5 (3971 m^2/g) [52], all of which demonstrate excellent gas storage potential. In contrast, MOFs with lower surface areas are often employed for aqueous-phase adsorption. For instance, MIL-101(Fe)/WO_3 (20.74 m^2/g) effectively adsorbs tetracycline-HCl from solution [53], MIL-101(Fe)/Bi_2MoO_6 (39.72 m^2/g) exhibits both photodegradation and photofixation capabilities toward tetracycline-HCl [54]. For GA-based metal complexes, surface area variations significantly influence adsorption performance. A nitrogen-grafted CuGA complex, despite its low surface area (2.00 m^2/g) [37]. In contrast, a non-grafted

CuGA complex, with a higher surface area (198.22 m^2/g), shows slightly lower adsorption efficiency for the same dye [36]. NiGA exhibits a surface area of 196 m^2/g, showing antimicrobial and anticancer activity [55].

These studies highlight the nuanced classification of GA-based metal complexes. While they often align with MPNs due to their amorphous nature, mild synthesis conditions, and spontaneous self-assembly, some may resemble MOFs. Additionally, GA's susceptibility to oxidation under high-temperature conditions limits its suitability as a ligand for MOF synthesis. Collectively, these factors underscore the advantages of MPNs, particularly for applications that demand flexible and low-energy synthesis.

4. Coordination Chemistry of GA-Based Metal Complexes

4.1. Protonation and Deprotonation of Ligands

Understanding the ionization behavior of ligands is critical for elucidating the formation of chelate complexes [56]. Phenolic compounds, as ligands, possess ionizable functional groups that can undergo protonation or deprotonation, processes that are strongly modulated by pH. In the case of gallic acid (GA), it contains four ionizable groups, that is, one carboxylic moiety and three hydroxy phenolic moieties [57,58]. The carboxylic moiety, being the most acidic group, deprotonates first as the pH increases, followed by the sequential deprotonation of the hydrogen ions (H^+) from the three hydroxy phenolic moieties. This stepwise deprotonation of GA can be expressed by the following equilibrium Equations (1)–(4) and is illustrated in Figure 3:

$$GAH_4 \rightleftharpoons H^+ + GAH_3^- \; [H^+][GAH_3^-] = Ka_1[GAH_4] \quad (1)$$

$$GAH_3^- \rightleftharpoons H^+ + GAH_2^{2-} \; [H^+]\left[GAH_2^{2-}\right] = Ka_2\left[GAH_3^-\right] \quad (2)$$

$$GAH_2^{2-} \rightleftharpoons H^+ + GAH^{3-} \; [H^+]\left[GAH^{3-}\right] = Ka_3\left[GAH_2^{2-}\right] \quad (3)$$

$$GAH^{3-} \rightleftharpoons H^+ + GA^{4-} \; [H^+]\left[GA^{4-}\right] = Ka_4\left[GAH^{3-}\right] \quad (4)$$

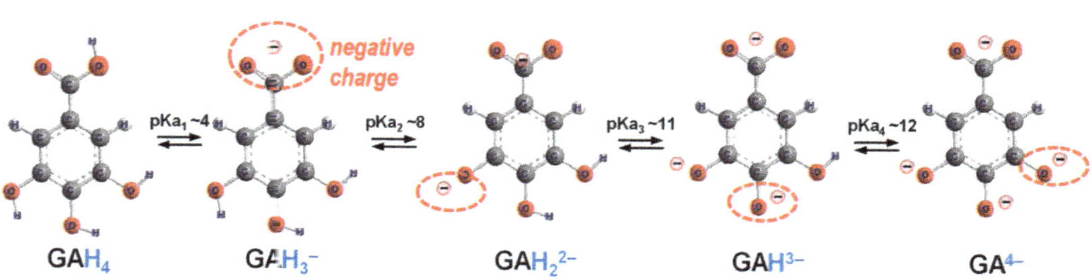

Figure 3. Stepwise ionization constants of gallic acid (GA). GAH_4 represents the fully protonated molecule. Each deprotonation step leads to the sequential formation of negatively charged species: $GAH_3{}^-$, $GAH_2{}^{2-}$, GAH^{3-}, and GA^{4-}. In the molecular structure, oxygen atoms are represented by red spheres, hydrogen atoms by white spheres (with larger spheres indicating non-ionizable hydrogen atoms and smaller spheres indicating ionizable hydrogen atoms), and carbon atoms by grey spheres. The red dashed circles highlight the sites of subsequent deprotonation.

The protonated ligand is designated by the general formula $AH_x{}^z$, where A denotes the ligand abbreviation (e.g., GA for gallic acid), H represents the proton, x is the number of protons associated with all ionizable groups, and z is the net charge of the species at its current state. The protonation constant (Ka, also referred to as acidity constant, ionization constant, or acid dissociation constant) reflects the relative capacity of a compound to donate a proton [59]. Its logarithmic value, pKa, indicates the pH at which protonation or deprotonation of the ionizable group occurs.

In its fully protonated state, GA possesses four protons, one at each ionizable group, and is denoted as GAH_4, indicating four protons and a neutral charge. The first deprotonation occurs at the carboxylic group, with a pKa_1 value approximately 4, forming the negatively charged species of GAH_3^-. The second deprotonation occurs at one of the hydroxy groups, with a pKa_2 value of approximately 8, forming GAH_2^{2-}. The third and fourth deprotonations, with pKa_3 and pKa_4 values of approximately 11 (GAH^{3-}) and 12 (GA^{4-}), respectively, require highly alkaline conditions.

Table 1 summarizes the reported pKa values of GA from various publications, ranging from older to more recent studies. The reported values are generally consistent for pKa_1 and pKa_2. However, significant variations in the reported pKa_3 and pKa_4 values is likely due to the extreme alkaline conditions required for accurate determination. Additionally, discrepancies may arise from differences in experimental conditions, such as the use of different salts and ionic strengths. For instance, salts such as KCl are highly reactive and can significantly influence ionization behavior.

Table 1. Ionization or dissociation constant of the gallic acid.

Condition	pKa_1	pKa_2	pKa_3	pKa_4	Ref.
I = 0.2 M [1] and T = 25 °C	4.22	8.69	11.19		[60]
I = 0.1 M [1] and T = 25 °C	4.4	8.6	11.2	12	[24]
I = 0.1 M $NaNO_3$ and T = 25 °C	4.10	8.38			[17]
I = 0.1 M $NaNO_3$ and T = 25 °C	4.12	8.32			[61]
I = 0.1 M KCl and T = 25 °C	3.75	7.50	9.50	10.50	[62]

[1] Salt source undefined.

4.2. *Metal Complex Formation and Stability Constant*

The interaction between ligands and metal ions is governed by the specific functional groups of the ligand, as well as the charge density and polarizability of the metal ion. According to Pearson's Hard and Soft Acids and Bases (HSAB) theory (see Supplementary Information Table S2), metal ions act as Lewis acids (electron acceptors), while ligands function as Lewis bases (electron donors) [63]. Hard acids, characterized by high charge density and low polarizability, preferentially bind to hard bases through ionic interactions. In contrast, soft acids, which have low charge density and high polarizability, form covalent bonds with soft bases. Gallic acid (GA), with its hydroxyl (–OH) functional groups, behaves as a hard base, forming stable coordination bonds with hard acids such as Fe^{3+} [17,27], a metal ion commonly used in metal-phenolic network (MPN) synthesis. While GA can also coordinate with borderline and soft acids, these complexes exhibit lower stability.

Crystal field theory (CFT) further explains metal ligand stability through ionic potential (charge-to-radius ratio), underpinning the Irving–Williams Series (IWS). IWS predicts the stability trend of first-row transition metal complexes (i.e., Mn^{2+} to Zn^{2+}) [64]. According to IWS, decreasing ionic radius from Mn^{2+} to Zn^{2+} leads to increased stabilization energy, enhancing complex stability. However, Cu^{2+} deviates from this trend due to the Jahn-Teller effect [65], which provides additional stabilization, making Cu^{2+} complexes more stable than those of Ni^{2+}. The general stability trend follows: $Mn^{2+} < Fe^{2+} < Co^{2+} < Ni^{2+} < Cu^{2+} > Zn^{2+}$.

To quantitatively assess metal ligand interaction strength, the stability constant (log$K$) is a key parameter for predicting complex formation and stability [66,67], serving as a fundamental metric for evaluating MPN stability. The stepwise stability constant (K_n) for each stage of metal ligand complexation is expressed as Equation (5).

$$ML_{n-1} + L \rightleftharpoons ML_n \; K_n = \frac{[ML_n]}{[ML_{n-1}][L]} \tag{5}$$

where M is the metal ion, L is the ligand, and n represents the number of ligands coordinated to the metal. The overall stability of the metal complex is represented by the cumulative stability constant (logβ), given by Equation (6).

$$M + nL \rightleftharpoons ML_n \; \beta_n = \frac{[ML_n]}{[M][L]^n} \tag{6}$$

$$\beta_n = K_1 \times K_2 \times K_3 \times \ldots \times K_n = \sum_{i=1}^{n} K_i$$

where β_n denotes the overall formation constant for a complex with n ligands. The higher the β_n value, the greater the stability of the resulting metal ligand complex, making it a crucial factor in determining the feasibility of MPN formation and adsorption efficiency.

Before complex formation, ligand deprotonation occurs, enhancing its electronegativity and facilitating metal ion attraction [68]. The metal-to-ligand ratio and pH are key factors affecting complex type and stability. Table 2 presents the log$K$ values for complex formation between GA and various transition metal ions, such as Cu^{2+}, Zn^{2+}, Ni^{2+}, Fe^{3+}, and Co^{2+}. Among these ions, Fe^{3+}, with its higher charge density and smaller ionic radius, facilitates strong binding with GA, leading to the formation of the most stable complex, aligning with HSAB theory. The observed GA-metal stability trend ($Co^{2+} < Ni^{2+} < Cu^{2+} > Zn^{2+}$) also aligns with the IWS prediction.

Table 2. Stability constant (log$K$) of complexes between GA and transition metal ions.

Metal	Complex Species M_pL_q		Condition	log$K$	Ref.
	p	q			
Cu^{2+}	1	1	I = 0.1 M $NaNO_3$, T = 25 °C	9.75	[61]
			I = 1 N $NaNO_3$, T = 27 °C	9.80	[26]
	1	2	I = 0.1 M $NaNO_3$, T = 25 °C	6.75	[61]
Zn^{2+}	1	1	I = 0.1 M $NaNO_3$, T = 25 °C	8.56	[61]
			I = 1 N $NaNO_3$, T = 27 °C	7.98	[26]
	1	2	I = 0.1 M $NaNO_3$, T = 25 °C	5.83	[61]
	2	1	I = 0.1 M $CaCl_2$, T = 25 °C, pH = 8	11.38	[24]
Ni^{2+}	1	1	I = 0.1 M $NaNO_3$, T = 25 °C	8.00	[61]
			I = 1 N $NaNO_3$, T = 27 °C	6.74	[26]
	1	2	I = 0.1 M $NaNO_3$, T = 25 °C	5.50	[61]
Fe^{3+}	1	1	I = 0.1 M $NaNO_3$, T = 25 °C	14.73	[17]
			I = 1 N $NaNO_3$, T = 27 °C	10.98	[26]
	1	2	I = 0.1 M $NaNO_3$, T = 25 °C	11.93	[61]
Co^{2+}	1	1	I = 0.1 M $NaNO_3$, T = 25 °C	7.25	[61]
			I = 1 N $NaNO_3$, T = 27 °C	7.13	[26]
	1	2	I = 0.1 M $NaNO_3$, T = 25 °C	4.75	[61]

Phenolic compounds, including GA, can form complexes with metal ions, but the stability of these complexes varies significantly and must be carefully evaluated. Not all phenolic compounds form highly stable complexes. For instance, Radalla (2015) reported log$K$ values for several phenolic acids complexed with transition metal ions, demonstrating that GA complexes exhibit the highest stability among phenolic acid ligands [61]. Similarly, Sursyakova et al. (2017) demonstrated that GA forms more stable complexes than succinic acid, with log$K$ values of 2.89 for the CuGA complex and 3.88 for $Cu(GA)_2$, with CuGA being the predominant species in the pH range of 3.5 to 8 [69]. The stability of metal ligand complexes is heavily influenced by the nature of the ligand. Some ligands form only sigma (σ) bonds with metal ions, whereas others, such as GA, can form both σ and pi (π) bonds. Ligands containing a benzene ring are particularly effective due to their ability to donate electrons through π-symmetric interactions, which not only influence the geometrical configuration but also significantly enhance the stability of the resulting complexes [70,71].

5. Analysis Techniques for Predicting the Metal Complex Formation

MPNs are (supra)molecular structures that form in aqueous solutions through the coordination of metal ions with organic ligands. The process begins with the deprotonation of the ligand, generating a negatively charged ligand field that strongly attracts positively charged metal ions. This deprotonation, involving the release of protons from the ligand, can be quantitatively analyzed using potentiometric titration, where each ionizable group's deprotonation appears as an inflection point in the titration curve. The presence of metal ions often facilitates proton release, shifting the titration curve toward lower pH values and providing valuable insights into the complexation process.

To characterize metal ligand complex formation at the molecular level, advanced analytical techniques are crucial. Spectroscopic methods such as Raman, electron paramagnetic resonance (EPR), UV-Vis, Fourier-transform infrared (FTIR), and nuclear magnetic resonance (NMR) spectroscopy are particularly effective in evaluating complex formation. These techniques offer detailed structural and electronic information, enabling a comprehensive understanding of the coordination chemistry governing MPN assembly.

5.1. Potentiometric Titration

Potentiometric titration is a robust and widely used analytical technique for predicting metal ligand complex formation across a broad pH range. This method provides critical insights into the likelihood of complexation by analyzing changes in protonation behavior and stability constants. Figure 4a illustrates the potentiometric titration curves for the titration of Fe^{3+} with GA (and/or glycine ligand). The protonation constant (pKa) of GA and the stability constant (log*K*) of its metal complex can be determined by analyzing the difference in the volume of NaOH required to achieve the same pH value in titration curves (*i*) and (*ii*) for pKa, and (*i*) and (*iv*) for log*K*. Similarly, for glycine, pKa and logK are derived from the differences in NaOH consumption between curves (*i*) and (*iii*) and curves (i) and (v), respectively, following the Irving and Rossotti methodology [17]. The formation of metal complexes is evidenced by the shift in the titration curve for the system containing Fe^{3+} and GA (curve *iv*) to lower pH values compared to the curve for GA only (curve *i*), as highlighted by the green shaded area in Figure 4a. This shift indicates the binding of Fe^{3+} to GA, which alters the protonation behavior of the ligand.

Using the open-access Hyperquad Simulation and Speciation 2009 (HySS2009) software [72], the log*K* values corresponding to the stability constants of the metal complexes can be stimulated, allowing for visualization of the distribution of metal complex species across various pH ranges [73]. Illustrated in Figure 4b is the species distribution of Fe^{3+} and GA complexes. FeGA begins forming at pH 5.0 and is predominantly abundant between pH 7.0 and highly alkaline pH. This observation aligns well with the shift observed in the titration curve in Figure 4a. In contrast, $Fe(GA)_2$ forms at lower abundance in the same pH range, due to its reduced stability (lower log*K*), which is attributed to weaker electrostatic interactions between Fe^{3+} and the second ligand molecules [74]. The structural representation of the typical Fe^{3+}-GA complexes at various metal-to-ligand ratios is presented in Figure 5.

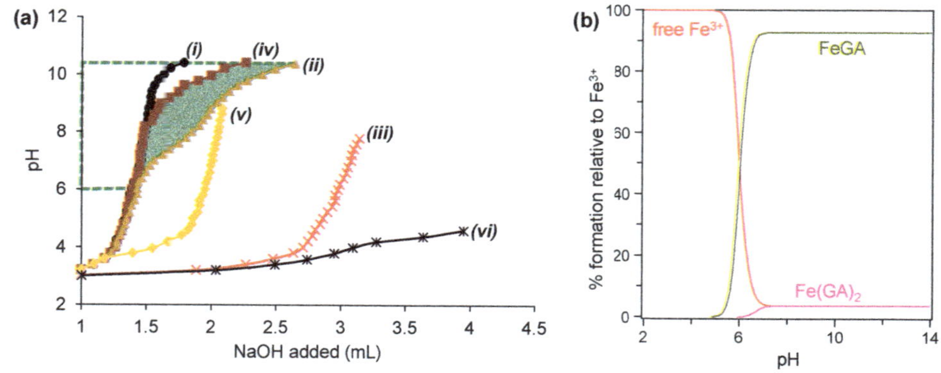

Figure 4. (**a**) Titration curves of a solution containing gallic acid, Fe^{3+}, and glycine at 0.1 M ionic strength: (*i*) 3 mM HNO_3, (*ii*) solution *i* + 1 mM gallic acid (ligand only), (*iii*) solution *ii* + 0.4 mM Fe^{3+}, (*iv*) solution *i* + 1 mM glycine, (*v*) solution *iv* + 0.4 mM Fe^{3+}, (*vi*) solution *i* + 1 mM gallic acid + 1 mM glycine + 0.4 mM Fe^{3+}. The green shaded area highlights the shift in the gallic acid curve resulting from the addition of Fe^{3+}. The figure was redrawn from Ref. [17] by extracting data points using Origin 2024(10.1) software. The data points for drawing these curves are provided in Supplementary Table S3. (**b**) Species distribution curve as a function of pH for GA-Fe^{3+} system, with percentages of formation relative to Fe^{3+}. The figure was created using HySS2009 software to model metal ligand complex distribution, with input pKa and logK values obtained from Ref. [17]. The data points for drawing the curve are provided in Supplementary Table S4.

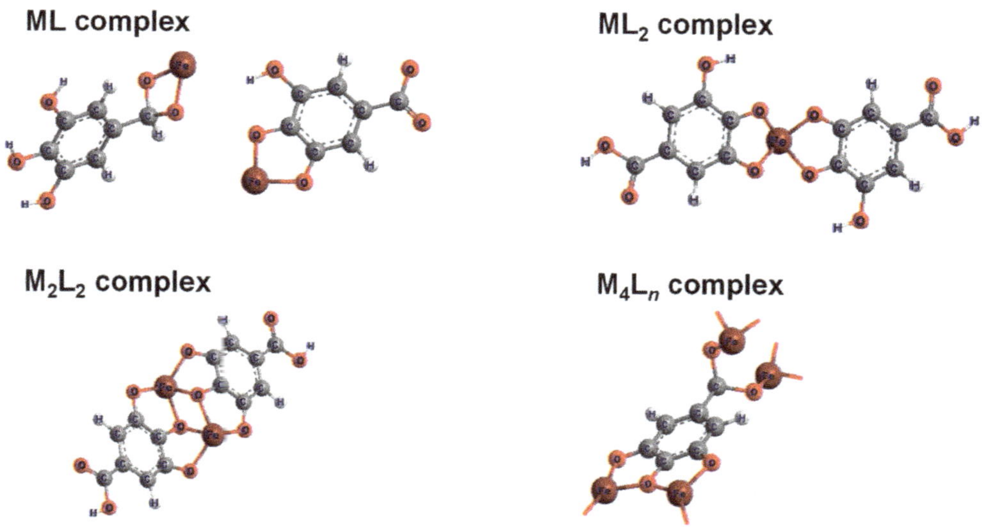

Figure 5. Structural representation of typical binary Fe^{3+}-GA complexes. Typical binary complexes are formed with metal-to-ligand ratios of 1:1 (FeGA), 1:2 ($Fe(GA)_2$), and ratios such as $Fe_2(GA)_2$ and $Fe_4(GA)_n$. Structural illustrations adapted and redrawn from Refs. [17,75]. In the molecular structure, oxygen atoms are represented by red spheres, hydrogen atoms by white spheres (with larger spheres indicating non-ionizable hydrogen atoms and smaller spheres indicating ionizable hydrogen atoms), carbon atoms by grey spheres, and iron atoms by the brown spheres.

The formation of the complexes is influenced by two primary factors: (i) pH, which dictates ligand deprotonation and the availability of deprotonated groups for coordination, and (ii) metal-to-ligand ratios, where higher metal or ligand concentrations can favor the formation of complexes with higher stoichiometry. At acidic pH (2.0–6.0), uncoordinated

Fe^{3+} and GA, dominate because the ligand remains mostly protonated, limiting strong electrostatic interactions. At pH > 6.0, abundant deprotonated ligand species readily coordinate with Fe^{3+}, explaining why MPN synthesis typically occurs under alkaline conditions. However, while FeGA persists under extreme alkaline pH, synthesis in such conditions is impractical due to the need for excessive alkaline compounds. High OH^- concentrations can lead to the formation or precipitation of metal hydroxide, especially in concentrated metal ligand solutions often used in MPN synthesis. This highlights the importance of optimizing synthesis conditions to promote effective complex formation.

5.2. Spectrophotometric-Based Analyses

Spectrophotometric techniques are well suited to analyze metal complexes because of their unique electronic structures and optical properties, which arise from the interactions between metal ions and ligands. Crystal field theory (CFT) and ligand field theory (LFT) provide the basic framework for understanding these properties, explaining how ligands create electric fields that partition d orbitals into different energy levels and how molecular orbital interactions influence electronic transitions [76,77]. Transition metal complexes, especially those involving d- and f-block elements, exhibit diverse oxidation states and electronic configurations, leading to characteristic transitions such as d-d, charge transfer, and ligand-centered transitions, all of which can be examined using spectrophotometry. Advanced computational methods, including molecular orbital (MO) and density functional theory (DFT), further enhance the ability to predict and interpret absorption and emission spectra, offering insights into stability, reactivity, and bonding interactions [76,78]. These capabilities make spectrophotometry a versatile and indispensable tool, playing a crucial role in advancing research in areas such as analytical chemistry, supramolecular chemistry, and the development of functional materials such as metal-organic frameworks (MOFs), molecular machines, including MPNs [76,77,79,80].

5.2.1. Raman Spectroscopy

Raman spectroscopy, particularly when coupled with Surface-Enhanced Raman Spectroscopy (SERS), serves as a highly sensitive and powerful tool for probing the functional groups of ligands involved in interactions with metal ions during the formation of metal complexes. These vibrational spectroscopy techniques enable the detection and identification of target analytes at the single-molecule level, offering unparalleled insights into molecular interactions [81]. For instance, Sánchez-Cortés and García-Ramos (2000) demonstrated the effectiveness of these techniques by examining the spectral changes in GA upon its interaction with Ag colloid [82]. In their study, the band corresponding to the carboxylic (–COOH) group, observed at 1690 cm^{-1} in the solid-state spectrum of GA (Figure 6(ai)), disappeared in the SERS spectrum of the GA dissolved in ethanol (Figure 6(aii), red-highlighted spectrum). This disappearance indicates the ionization of the –COOH group at neutral pH. Additionally, the SERS spectrum of GA on Ag colloid (Figure 6(aiii), green-highlighted spectrum) revealed the emergence of new spectral features, reflecting chemical changes in GA induced by the presence of the Ag colloid. These spectral alterations are consistent with the formation of metal ligand coordination complexes, highlighting the utility of SERS in elucidating such interactions.

Similarly, Espina et al. (2022) investigated the Raman spectra of iron gall ink species, specifically the Fe-GA complex, and observed significant restructuring of the vibrational spectra between free GA and metal-complexed GA (Figure 6(bi,bii) compared to Figure 6(biii,biv)) [83]. Three prominent bands at 1470 cm^{-1}, 1322 cm^{-1}, and 576 cm^{-1} were identified as characteristic of iron gall ink species, with these bands being widely documented in prior studies [84–86]. Density functional theory (DFT) calculations on the

Fe-GA complex revealed that the band at 1470 cm^{-1} arises from benzene ring vibrations coupled to C–O stretching and C–H bending. This band intensifies in the spectrum of metal-complexed GA due to changes in polarizability induced by metal coordination. The band at 1322 cm^{-1} corresponds to ring stretching coupled to C–O stretching and C–H bending, while the band at 576 cm^{-1}, exclusive to the metal-complexed GA, is attributed to Fe–O stretching, confirming metal coordination.

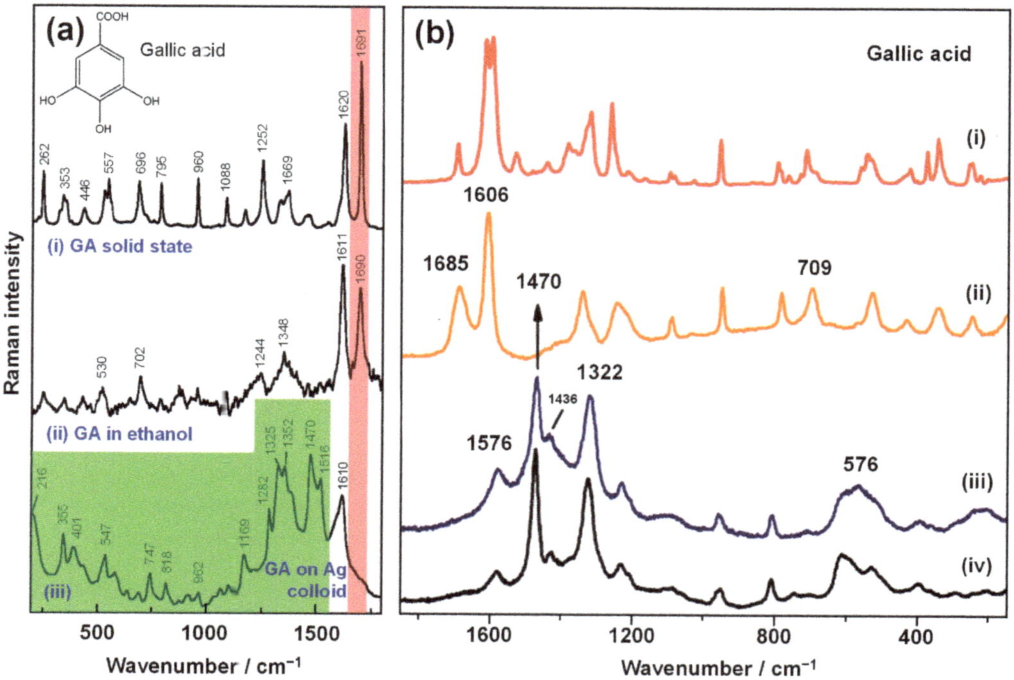

Figure 6. (**a**) Raman spectra illustrating the interaction of gallic acid (GA) with silver (Ag) in a colloidal system: (i) solid GA, (ii) 60 mM GA in ethanol, and (iii) 5 mM GA on Ag colloid at pH 3.9. The red-highlighted region corresponds to the vibrational modes of the carboxylic group in GA, which disappear due to deprotonation upon dissolution in ethanol. The green-highlighted region represents new spectral bands in GA induced by its interaction with the Ag colloid. Image adapted from Sánchez-Cortés and García-Ramos (2000) [82], with permission from Academic Press. All rights reserved. (**b**) Raman spectra of the metal complex between gallic acid (GA) and Fe^{3+}: (i) solid GA, (ii) 20 mM GA in aqueous solution, (iii) Fe-GA complex on paper, and (iv) Fe-GA complex in solution at a ligand-to-metal molar ratio of 1:3 and pH of 7. Image adapted from Espina et al. (2022) [83], licensed under CC-BY-NC-ND 4.0.

These findings underscore the exceptional capability of SERS to provide detailed insights into molecular interactions and structural transformations at the nanoscale. By capturing subtle spectral changes and correlating them with specific molecular vibrations, SERS enables a deeper understanding of the mechanisms underlying metal ligand complexation and related chemical processes.

5.2.2. Fourier Transform Infrared (FTIR) Spectroscopy

FTIR spectroscopy is a valuable tool for confirming the successful formation of metal ligand complexes by comparing the spectra of the complex with those of the original ligand. It is particularly effective in detecting prominent vibrations of oxygen-containing functional groups. For example, Santoso et al. (2021) demonstrated the use of FTIR to confirm the formation of the metal complex N_g-CuGA from GA and Cu^{2+}, with additional nitrogen group functionalization via ternary complexation with glycine [37]. The complexation of GA with Cu^{2+} induces vibrational band shifts in the galloyl group, specifically in the –OH groups, as indicated by the shifting of the bands between the wavenumbers 3282 and 3496 cm^{-1} (Figure 7(aiii)). Furthermore, nitrogen functionalization is confirmed by the presence of a band at 3161 cm^{-1} in the N_g-CuGA spectra, indicating the –NH group. The observed shift in this amine group's spectral position, compared to glycine alone, suggests its involvement in metal ion coordination during complex formation.

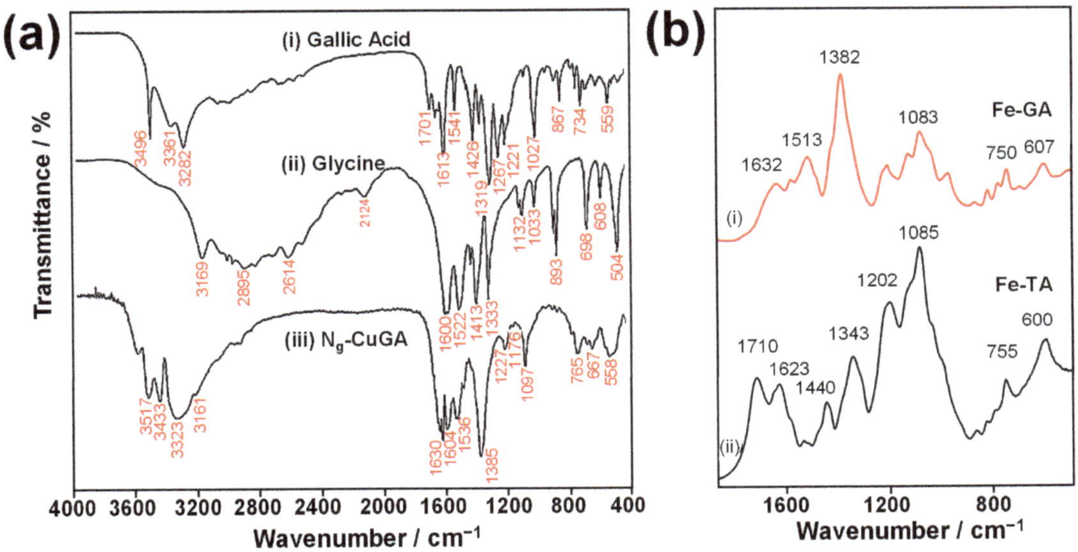

Figure 7. (**a**) FTIR spectra of the solid-state metal complexes formed by the coordination of Cu^{2+} with gallic acid (GA): (i) solid GA, (ii) solid glycine, and (iii) solid ternary metal complex. Image adapted from Santoso et al. (2021) [37], licensed under CC-BY. (**b**) FTIR spectra of solid-phase metal complexes of Fe^{3+} with gallic acid (GA) or tannic acid (TA): (i) Fe-GA complex and (ii) Fe-TA complex. Image adapted from Espina et al. (2022) [83], licensed under CC-BY-NC-ND 4.0.

Similarly, Espina et al. (2022) highlighted the robustness of FTIR in identifying functional groups involved in metal chelation [78]. Upon complexation with Fe^{3+}, the band corresponding to C–O stretching (1309 cm^{-1}) and C–OH bending (1014 cm^{-1}) decreases in intensity compared to GA alone, while a strong band at 1083 cm^{-1} appears instead (Figure 7(bi)), indicating Fe^{3+} coordination with the –OH of the phenolic group. Additionally, the appearance of a strong band at 1382 cm^{-1} has been attributed to the interaction between Fe^{3+} and the –COOH group. The Fe–O coordination is observed at 607 cm^{-1} for the FeGA complex and at 600 cm^{-1} for the FeTA complex (Figure 7(bii)).

Several other works also reported changes in the spectra as induced by the complexation. Lunardi et al. (2023) reported a reduction from a doublet to a single peak near ~2900 cm^{-1} [6], while Liu et al. (2021) observed the disappearance of the O–H vibration peak at ~3300 cm^{-1} [87]. Espina et al. (2022) documented alterations in peaks within the 1300–1000 cm^{-1} range [83]. These spectral changes provide critical insights into the coordination environment and structural transformations of the ligand upon complexation.

5.2.3. Electron Paramagnetic Resonance (EPR) Spectroscopy

EPR spectroscopy is a powerful technique for characterizing ligand fields (the negative point charges of ligands) and determining the oxidation states of metal complexes. It provides valuable insights into the reaction chemistry between metals and ligands across a wide range of experimental conditions [88]. For instance, Pirker et al. (2012) successfully employed EPR spectroscopy to investigate the formation of mononuclear Cu^{2+} complexes with GA [89]. Their study identified three distinct Cu-GA complex species through S-band and X-band EPR spectra, denoted as Complexes I, II, and III.

Complexes I and II were assigned as mono- and bis-Cu^{2+} GA complexes, respectively, with Cu^{2+} coordinating to two hydroxy-phenolic groups of GA, as illustrated in Figure 5 for ML and ML_2 complexes. Complex III, on the other hand, was hypothesized to form through additional coordination of glycerol. The S-band EPR spectra at an acidic pH of 5 revealed the presence of hydrated, uncomplexed Cu^{2+}, $[Cu(H_2O)6]^{2+}$. While Complex I was hardly detectable in the S-band spectra (Figure 8a), it was clearly observed in the X-band spectra (Figure 8b). At pH 7, the S-band spectra exhibited overlapping signals from Complexes II and III, whereas only Complex III was detected at higher pH values.

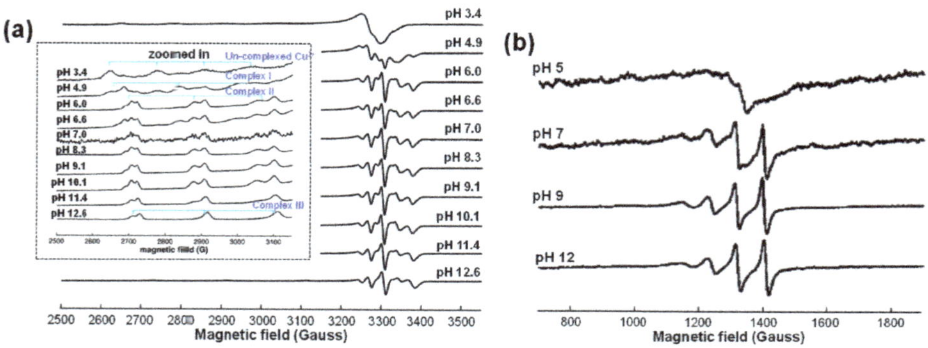

Figure 8. EPR spectra of gallic acid (GA) complexed with Cu^{2+} metal ions at metal-to-ligand molar ratio of 1:5 at different pH values: (**a**) S-band spectra and (**b**) X-band spectra. Images were reproduced from Pirker et al. (2021) [89], with permission from Elsevier Inc.

5.2.4. UV-Vis Spectroscopy

The stability of metal ligand complexes in aqueous solutions is commonly quantified using the stability constant (log*K*), while their structural characteristics are elucidated through spectroscopic techniques. These methods are crucial for confirming the successful formation of metal phenolic networks (MPNs) or modifications induced by MPNs on substrates. The complexation of metal ions with polyphenols under varying pH conditions often results in distinct visible color changes, which arise from different modes of metal ligand coordination and the formation of diverse complex species. UV-visible absorption spectroscopy provides a straightforward means to analyze ligand-to-metal charge transfer (LMCT), which is responsible for these color changes [27,83,90–92].

Free GA molecules exhibit two characteristic absorption bands at ~210 nm and ~260 nm, corresponding to the ^1L_a and ^1L_b singlet excited states, respectively, associated with $\pi \rightarrow \pi^*$ transitions. Upon complexation with metal ions at pH 7, these bands significantly decrease in intensity and may undergo a redshift. Additionally, a new, broad absorption band emerges in the visible region at ~600 nm (Figure 9(aii), indicated by the sky-blue colored bar), attributed to LMCT between the metal and the ligand [83]. This band is primarily responsible for the pronounced color changes observed during complexation. For metal complexes formed at pH 11, the band at ~600 nm undergoes a blueshift, which is attributed to the auto-oxidation of GA to quinone structures. Similar observations have been reported by Yadav et al. [49] and Masoud et al. (2014) [27], Figure 9b,c, where complexation with Fe^{3+} results in the appearance of a band in the visible range with λ_{max}~600 nm.

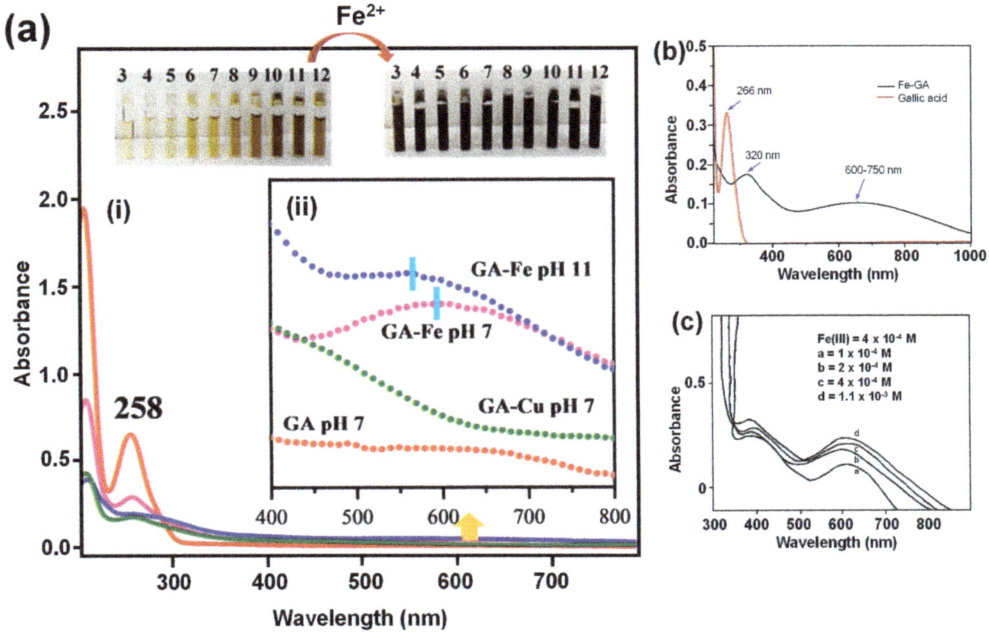

Figure 9. (**a**) UV-Vis absorption spectra of the complex between Fe^{3+} and gallic acid (GA): (i) 0.066 mM GA solution and (ii) metal complex at a metal-to-ligand molar ratio of 3:1. The red line represents the GA solution at pH 7, the green line represents the GA-Cu^{2+} complex at pH 7, the pink line represents the GA-Fe^{3+} complex at pH 7, and the blue line represents the GA-Fe^{3+} complex at pH 11. Image reproduced from Espina et al. (2022) [83], licensed under CC-BY-NC-ND 4.0. (**b**) Electronic absorption spectra of GA and its corresponding iron gallate nanocomplex. Image reproduced from Yadav et al. (2019) [49], with permission from Elsevier B.V. All rights reserved. (**c**) UV-Vis spectra of GA complexes with Fe^{3+} at various molar concentrations. The image was reproduced from Masoud et al. (2014) [27], with permission from Elsevier B.V. All rights reserved.

UV-Vis spectrophotometry is a simple yet effective technique for confirming the successful coordination of metals and ligands in the formation of MPNs. This technique is applicable to various types of ligands. For instance, Mazaheri et al. (2022) [93] demonstrated the formation of Fe^{3+}-tannic acid (TA) MPN through the absorption band within the range 450–650 nm with λ_{max} at 475 nm, corresponding to the formation of the mono Fe^{3+}-TA complexes. This study also highlighted the stability of MPNs over prolonged aging times, noting a decrease in absorption intensity with increasing aging time due to complexation with acetonitrile, which was used as the solvent in their work.

5.2.5. Nuclear Magnetic Resonance (NMR) Spectroscopy

Nuclear magnetic resonance (NMR) spectroscopy is a powerful, non-destructive analytical technique widely employed for molecular structure identification. It operates by exploiting the magnetic properties of atomic nuclei, providing detailed insights into molecular environments and interactions. In the context of metal complex characterization, NMR is particularly valuable for discerning oxidation states, coordination environments, and ligand-metal interactions. To elucidate comprehensive structural information, NMR is often combined with complementary techniques such as polarization transfer, relaxometry, and multidimensional NMR spectroscopy [80,94].

Fazary et al. (2009) have demonstrated the use of ^{1}H and ^{13}C NMR spectroscopy to characterize the isolated solid FeGA complex. Their analysis revealed the presence of functional ionizable groups in the ligand; however, the data did not provide conclusive evidence for the complexation between the ligand and the metal ion. In a more recent study, Etou et al. (2024) demonstrated the efficacy of liquid-state NMR spectroscopy, coupled with density functional theory (DFT) calculations, for elucidating the structure of metal complexes involving gallic acid (GA) and Al^{3+} [95]. By employing ^{27}Al NMR, they compared the spectra of a solution containing only Al-salt with that of an Al^{3+}-GA mixed solution. While a single peak at 0 ppm was observed for the Al-salt alone, two distinct peaks at 1.17 ppm and 15.7 ppm were detected in the Al^{3+}-GA system. These peaks were attributed to the formation of monodentate and bidentate AlGA complexes, respectively, with the former arising from coordination between Al^{3+} and the –COOH group of GA, and the latter resulting from coordination with two –OH groups of GA. Further confirmation was provided by ^{1}H NMR analysis, which compared the spectra of GA alone with the Al^{3+}-GA mixed solution. A peak at 7.14 ppm, corresponding to GA alone, was accompanied by two additional peaks at 6.95 ppm and 7.04 ppm in the Al^{3+}-GA system, consistent with the formation of the bidentate complex. This study highlights the utility of NMR spectroscopy, particularly when integrated with computational methods, for unraveling the structural intricacies of metal ligand interactions.

6. Versatility of GA-Based MPNs as Adsorbents

MPNs stand out as unique materials characterized by their amorphous structure, achieved through low-energy synthesis. Unlike other frameworks, MPNs are synthesized at room temperature, within short durations, and under moderate concentrations of metals and ligands. This mild synthesis process preserves the intrinsic properties of (poly)phenols, enabling strong interactions with various macromolecular substrates and metal ions. These interactions grant MPNs exceptional adhesive properties, making them highly versatile for surface engineering and adsorbent applications [96].

6.1. MPN as Standalone Adsorbent

MPNs have been extensively utilized as surface modifiers due to their strong adhesive properties. However, this focus has often overshadowed the exploration of MPNs as standalone materials. MPNs can be isolated as solid particles by carefully controlling parameters in the reaction, including the use of minimal amounts of solvent based on the solubility of the ligand, controlling the pH, and regulating the concentrations of metal ions and ligands during synthesis. This well-established method has been employed for decades. For instance, Patel et al. (1971) successfully isolated Ni^{2+} complexes with ethylenediamine (en) ligands, such as catechol and pyrogallol, in 1:2 and 1:3 ratios, forming $Ni(en)_2$ and $Ni(en)_3$. These complexes precipitated as the solution pH reached 7 upon ligand addition [60]. $Ni(en)_2$ required less water and ligand, while $Ni(en)_3$ necessitated larger quantities of both. Heating is sometimes applied to accelerate the formation of solid precipitates, though the temperatures required are significantly lower than those used in the synthesis of MOFs. Figure 10 illustrates the assembly of the metal complexes into MPN supramolecule. The formation of solid metal complexes resembles the agglomeration of coordinated metal ligand molecular compounds, which combine to form large clusters of supramolecular complexes. This clustering is driven by charge neutralization, which occurs as metal ligand interactions are prolonged or heating is applied. The neutralization process may also be influenced by attached water molecules to the metal ions.

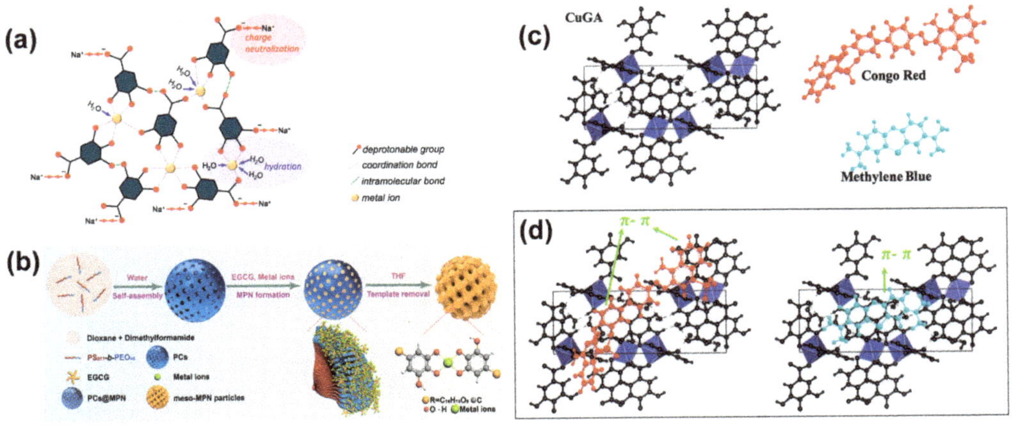

Figure 10. (**a**) Schematic representation of the molecular assembly process of metal phenolic networks as a standalone material. (**b**) Schematic representation of the preparation process of ordered mesoporous MPN. Adapted with permission from Lin et al. (2019) [97], Copyright © 2019, American Chemical Society. (**c**,**d**) Schematic illustration of the molecular structure of Cu^{2+}-gallic acid (CuGA) complexes and their subsequent interaction mode toward Congo red and methylene blue adsorbates. Image reproduced from Azhar et al. (2020) [36], licensed under CC-BY.

A major challenge in this process is the potential decomposition of the ligand during cluster formation, particularly due to oxygen exposure. Ligand decomposition can lead to the formation of metal oxides or hydroxides instead of the desired metal ligand complex. Morales et al. (2023) reported such phenomena, noting that ligand decomposition is more likely to occur when metal acetate salts are used as the metal source [98]. The acetate ions can induce the oxidation of gallic acid (GA), leading to its decomposition into phenols, quinones, and hydroxyquinones. To prevent ligand decomposition, metal chloride salts can be used instead of metal acetates, as chloride ions do not promote oxidation.

Santoso et al. (2021) demonstrated a technique for isolating MPNs as solid particles to prepare an adsorbent [37]. The MPN adsorbent was synthesized by reacting an equimolar concentration of GA with $CuCl_2{\cdot}5H_2O$ in a minimal amount of water. The solid MPN precipitate was isolated by adjusting the reaction pH to 8 with NaOH. To enhance the adsorption performance, glycine was introduced, providing nitrogen (N)-containing adsorption sites. The resulting solid MPN can be employed in the adsorption system at various pH levels, starting from a pH of 2 to 10, indicating the material's capability to withstand various environmental pH levels. The MPN has excellent adsorption efficiency for methylene blue, with an adsorption capacity of 190.81 mg/g. This efficiency was attributed to interactions such as electrostatic attraction, hydrogen bonding, dipole-dipole interactions, and n–π stacking.

Solution pH plays a critical role in the formation of MPN complexes, which directly affects their adsorption performance. Azhar et al. [36] demonstrated the impact of pH on the physicochemical properties of complexes formed between Cu^{2+} and GA, as well as their adsorption efficiency toward methylene blue and Congo red. At low pH (limited NaOH addition), insufficient hydroxide ions hinder the complete deprotonation of GA, limiting its coordination with Cu^{2+}. Conversely, excessive NaOH addition can result in the oxidation of Cu^{2+}, forming Cu_2O instead of the desired complexes. Moreover, the synthesis temperature was also shown to influence the formation of the complexes. The CuGA complexes with minimal Cu_2O impurities were achieved at a Cu:GA:NaOH molar ratio of 1:1:2.2, resulting in adsorption capacities of 124.64 mg/g for methylene blue and 344.54 mg/g for Congo red, where the molecular interaction between the adsorbent and adsorbate is shown in Figure 10c,d. In another study, Azhar et al. also demonstrated a high adsorption capacity for basic red 9 (115.08 mg/g) [41].

Isolated MPN complexes typically do not exhibit well-defined porosity. To overcome this, Lin et al. (2019) [97] developed a method to create highly ordered mesoporous MPN particles using a sacrificial templating technique. In this approach, cubosome templates were created from PS_{217}-b-PEO_{45} block copolymers (PC) via a cosolvent method. An MPN precursor solution containing GA or epigallocatechin (EGCG) and $FeCl_3.6H_2O$ was added to the cubosome-containing solution, and the pH was adjusted to 6.5 to drive coordination-driven assembly. The stability of Fe-ligand bonds was demonstrated under physiological conditions (pH 7.4), with Fe^{3+} ions detaching at acidic pH levels. This pH-dependent disassembly behavior is also confirmed in other works [99–101]. These mesoporous MPN materials exhibited exceptional protein-loading capacities (362 mg/g for glucose oxidase and 486 mg/g for horseradish peroxidase), outperforming commercial mesoporous SiO_2 particles.

Despite the promising capabilities of GA-derived metal phenolic networks (MPNs), their adsorption potential remains underexplored, presenting an exciting avenue for future research. For instance, FeGA MPNs have demonstrated electrochemical and redox properties [34] which could be harnessed not only for metal adsorption but also for the removal of hazardous compounds through oxidation pathways. Fe-containing adsorbents have shown the ability to facilitate Fenton-like oxidation, further enhancing the degradation of organic contaminants in water. Additionally, the presence of GA in the Fenton process can enhance the efficiency of the process [102].

6.2. MPN for Surface Modification of Adsorbent

The surface modification of adsorbents with MPNs significantly enhances their adsorption performance by improving binding affinity and capacity. Additionally, MPN moieties impart pH-responsive behavior to the modified adsorbents. For instance, Zeng et al. [103] developed the development of Fe^{3+}-GA (FeGA) MPNs that are stable at pH 5 and non-cytotoxic to mouse cells. These FeGA MPNs also exhibit photon absorption properties, making them potential photothermal guidance agents. At pH 7, FeGA undergoes decomposition due to the metabolic activity of the tested mouse cells with decomposition products efficiently excreted without causing cytotoxicity. The non-cytotoxic nature of FeGA is advantageous in applications targeting aquatic ecosystems, as it minimizes adverse effects on surrounding biota.

Figure 11a illustrates the self-assembly process of MPN formation as a film coating on a substrate. The process begins with the deposition of metal ions onto the substrate surface. Using a substrate that contains functional groups capable of binding metal ions, such as carboxylic or amino groups, can enhance the stability of the resulting MPN coating [104]. Upon the addition of phenolic acids (e.g., GA) under suitable pH conditions, iron complexes form spontaneously, leading to the development of a supramolecular MPN coating on the substrate surface. For example, the modification of cellulose using MPN via the coordination of GA and Fe^{3+} has been successfully demonstrated by Lunardi et al. (2024) [6]. This process involved immersing the cellulose substrate in a Fe^{3+}-containing solution, followed by the addition of a GA solution. MPN self-assembly was achieved by adjusting the solution pH to 8 by adding NaOH. Although the microscale size of the MPN particles hindered direct morphological observation, EDX analysis confirmed the presence of Fe, indicating successful MPN coating (Figure 11b). The modification was further evidenced by a color change on the cellulose surface, with distinct colors observed for different metal-to-ligand ratios (Figure 11c). This characteristic coloration of Fe-GA MPN has also been utilized in the preparation of dyes and pigments, such as iron gall inks [75,104].

Cellulose modified with MPN at different metal-to-ligand ratios exhibited varying removal efficiencies. The highest removal efficiency (~90%) was achieved at a 1:3 ratio, while lower efficiencies were observed at ratios of 1:1 and 3:1. The maximum adsorption capacity of MPN-modified cellulose for Cr^{6+} was 73.63 mg/g, approximately 1.6 times higher than that of unmodified cellulose. Increasing the temperature further enhanced the adsorption capacity to 116.8 mg/g.

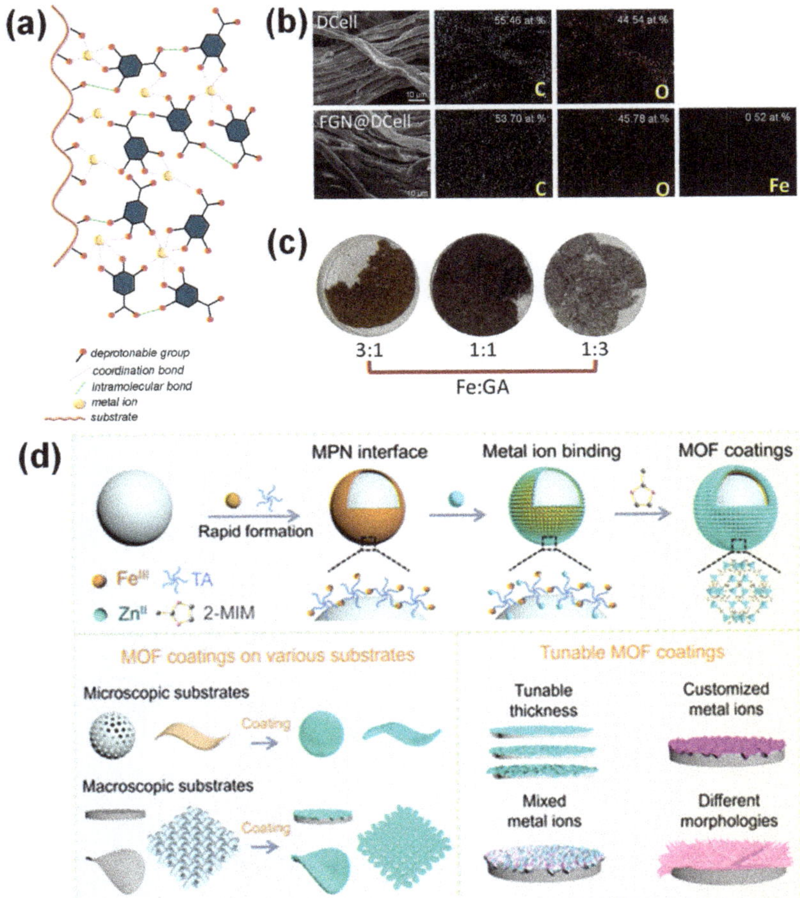

Figure 11. (**a**) Schematic illustration of MPN assembly on cellulose substrates. (**b**,**c**) SEM-EDX image and visual appearance of adsorbent composite prepared from coating of cellulose (DCell) with Fe-gallic acid MPN. Adapted with permission from Lunardi et al. (2024) [6], Copyright © 2023. Elsevier B.V. All rights reserved. (**d**) MPN-mediated MOF coatings: schematic representation of tunable MOF coating thickness on substrates. Adapted with permission from Wang et al. (2024) [96], Copyright © 2024 John Wiley and Sons, Inc.

The combination of MPNs and MOFs has become a widely adopted strategy to enhance the functionality of MOFs. FeGA MPNs, for example, offer photodynamic activity for theragnostic applications, antibacterial and antifungal properties ideal for surface coatings, and pH-responsive behavior suitable for adsorption and controlled drug release. The integration of MPNs with MOFs often involves bridging bonds, such as the Fe–O–Cu bond observed in the combination of Cu^{2+}-GA (CuGA) MPN with NH_2-MIL-88B MOF, as reported by Li et al. [105]. In this system, GA served as a chelating agent, binding

Fe-metal clusters in the MOF while simultaneously coordinating with free Cu^{2+} ions. This adhesion, resulting in the formation of CuGA@NH_2-MIL-88B, enhances the photodynamic behavior of the MOF by facilitating electron transfer from photons to the MOF through the bridging bond. Additionally, this composite demonstrated significant adsorption capability as demonstrated by a significant reduction in glutathione absorbance in the solution within 30 min of contact.

The weak adhesive nature of MOFs poses a challenge for their integration with various substrates, as direct growth of MOFs on substrate surfaces often results in particle separation due to insufficient bonding. In this context, MPNs can serve as effective bridging molecules, facilitating MOF adhesion to substrates. This process involves coating the substrate with MPNs under mild conditions, such as room temperature in aqueous solvents. The adhered MPN molecules provide active coordination sites, facilitating interactions with MOF precursors and promoting MOF growth on the substrate surface (Figure 11d). This approach has been successfully used to prepare self-standing hollow MPN@ZIF-8 structures, which function as effective H_2/CH_4 gas adsorbents [96]. Additionally, Luo et al. (2019) demonstrated the ease of MPN coating by applying it to a biomass-derived microporous membrane, enhancing its ability to effectively capture uranium from seawater [106].

Rahim et al. [107] demonstrated the grafting of GA onto Fe_3O_4 magnetic nanoparticles (FMNPs) using alkoxysilanes and the amine-containing compound 1,2-ethylenediamine (1,2-en) as mediators. In this process, the oxygen-containing group of the silane compound forms a chelation-like coordination bond with FMNPs, while the silicon moiety reacts with the ethyl group of 1,2-en via a nucleophilic-driven reaction. The amine group of 1,2-en interacts with GA through hydrogen bonding and electrostatic interactions. This composite exhibited a high adsorption capacity, binding 4.23–4.98 mmol of Fe^{3+} ions per gram of adsorbent.

Pirozzi et al. [108] explored a reverse adsorption strategy leveraging coordination chemistry between GA and metal ions. In their study, GA molecules were adsorbed onto a magnetic metal-ceramic nanocomposite by coordinating with active Lewis acid sites on the nanocomposite surface. The highest GA uptake was observed at pH 5, where the surface charge of the nanocomposite is nearly neutral, minimizing electrostatic repulsion with the moderately protonated GA. This optimal pH balanced the nanocomposite's surface charge, GA protonation state, and Lewis acidic sites availability of for interaction.

In a recent study by Nilnit et al. [109], an in-situ strategy was developed for synthesizing a magnetic adsorbent coated with a phenolic compound. In this method, phenolics extracted from the *Hevea brasiliensis* Muell. Arg. bark was directly reacted with $FeSO_4.7H_2O$ at 65 °C under sonication. The phenolic acted as ligand, chelating Fe^{2+} ions to form Fe-phenolic complexes. Sodium hydroxide was then added to the reaction mixture to precipitate $Fe(OH)_2$-phenolic, which subsequently oxidized in water to produce magnetite Fe_3O_4-phenolic particles. This strategy resembles conventional MPN synthesis but differs in the absence of an exterior substrate, with Fe_3O_4 serving as both a substrate and a metal cluster. The phenolic-coated Fe_3O_4 selectively adsorbed tetracycline residue in honey via a solid-phase extraction process, facilitating the preconcentration of the residues before HPLC-UV analysis. The adsorbed tetracycline residues ranged from 12 to 127.2 mg/g. However, the phenolic-coated Fe_3O_4 particles were prone to thermal and chemical degradation, rendering them unsuitable for reuse.

While GA-MPN adsorbents offer versatile functionalities, their stability under acidic conditions remains a challenge due to protonation and the subsequent weakening of metal ligand coordination bonds, requiring further research to address this issue. One promising approach is hybridization with inorganic materials, such as metal oxides and MOFs, which can enhance structural rigidity and protect coordination sites from protonation.

Additionally, complexation with metal ions that results in a higher stability constant can provide greater resistance to acidic environments. By optimizing metal ligand interactions, GA-MPNs can potentially be tailored for improved durability in low-pH applications.

6.3. Other MPNs-Based Adsorbent

Tannic acid (TA)—a natural biopolymer composed of eight or more gallic acid (GA) molecules—was the first phenolic compound introduced for metal-phenolic network (MPN) synthesis, as reported in the pioneering work of Ejima et al. (2013) [14]. As a structurally undefined dendritic polyphenol, it is widely used as a ligand in MPN preparation due to its abundant carboxylic and phenolic hydroxyl groups. These functional groups contribute to the complexity of determining its pKa value [110]. As pH increases, TA undergoes deprotonation, generating negatively charged ligand species that readily donate electron pairs to metal ions, forming stable complexes [83].

FeTA MPNs are particularly notable for their antifungal activity. Hou et al. demonstrated that FeTA coatings on Ag-based MOFs exhibit a synergistic antifungal effect, effectively inhibiting the growth of *Fusarium oxysporum* [111]. The FeTA coating not only provides active sites for myclobutanil loading but also enables controlled, pH-responsive release, enhancing the efficiency and precision of antifungal delivery. Furthermore, FeTA MPNs are non-cytotoxic and ecologically compatible, as evidenced by their lack of inhibitory effects on Pakchoi seed germination.

Rahim et al. (2019) explored the chelating ability of phenolic ligands for metal ion sequestration, developing a metal-phenolic sorbent (MPS) via a straightforward sol-gel process by mixing TA and Zr^{4+} in a 1:1.2 molar ratio at 85 °C for 3 min, followed by a 10 min gelation period at room temperature [112]. The resulting material exhibited nearly 100% sequestration efficiency across a broad range of metals, attributed to the active catechol and gallol groups in TA that remained unbound to Zr^{4+}. Investigating GA as a ligand for MPS synthesis could provide further insights into optimizing metal sequestration performance.

MPNs also show promise as controlled-release agents for nutrients, offering a practical alternative to post-application urea removal from soil or water. Mazaheri et al. (2022) demonstrated FeTA MPN self-assembly in acetonitrile for urea coating, where complexation occurred immediately upon urea addition, leading to uniform deposition [93]. A solvent-free approach was later developed via mechanochemical grinding of MPN and urea granules, followed by aging, which enhanced complexation and coating stability [113]. The most recent work by Mazaheri et al. (2024) further improved nutrient release control by incorporating silicate into MPN-encapsulated urea [114]. Applying similar strategies to GA-based MPNs could unlock new opportunities for sustainable nutrient delivery, leveraging GA's unique binding properties for enhanced stability and controlled release.

7. Conclusions and Future Perspective

7.1. Concluding Remark

Gallic acid (GA)-based phenolic-metal networks (GA-MPNs) have emerged as a versatile class of functional materials with broad applications in adsorption and surface modification. Their environmentally benign nature, universal adhesion properties, and tunable coordination chemistry make them highly effective for enhancing adsorbent performance. This review has provided a comprehensive overview of the fundamental chemistry governing the formation of GA-MPNs, their structural and chemical characterization, and their roles in various adsorbent architectures, including stand-alone MPNs, MPN-mediated mesoporous materials, MPN-MOF composites, and MPN-coated substrates. By consolidating recent advances in GA-based MPNs, this work contributes to the fundamental understanding required for the rational design of next-generation MPN-based adsorbents.

7.2. Future Research Directions

To unlock the full potential of GA-based MPNs, future research should focus on three important aspects: (1) advancing their synthesis, (2) expanding their applications in emerging fields, and (3) improving their sustainability through biomass-derived ligands.

1. Advancing MPN Synthesis and Structural Control: While GA-based MPNs have demonstrated promising adsorption properties, their amorphous nature presents challenges in precisely tuning pore size, morphology, and surface area. Future efforts should focus on:
    - Template-Assisted Synthesis: Using sacrificial templates or structure-directing agents to achieve better control over pore architecture and surface area.
    - Ligand-to-Metal Ratio Optimization: Fine-tuning coordination chemistry to enhance stability, redox properties, and adsorption performance.
    - Post-Synthesis Modifications: Functionalizing MPNs with catalytic sites, redox-active moieties, or hybrid nanomaterials to broaden their utility.

    Additionally, exploring a wider range of metal centers with tailored functionalities—such as enhanced redox activity, photodegradability, or photocatalytic properties—could expand MPN applications in energy and environmental fields.
2. Expanding Applications in Emerging Fields: Beyond adsorption and surface modification, GA-MPNs hold potential for various high-impact applications:
    - Biomedical Applications: GA's intrinsic bioactivity, combined with metal coordination, enables potential use in antimicrobial coatings, drug delivery systems, and biosensors. Investigating MPNs as biodegradable, metal-coordinated drug carriers or bioadhesives could open new biomedical frontiers.
    - Environmental Remediation: Functionalized GA-MPNs could be engineered for targeted pollutant removal, photocatalytic degradation of contaminants, and recovery of critical metals from wastewater. Additionally, integrating MPNs with membranes or composite materials could improve their practicality in filtration technologies.
    - Energy Storage and Catalysis: MPNs with redox-active metals may serve as electrode materials in supercapacitors or electrocatalysts for water splitting and CO_2 reduction.
3. Sustainable Development Using Biomass-Derived Ligands: A promising avenue for cost-effective and eco-friendly MPN development is the use of crude biomass extracts as sources of phenolic ligand instead of purified GA. However, key challenges remain:
    - Extraction Optimization: Developing efficient, scalable methods to obtain high-phenolic-content extracts with minimal impurities.
    - Ligand Purity Control: Understanding the impact of natural extract variability on MPN formation and performance.
    - Complexation Efficiency: Investigating how mixed phenolic compounds in crude extracts influence coordination chemistry and material stability.

By addressing these challenges, GA-based MPNs could become a sustainable and economically viable alternative for adsorption, catalysis, and biomedicine. Continued interdisciplinary research integrating chemistry, materials science, and engineering will be essential to fully realize their potential in both scientific and industrial domains.

Supplementary Materials: The following supporting information can be downloaded at: https://www.mdpi.com/article/10.3390/molecules30061218/s1, Table S1: Gallic acid-based metal complexes; Table S2: Hard-Soft Acid-Base List; Table S3: Data point of titration curves of solution containing gallic acid, Fe^{3+}, and glycine at 0.1 M ionic strength; Table S4: HySS2009-Concentration table. References [115–119] are cited in the supplementary materials.

Author Contributions: Writing—original draft preparation, S.P.S., A.E.A. and K.-C.C.; writing—review and editing, S.-P.L., C.-W.H. and O.S.; visualization, S.P.S., H.-Y.H. and A.R.; supervision, S.I. All authors have read and agreed to the published version of the manuscript.

Funding: This research received no external funding.

Institutional Review Board Statement: Not applicable.

Informed Consent Statement: Not applicable.

Data Availability Statement: No primary research results have been included and no new data were generated or analyzed as part of this review.

Conflicts of Interest: The authors declare no conflicts of interest.

Abbreviations

The following abbreviations are used in this manuscript:

MPNs	Metal-phenolic networks
MOFs	Metal-organic frameworks
GA	Gallic acid
TA	Tannic acid
HSAB	Hard-Soft Acid-Base
IWS	Irving Williams Series
SEM	Scanning electron microscopy
EDX	Energy dispersive X-ray
SERS	Surface-Enhanced Raman Spectroscopy
FTIR	Fourier transform infrared spectroscopy
EPR	Electron Paramagnetic Resonance
NMR	Nuclear Magnetic Resonance
TEM	Transmission electron microscopy
HPLC	High performance liquid chromatography
DFT	Density functional theory
SSA	Surface specific area
PC	PS_{217}-b-PEO_{45} block copolymers
EGCG	Epigallocatechin
en	Ethylenediamine
FMNPs	Fe_3O_4 magnetic nanoparticles
LMCT	Ligand-to-metal charge transfer

References

1. Manyangadze, M.; Chikuruwo, N.H.M.; Narsaiah, T.B.; Chakra, C.S.; Radhakumari, M.; Danha, G. Enhancing adsorption capacity of nano-adsorbents via surface modification: A review. *S. Afr. J. Chem. Eng.* **2020**, *31*, 25–32. [CrossRef]
2. Jha, M.K.; Joshi, S.; Sharma, R.K.; Kim, A.A.; Pant, B.; Park, M.; Pant, H.R. Surface Modified Activated Carbons: Sustainable Bio-Based Materials for Environmental Remediation. *Nanomaterials* **2021**, *11*, 3140. [CrossRef]
3. Petrovic, B.; Gorbounov, M.; Soltani, S.M. Influence of surface modification on selective CO_2 adsorption: A technical review on mechanisms and methods. *Microporous Mesoporous Mater.* **2021**, *312*, 110751. [CrossRef]
4. Abegunde, S.M.; Idowu, K.S.; Adejuwon, O.M.; Adeyemi-Adejolu, T. A review on the influence of chemical modification on the performance of adsorbents. *Resour. Environ. Sustain.* **2020**, *1*, 100001. [CrossRef]
5. Liang, B.; Zhu, P.; Gu, J.; Yuan, W.; Xiao, B.; Hu, H.; Rao, M. Advancing Adsorption and Separation with Modified SBA-15: A Comprehensive Review and Future Perspectives. *Molecules* **2024**, *29*, 3543. [CrossRef]

6. Lunardi, V.B.; Cheng, K.-C.; Lin, S.-P.; Angkawijaya, A.E.; Go, A.W.; Soetaredjo, F.E.; Ismadji, S.; Hsu, H.-Y.; Hsieh, C.-W.; Santoso, S.P. Modification of cellulosic adsorbent via iron-based metal phenolic networks coating for efficient removal of chromium ion. *J. Hazard. Mater.* **2024**, *465*, 132973. [CrossRef]
7. Zhang, Y.; Shen, L.; Zhong, Q.-Z.; Li, J. Metal-phenolic network coatings for engineering bioactive Interfaces. *Colloids Surf. B Biointerfaces* **2021**, *205*, 111851. [CrossRef]
8. Cheng, X.; Zhu, Y.; Tang, S.; Lu, R.; Zhang, X.; Li, N.; Zan, X. Material priority engineered metal-polyphenol networks: Mechanism and platform for multifunctionalities. *J. Nanobiotechnol.* **2022**, *20*, 255. [CrossRef]
9. Wen, Y.; Yang, X.; Li, Y.; Yan, L.; Zhao, Y.; Shao, L. Progress reports of metal-phenolic network engineered membranes for water treatment. *Sep. Purif. Technol.* **2023**, *320*, 124225. [CrossRef]
10. Zhong, Q.-Z.; Pan, S.; Rahim, M.A.; Yun, G.; Li, J.; Ju, Y.; Lin, Z.; Han, Y.; Ma, Y.; Richardson, J.J.; et al. Spray Assembly of Metal–Phenolic Networks: Formation, Growth, and Applications. *ACS Appl. Mater. Interfaces* **2018**, *10*, 33721–33729. [CrossRef]
11. Välimets, A.; Koort, K.; Mortimer, M. Biocompatibility of Metal–Phenolic Network-Coated Nanoparticles. *Proceedings* **2023**, *92*, 33. [CrossRef]
12. Fan, G.; Cottet, J.; Rodriguez-Otero, M.R.; Wasuwanich, P.; Furst, A.L. Metal–Phenolic Networks as Versatile Coating Materials for Biomedical Applications. *ACS Appl. Bio Mater.* **2022**, *5*, 4687–4695. [CrossRef]
13. Geng, H.; Zhong, Q.-Z.; Li, J.; Lin, Z.; Cui, J.; Caruso, F.; Hao, J. Metal Ion-Directed Functional Metal–Phenolic Materials. *Chem. Rev.* **2022**, *122*, 11432–11473. [CrossRef]
14. Ejima, H.; Richardson, J.J.; Liang, K.; Best, J.P.; Koeverden, M.P.V.; Such, G.K.; Cui, J.; Caruso, F. One-Step Assembly of Coordination Complexes for Versatile Film and Particle Engineering. *Science* **2013**, *341*, 154–157. [CrossRef] [PubMed]
15. Wianowska, D.; Olszowy-Tomczyk, M. A Concise Profile of Gallic Acid—From Its Natural Sources through Biological Properties and Chemical Methods of Determination. *Molecules* **2023**, *28*, 1186. [CrossRef]
16. Andjelković, M.; Camp, J.V.; Meulenaer, B.D.; Depaemelaere, G.; Socaciu, C.; Verloo, M.; Verhe, R. Iron-chelation properties of phenolic acids bearing catechol and galloyl groups. *Food Chem.* **2006**, *98*, 23–31. [CrossRef]
17. Fazary, A.E.; Taha, M.; Ju, Y.-H. Iron Complexation Studies of Gallic Acid. *J. Chem. Eng. Data* **2008**, *54*, 35–42. [CrossRef]
18. Rahim, M.A.; Kempe, K.; Müllner, M.; Ejima, H.; Ju, Y.; Koeverden, M.P.v.; Suma, T.; Braunger, J.A.; Leeming, M.G.; Abrahams, B.F.; et al. Surface-Confined Amorphous Films from Metal-Coordinated Simple Phenolic Ligands. *Chem. Mater.* **2015**, *27*, 5825–5832. [CrossRef]
19. Lin, Z.; Liu, H.; Richardson, J.J.; Xu, W.; Chen, J.; Zhou, J.; Caruso, F. Metal–phenolic network composites: From fundamentals to applications. *Chem. Soc. Rev.* **2024**, *53*, 10800–10826. [CrossRef]
20. Venkateswarlu, C.; Das, M.S.; Athavale, V.T. Studies of gallic acid complexes with metals and their analytical applications. Part I. Spectrophotometric Investigation. *Proc. Indian Acad. Sci. Sect. A* **1954**, *40*, 260–269. [CrossRef]
21. Venkateswarlu, C.; Das, M.S.; Athavale, V.T. Studies of gallic acid complexes with metals and their analytical applications—Part III. Spectrophotometric estimation of gallic acid. *Proc. Indian Acad. Sci. Sect. A* **1956**, *44*, 241–246. [CrossRef]
22. Varde, M.S.; Athavale, V.T. Studies of gallic acid complexes with metals and their analytical applications—Part II. A Spectrophotometric study of molybdenum complexes. *Proc. Indian Acad. Sci. Sect. A* **1956**, *44*, 228–240. [CrossRef]
23. Powell, H.K.J.; Taylor, M.C. Interactions of Iron(II) and Iron(III) with Gallic Acid and its Homologues: A Potentiometric and Spectrophotometric Study. *Aust. J. Chem.* **1982**, *35*, 739–756. [CrossRef]
24. Sandmann, B.J.; Chien, M.H.; Sandmann, R.A. Stability Constants of Calcium, Magnesium and Zinc Gallate Using a Divalent Ion-Selective Electrode. *Anal. Lett.* **1985**, *18*, 149–159. [CrossRef]
25. Elinany, G.A.; Ebeid, F.M.; Zahra, A.M.; Ziedan, F.I. Polarography of Metal-Gallic Complexes. *J. Electroanal. Chem.* **1976**, *72*, 363–369. [CrossRef]
26. Rahim, S.A.; Hussain, S.; Farooqui, M. Protonation Equilibria of Gallic Acid (GA) and Stability Constants of Its Complexes with Transition Metal Ions in Aqueous Media. *J. Chem. Biol. Phys. Sci.* **2017**, *7*, 267–273.
27. Masoud, M.S.; Ali, A.E.; Haggag, S.S.; Nasr, N.M. Spectroscopic studies on gallic acid and its azo derivatives and their iron(III) complexes. *Spectrochim. Acta Part A Mol. Biomol. Spectrosc.* **2014**, *120*, 505–511. [CrossRef]
28. Zaccaron, S.; Ganzerla, R.; Bortoluzzi, M. Iron complexes with gallic acid: A computational study on coordination compounds of interest for the preservation of cultural heritage. *J. Coord. Chem.* **2012**, *66*, 1709–1719. [CrossRef]
29. Gao, F.; Zheng, D.; Tanaka, H.; Zhan, F.; Yuan, X.; Gao, F.; Wang, Q. An electrochemical sensor for gallic acid based on Fe_2O_3/electro-reduced graphene oxide composite: Estimation for the antioxidant capacity index of wines. *Mater. Sci. Eng. C-Mater. Biol. Appl.* **2015**, *57*, 279–287. [CrossRef]
30. Mishra, B.; Chandra, M. Evaluation of phytoremediation potential of aromatic plants: A systematic review. *J. Appl. Res. Med. Aromat. Plants* **2022**, *31*, 100405. [CrossRef]
31. Daduang, J.; Palasap, A.; Daduang, S.; Boonsiri, P.; Suwannalert, P.; Limpaiboon, T. Gallic acid enhancement of gold nanoparticle anticancer activity in cervical cancer cells. *Asian Pac. J. Cancer Prev.* **2015**, *16*, 169–174. [CrossRef] [PubMed]

32. Khan, A.N.; Hassan, M.N.; Khan, R.H. Gallic acid: A naturally occurring bifunctional inhibitor of amyloid and metal induced aggregation with possible implication in metal-based therapy. *J. Mol. Liq.* **2019**, *285*, 27–37. [CrossRef]
33. Sharma, S.; Mittal, D.; Verma, A.K.; Roy, I. Copper-Gallic Acid Nanoscale Metal–Organic Framework for Combined Drug Delivery and Photodynamic Therapy. *ACS Appl. Bio Mater.* **2019**, *2*, 2092–2101. [CrossRef]
34. Cherepanov, P.V.; Rahim, M.A.; Bertleff-Zieschang, N.; Sayeed, M.A.; O'Mullane, A.P.; Moulton, S.E.; Caruso, F. Electrochemical Behavior and Redox-Dependent Disassembly of Gallic Acid/Fe III Metal–Phenolic Networks. *ACS Appl. Mater. Interfaces* **2018**, *10*, 5828–5834. [CrossRef] [PubMed]
35. Wu, J.; Ouyang, D.; He, Y.; Su, H.; Yang, B.; Li, J.; Sun, Q.; Lin, Z.; Cai, Z. Synergistic Effect of Metal–Organic Framework/Gallic Acid in Enhanced Laser Desorption/Ionization Mass Spectrometry. *ACS Appl. Mater. Interfaces* **2019**, *11*, 38255–38264. [CrossRef]
36. Azhar, B.; Angkawijaya, A.E.; Santoso, S.P.; Gunarto, C.; Ayucitra, A.; Go, A.W.; Tran-Nguyen, P.L.; Ismadji, S.; Ju, Y.-H. Aqueous synthesis of highly adsorptive copper–gallic acid metal–organic framework. *Sci. Rep.* **2020**, *10*, 19212. [CrossRef]
37. Santoso, S.P.; Bundjaja, V.; Angkawijaya, A.E.; Gunarto, C.; Go, A.W.; Yuliana, M.; Tran-Nguyen, P.L.; Hsieh, C.-W.; Ju, Y.-H. One-step synthesis of nitrogen-grafted copper-gallic acid for enhanced methylene blue removal. *Sci. Rep.* **2021**, *11*, 12021. [CrossRef]
38. Hou, X.; Zhang, L.; Chen, Y.; Liu, Z.; Zhao, X.; Lu, B.; Luo, Y.; Qu, X.; Musskaya, O.; Glazov, I.; et al. Photothermal switch by gallic acid-calcium grafts synthesized by coordination chemistry for sequential treatment of bone tumor and regeneration. *Biomaterials* **2025**, *312*, 122724. [CrossRef]
39. Gu, L.; Li, X.; Chen, G.; Yang, H.; Qian, H.; Pan, J.; Miao, Y.; Li, Y. A glutathione-activated bismuth-gallic acid metal-organic framework nano-prodrug for enhanced sonodynamic therapy of breast tumor. *J. Colloid Interface Sci.* **2025**, *679*, 214–223. [CrossRef]
40. Wasuwanich, P.; Fan, G.; Burke, B.; Furst, A.L. Metal-phenolic networks as tuneable spore coat mimetics. *J. Mater. Chem. B* **2022**, *10*, 7600–7606. [CrossRef]
41. Azhar, B.; Avian, C.; Tiwikrama, A.H. Green synthesis optimization with artificial intelligence studies of copper–gallic acid metal–organic framework and its application in dye removal from wastewater. *J. Mol. Liq.* **2023**, *389*, 122844. [CrossRef]
42. Zhou, H.-C.; Long, J.R.; Yaghi, O.M. Introduction to Metal–Organic Frameworks. *Chem. Rev.* **2012**, *112*, 673–674. [CrossRef] [PubMed]
43. Lin, K.-S.; Adhikari, A.K.; Ku, C.-N.; Chiang, C.-L.; Kuo, H. Synthesis and characterization of porous HKUST-1 metal organic frameworks for hydrogen storage. *Int. J. Hydrogen Energy* **2012**, *37*, 13865–13871. [CrossRef]
44. Mahmoodi, N.M.; Abdi, J.; Oveisi, M.; Asli, M.A.; Vossoughi, M. Metal-organic framework (MIL-100 (Fe)): Synthesis, detailed photocatalytic dye degradation ability in colored textile wastewater and recycling. *Mater. Res. Bull.* **2018**, *100*, 357–366. [CrossRef]
45. Zhao, H.; Li, Q.; Wang, Z.; Wu, T.; Zhang, M. Synthesis of MIL-101(Cr) and its water adsorption performance. *Microporous Mesoporous Mater.* **2020**, *2020*, 110044. [CrossRef]
46. Mu, X.; Chen, Y.; Lester, E.; Wu, T. Optimized synthesis of nano-scale high quality HKUST-1 under mild conditions and its application in CO_2 capture. *Microporous Mesoporous Mater.* **2018**, *270*, 249–257. [CrossRef]
47. Zou, M.; Dong, M.; Zhao, T. Advances in Metal-Organic Frameworks MIL-101(Cr). *Int. J. Mol. Sci.* **2022**, *23*, 9396. [CrossRef]
48. Yilmaz, B.S. Antimicrobial and Anticancer Activity of Gallic Acid–Cu(II) Hybrid Nanoflowers and Gallic Acid–Zn(II) Hybrid Nanoflowers. *J. Inorg. Organomet. Polym. Mater.* **2024**, *34*, 5329–5341. [CrossRef]
49. Yadav, A.; Sharma, A.; Sharma, R.K. Mesoporous iron gallate nanocomplex for adsorption and degradation of organic dyes. *Colloids Surf. A Physicochem. Eng. Asp.* **2019**, *579*, 123694. [CrossRef]
50. Wang, T.C.; Bury, W.; Gómez-Gualdrón, D.A.; Vermeulen, N.A.; Mondloch, J.E.; Deria, P.; Zhang, K.; Moghadam, P.Z.; Sarjeant, A.A.; Snurr, R.Q.; et al. Ultrahigh Surface Area Zirconium MOFs and Insights into the Applicability of the BET Theory. *J. Am. Chem. Soc.* **2015**, *137*, 3585–3591. [CrossRef]
51. Peng, Y.; Krungleviciute, V.; Eryazici, I.; Hupp, J.T.; Farha, O.K.; Yildirim, T. Methane storage in metal-organic frameworks: Current records, surprise findings, and challenges. *J. Am. Chem. Soc.* **2013**, *135*, 118877–118894. [CrossRef] [PubMed]
52. Spanopoulos, I.; Tsangarakis, C.; Klontzas, E.; Tylianakis, E.; Froudakis, G.; Adil, K.; Belmabkhout, Y.; Eddaoudi, M.; Trikalitis, P.N. Reticular Synthesis of HKUST-like tbo-MOFs with Enhanced CH_4 Storage. *J. Am. Chem. Soc.* **2015**, *138*, 1568–1574. [CrossRef]
53. Yang, H.; Zhao, Z.-C.; Yang, Y.-P.; Zhang, Z.; Chen, W.; Yan, R.-Q.; Jin, Y.; Zhang, J. Defective WO_3 nanoplates controllably decorated with MIL-101(Fe) nanoparticles to efficiently remove tetracycline hydrochloride by S-scheme mechanism. *Sep. Purif. Technol.* **2022**, *300*, 121846. [CrossRef]
54. Zhao, Z.-C.; Wang, K.; Chang, L.; Yan, R.-Q.; Zhang, J.; Zhang, M.; Wang, L.; Chen, W.; Huang, G.-B. Construction of S-scheme MIL-101(Fe)/Bi_2MoO_6 heterostructures for enhanced catalytic activities towards tetracycline hydrochloride photodegradation and nitrogen photofixation. *Sol. Energy* **2023**, *264*, 112042. [CrossRef]
55. El-Shahawy, A.A.G.; Dief, E.M.; El-Dek, S.I.; Farghali, A.A.; El-Ela, F.I.A. Nickel-gallate metal–organic framework as an efficient antimicrobial and anticancer agent: In vitro study. *Cancer Nanotechnol.* **2023**, *14*, 60. [CrossRef]

56. Chen, Z.; Świsłocka, R.; Choińska, R.; Marszałek, K.; Dąbrowska, A.; Lewandowski, W.; Lewandowska, H. Exploring the Correlation Between the Molecular Structure and Biological Activities of Metal–Phenolic Compound Complexes: Research and Description of the Role of Metal Ions in Improving the Antioxidant Activities of Phenolic Compounds. *Int. J. Mol. Sci.* **2024**, *25*, 11775. [CrossRef]
57. Molski, M. Computation of the pKa Values of Gallic Acid and Its Anionic Forms in Aqueous Solution: A Self-Similar Transformation Approach for Accurate Proton Hydration Free Energy Estimation. *Molecules* **2025**, *30*, 742. [CrossRef]
58. Jabbari, M. Solvent dependence of protonation equilibria for gallic acid in water and different acetonitrile–water cosolvent systems. *J. Mol. Liq.* **2015**, *208*, 5–10. [CrossRef]
59. Roy, K.; Kar, S.; Das, R.N. Chemical Information and Descriptors. In *Understanding the Basics of QSAR for Applications in Pharmaceutical Sciences and Risk Assessment*, 1st ed.; Academic Press: Cambridge, MA, USA; Elsevier Inc.: Amsterdam, The Netherlands, 2015; pp. 47–80. [CrossRef]
60. Patel, D.C.; Bhattacharya, P.K. A study of ligand exchange in some nickel complexes. *J. Inorg. Nucl. Chem.* **1971**, *33*, 529–533. [CrossRef]
61. Radalla, A.M. Potentiometric studies on ternary complexes involving some divalent transition metal ions, gallic acid and biologically abundant aliphatic dicarboxylic acids in aqueous solutions. *Beni-Suef Univ. J. Basic Appl. Sci.* **2015**, *4*, 174–182. [CrossRef]
62. Şişmanoğlu, T.; İçhedef, Ç.; Akdut, G.; Soylu, G.S.P.; Medine, E.İ.; Teksöz, S. Complexation of gallic acid involving La^{3+}, Sm^{3+}, Th^{4+} and $UO_2{}^{2+}$ ions in aqueous solution by potentiometry at various temperatures. *J. Radioanal. Nucl. Chem.* **2024**, *334*, 623–636. [CrossRef]
63. Soldatović, T. Correlation between HSAB Principle and Substitution Reactions in Bioinorganic Reactions. In *Photophysics, Photochemical and Substitution Reactions—Recent Advances*; Saha, S., Kanaparthi, R.K., Soldatovic, T., Eds.; IntechOpen: London, UK, 2020.
64. Miličević, A.; Branica, G.; Raos, N. Irving-Williams Order in the Framework of Connectivity Index 3χv Enables Simultaneous Prediction of Stability Constants of Bivalent Transition Metal Complexes. *Molecules* **2011**, *16*, 1103–1112. [CrossRef] [PubMed]
65. Moreno, O.P.; Araiza, O.R.P.; Portillo, M.C.; Téllez, V.C.; Garrido, M.A.V. Jahn-Teller effect analysis at coordination complex $[Cu(NH_3)_4]^{2+}$ ion, growth by green synthesis in CuS nanocrystals. *Optik* **2022**, *251*, 168470. [CrossRef]
66. Singh, J.; Srivastav, A.N.; Singh, N.; Singh, A. Stability Constants of Metal Complexes in Solution. In *Stability and Applications of Coordination Compounds*; Srivastva, A.N., Ed.; IntechOpen: London, UK, 2020. [CrossRef]
67. Sóvágó, I.; Kállay, C.; Várnagy, K. Peptides as complexing agents: Factors influencing the structure and thermodynamic stability of peptide complexes. *Coord. Chem. Rev.* **2012**, *256*, 2225–2233. [CrossRef]
68. Lihi, N.; Lukács, M.; Raics, M.; Szunyog, G.; Várnagy, K.; Kállay, C. The effect of carboxylate groups on the complexation of metal ion with oligopeptides—Potentiometric investigation. *Inorganica Chim. Acta* **2018**, *472*, 165–173. [CrossRef]
69. Sursyakova, V.V.; Burmakina, G.V.; Rubaylo, A.I. Composition and stability constants of copper(II) complexes with succinic acid determined by capillary electrophoresis. *J. Coord. Chem.* **2017**, *70*, 431–440. [CrossRef]
70. El-Megharbel, S.M.; Hamza, R.Z. Synthesis, spectroscopic characterizations, conductometric titration and investigation of potent antioxidant activities of gallic acid complexes with Ca (II), Cu (II), Zn(III), Cr(III) and Se (IV) metal ions. *J. Mol. Liq.* **2022**, *358*, 119196. [CrossRef]
71. Santoso, S.P.; Angkawijaya, A.E.; Ju, Y.-H.; Soetaredjo, F.E.; Ismadji, S.; Ayucitra, A. Synthesis, characterization, thermodynamics and biological studies of binary and ternary complexes including some divalent metal ions, 2, 3-dihydroxybenzoic acid and N -acetylcysteine. *J. Taiwan Inst. Chem. Eng.* **2016**, *68*, 23–30. [CrossRef]
72. HySS2009, Hyperquad Simulation and Speciation. Available online: http://www.hyperquad.co.uk/ (accessed on 3 January 2025).
73. Alderighi, L.; Gans, P.; Ienco, A.; Peters, D.; Sabatini, A.; Vacca, A. Hyperquad simulation and speciation (HySS): A utility program for the investigation of equilibria involving soluble and partially soluble species. *Coord. Chem. Rev.* **1999**, *184*, 311–318. [CrossRef]
74. Santoso, S.P.; Chandra, I.K.; Soetaredjo, F.E.; Angkawijaya, A.E.; Ju, Y.-H. Equilibrium Studies of Complexes between *N*-Acetylcysteine and Divalent Metal Ions in Aqueous Solutions. *J. Chem. Eng. Data* **2014**, *59*, 1661–1666. [CrossRef]
75. Boyatzis, S.C.; Velivasaki, G.; Malea, E. A study of the deterioration of aged parchment marked with laboratory iron gall inks using FTIR-ATR spectroscopy and micro hot table. *Herit. Sci.* **2016**, *4*, 13. [CrossRef]
76. Tanase, T. *Introduction: What Is a Metal Complex?* ; Tanase, T., Ishii, Y., Eds.; Royal Society of Chemistry: London, UK, 2024; pp. 1–12.
77. Elattar, R.H.; El-Malla, S.F.; Kamal, A.H.; Mansour, F.R. Applications of metal complexes in analytical chemistry: A review article. *Coord. Chem. Rev.* **2024**, *501*, 215568. [CrossRef]
78. Frešer, F.; Hostnik, G.; Tošović, J.; Bren, U. Dependence of the Fe(II)-Gallic Acid Coordination Compound Formation Constant on the pH. *Foods* **2021**, *10*, 2689. [CrossRef]

79. Ye, H.; Jiang, S.; Yan, Y.; Zhao, B.; Grant, E.R.; Kitts, D.D.; Yada, R.Y.; Pratap-Singh, A.; Baldelli, A.; Yang, T. Integrating Metal–Phenolic Networks-Mediated Separation and Machine Learning-Aided Surface-Enhanced Raman Spectroscopy for Accurate Nanoplastics Quantification and Classification. *ACS Nano* **2024**, *18*, 26281–26296. [CrossRef]
80. Shin, J.; Lim, M.H.; Han, J. NMR spectroscopic investigations of transition metal complexes in organometallic and bioinorganic chemistry. *Bull. Korean Chem. Soc.* **2024**, *45*, 593–613. [CrossRef]
81. Pérez-Jiménez, A.I.; Lyu, D.; Lu, Z.; Liu, G.; Ren, B. Surface-enhanced Raman spectroscopy: Benefits, trade-offs and future developments. *Chem. Sci.* **2020**, *11*, 4563–4577. [CrossRef]
82. Sánchez-Cortés, S.; García-Ramos, J.V. Adsorption and Chemical Modification of Phenols on a Silver Surface. *J. Colloid Interface Sci.* **2000**, *231*, 98–106. [CrossRef]
83. Espina, A.; Cañamares, M.V.; Jurašeková, Z.; Sanchez-Cortes, S. Analysis of Iron Complexes of Tannic Acid and Other Related Polyphenols as Revealed by Spectroscopic Techniques: Implications in the Identification and Characterization of Iron Gall Inks in Historical Manuscripts. *ACS Omega* **2022**, *7*, 27937–27949. [CrossRef]
84. Carter, E.A.; Perez, F.R.; Garcia, J.M.; Edwards, H.G.M. Raman Spectroscopic Analysis of an Important Visigothic Historiated Manuscript. *Philos. Transcations R. Soc. A Math. Phys. Eng. Sci.* **2016**, *2082*, 20160041. [CrossRef]
85. Nastova, I.; Grupče, O.; Minčeva-Šukarova, B.; Turan, S.; Yaygingol, M.; Ozcatal, M.; Martinovska, V.; Jakovlevska-Spirovska, Z. Micro-Raman spectroscopic analysis of inks and pigments in illuminated medieval old-Slavonic manuscripts. *J. Raman Spectrosc.* **2012**, *43*, 1729–1736. [CrossRef]
86. Burgio, L.; Clark, R.J.H.; Hark, R.R. Raman microscopy and x-ray fluorescence analysis of pigments on medieval and Renaissance Italian manuscript cuttings. *Proc. Natl. Acad. Sci. USA* **2010**, *107*, 5726–5731. [CrossRef] [PubMed]
87. Liu, C.; Li, C.; Jiang, S.; Zhang, C.; Tian, Y. pH-responsive hollow Fe–gallic acid coordination polymer for multimodal synergistic-therapy and MRI of cancer. *Nanoscale Adv.* **2021**, *4*, 173–181. [CrossRef] [PubMed]
88. Severino, J.F.; Goodman, B.A.; Reichenauer, T.G.; Pirker, K.F. Is there a redox reaction between Cu(II) and gallic acid? *Free Radic. Res.* **2011**, *45*, 123–132. [CrossRef] [PubMed]
89. Pirker, K.F.; Baratto, M.C.; Basosi, R.; Goodman, B.A. Influence of pH on the speciation of copper(II) in reactions with the green tea polyphenols, epigallocatechin gallate and gallic acid. *J. Inorg. Biochem.* **2012**, *112*, 10–16. [CrossRef]
90. Santoso, S.P.; Angkawijaya, A.E.; Soetaredjo, F.E.; Ismadji, S.; Ju, Y.-H. Complex equilibrium study of some hydroxy aromatic ligands with beryllium ion. *J. Mol. Liq.* **2015**, *212*, 524–531. [CrossRef]
91. Sherwani, I.A.H.A.; Köse, A.; Güngör, Ö.; Kırpık, H.; Güngör, S.A.; Köse, M. Synthesis, characterization and investigation of photophysical and biological properties of Cu(II) and Zn(II) complexes of benzimidazole ligands. *Appl. Organomet. Chem.* **2022**, *36*, e6585. [CrossRef]
92. Berto, S.; Alladio, E. Application of Chemometrics Tools to the Study of the Fe(III)–Tannic Acid Interaction. *Front. Chem.* **2020**, *8*, 614171. [CrossRef]
93. Mazaheri, O.; Alivand, M.S.; Zavabeti, A.; Spoljaric, S.; Pan, S.; Chen, D.; Caruso, F.; Suter, H.C.; Mumford, K.A. Assembly of Metal–Phenolic Networks on Water-Soluble Substrates in Nonaqueous Media. *Adv. Funct. Mater.* **2022**, *32*, 2111942. [CrossRef]
94. Novotny, J.; Komorovsky, S.; Marek, R. Paramagnetic Effects in NMR Spectroscopy of Transition-Metal Complexes: Principles and Chemical Concepts. *Acc. Chem. Res.* **2024**, *57*, 1467–1477. [CrossRef]
95. Etou, M.; Yoshida, M.; Tsuji, Y.; Murakami, M.; Inoue, T. Gallic acid complexation with Al^{3+} under acidic condition—^{27}Al NMR and DFT study. *Inorganica Chim. Acta* **2024**, *571*, 122229. [CrossRef]
96. Wang, T.; Lin, Z.; Mazaheri, O.; Chen, J.; Xu, W.; Pan, S.; Kim, C.-J.; Zhou, J.; Richardson, J.J.; Caruso, F. Crystalline Metal–Organic Framework Coatings Engineered via Metal–Phenolic Network Interfaces. *Angew. Chem. Int. Ed.* **2024**, *63*, e202410043. [CrossRef] [PubMed]
97. Lin, Z.; Zhou, J.; Cortez-Jugo, C.; Han, Y.; Ma, Y.; Pan, S.; Hanssen, E.; Richardson, J.J.; Caruso, F. Ordered Mesoporous Metal–Phenolic Network Particles. *J. Am. Chem. Soc.* **2019**, *142*, 335–341. [CrossRef] [PubMed]
98. Moralles, V.A.; Davolos, M.R.; Cebim, M.A. Europium(III) and gallic acid ligand coordination compounds: Synthesis, characterization, photophysical processes, and optimization of luminescent properties. *Inorganica Chim. Acta* **2023**, *558*, 121748. [CrossRef]
99. Kim, N.; Lee, I.-S.; Choi, Y.; Ryu, J. Molecular design of heterogeneous electrocatalysts using tannic acid-derived metal-phenolic networks. *Nanoscale* **2021**, *13*, 20374–20386. [CrossRef]
100. Wei, Y.; Wei, Z.; Luo, P.; Wei, W.; Liu, S. pH-sensitive metal-phenolic network capsules for targeted photodynamic therapy against cancer cells. *Artif. Cells Nanomed. Biotechnol.* **2018**, *46*, 1552–1561. [CrossRef]
101. Tomasetig, D.; Wang, C.; Hondl, N.; Friedl, A.; Ejima, H. Exploring Caffeic Acid and Lignosulfonate as Key Phenolic Ligands for Metal-Phenolic Network Assembly. *ACS Omega* **2024**, *9*, 20444–20453. [CrossRef]
102. Pereira, J.P.; Borges, C.H.; Tabelini; Aguiar, A. A Review of Gallic Acid-Mediated Fenton Processes for Degrading Emerging Pollutants and Dyes. *Molecules* **2023**, *28*, 1166. [CrossRef]

103. Zeng, J.; Cheng, M.; Wang, Y.; Wen, L.; Chen, L.; Li, Z.; Wu, Y.; Gao, M.; Chai, Z. pH-Responsive Fe(III)–Gallic Acid Nanoparticles for In Vivo Photoacoustic-Imaging-Guided Photothermal Therapy. *Adv. Healthc. Mater.* **2016**, *5*, 772–780. [CrossRef]
104. Fitz-Binder, C.; Manian, A.P.; Lenninger, M.; Ortlieb, S.; Bechtold, T.; Pham, T. Dyeing behaviour of iron(III)-gallic acid complexes on wool as function of pH-dependent iron(III)-complex stoichiometry. *Dye. Pigment.* **2025**, *233*, 112502. [CrossRef]
105. Li, Y.; Li, C.; Liu, S.; Wang, Q.; Tang, Z.; Qu, J.; Ye, J.; Lu, Y.; Wang, J.; Zhang, K.; et al. Nano-photosensitizers with gallic acid-involved Fe–O–Cu "electronic storage station" bridging ligand-to-metal charge transfer for efficient catalytic theranostics. *J. Colloid Interface Sci.* **2024**, *676*, 974–988. [CrossRef]
106. Luo, W.; Xiao, G.; Tian, F.; Richardson, J.J.; Wang, Y.; Zhou, J.; Guo, J.; Liao, X.; Shi, B. Engineering robust metal–phenolic network membranes for uranium extraction from seawater. *Energy Environ. Sci.* **2019**, *12*, 607–614. [CrossRef]
107. Rahim, A.M.A.; Ahmed, S.A.; Soliman, E.M. Adsorptive removal of Fe(III) using gallic acid anchored iron magnetic nano-adsorbents synthesized via two different routes under microwave irradiation. *Indian J. Chem.* **2020**, *59A*, 9–20.
108. Pirozzi, D.; Pansini, M.; Marocco, A.; Esposito, S.; Barrera, G.; Tiberto, P.; Allia, P.; Sannino, F. Adsorption of gallic acid by tailor-made magnetic metal-ceramic nanocomposites. *J. Mol. Liq.* **2023**, *371*, 121083. [CrossRef]
109. Nilnit, T.; Supharoek, S.-A.; Siriangkhawut, W.; Vichapong, J.; Ponhong, K. Ultrasound-assisted continuous flow synthesis of natural phenolic-coated Fe_3O_4 for magnetic solid phase extraction of tetracycline residues in honey. *Food Chem.* **2024**, *464*, 141642. [CrossRef]
110. Yan, W.; Shi, M.; Dong, C.; Liu, L.; Gao, C. Applications of tannic acid in membrane technologies: A review. *Adv. Colloid Interface Sci.* **2020**, *284*, 102267. [CrossRef]
111. Hou, Y.; Zhang, Y.; Huang, Y.; Zhou, A.; Han, J.; Yang, K.; Zhao, Y.; Zhou, J.; Wang, J.; Chen, G.; et al. A pH-responsive MOFs@MPN nanocarrier with enhancing antifungal activity for sustainable controlling myclobutanil release. *Chem. Eng. J.* **2024**, *497*, 155713. [CrossRef]
112. Rahim, M.A.; Lin, G.; Tomanin, P.P.; Ju, Y.; Barlow, A.; Bjonmalm, M.; Caruso, F. Self-Assembly of a Metal−Phenolic Sorbent for Broad-Spectrum Metal Sequestration. *ACS Appl. Mater. Interfaces* **2020**, *12*, 3746–3754. [CrossRef]
113. Mazaheri, O.; Zavabeti, A.; McQuillan, R.V.; Lin, Z.; Alivand, M.S.; Gaspera, E.D.; Chen, D.; Caruso, F.; Suter, H.; Mumford, K.A. Solid-State Encapsulation of Urea via Mechanochemistry-Driven Engineering of Metal–Phenolic Networks. *Chem. Mater.* **2023**, *35*, 7800–7813. [CrossRef]
114. Mazaheri, O.; Lin, Z.; Xu, W.; Mohankumar, M.; Wang, T.; Zavabeti, A.; McQuillan, R.V.; Chen, J.; Richardson, J.J.; Mumford, K.A.; et al. Assembly of Silicate–Phenolic Network Coatings with Tunable Properties for Controlled Release of Small Molecules. *Adv. Mater.* **2024**, *36*, 2413349. [CrossRef]
115. Wang, Y.; Zhang, J.; Zhang, C.; Li, B.; Wang, J.; Zhang, X.; Li, D.; Sun, S.-K. Functional-Protein-Assisted Fabrication of Fe–Gallic Acid Coordination Polymer Nanonetworks for Localized Photothermal Therapy. *ACS Sustain. Chem. Eng.* **2018**, *7*, 994–1005. [CrossRef]
116. Jing, Z.; Li, M.; Wang, H.; Yang, Z.; Zhou, S.; Ma, J.; Meng, E.; Zhang, H.; Liang, W.; Hu, W.; et al. Gallic acid-gold nanoparticles enhance radiation-induced cell death of human glioma U251 cells. *IUBMB Life* **2021**, *73*, 398–407. [CrossRef] [PubMed]
117. Novaković, T.B.; Pavlović, S.M.; Pagnacco, M.C.; Banković, P.T.; Mojović, Z.D. The Application of Alumina for Electroanalytical Determination of Gallic Acid. *Electrocatalysis* **2023**, *14*, 18–28. [CrossRef]
118. Goman, D.; Stanković, A.; Galović, O.; Džakula, B.N.; Kontrec, J.; Medvidović-Kosanović, M. Complexation of gallic acid with calcium: Electrochemical, potentiometric, and UV-VIS studies. *Anal. Methods* **2024**, *16*, 391–395. [CrossRef] [PubMed]
119. Hamedi, H.; Javanbakht, S.; Mohammadi, R. In-situ synthesis of copper-gallic acid metal–organic framework into the gentamicin-loaded chitosan hydrogel bead: A synergistic enhancement of antibacterial properties. *J. Ind. Eng. Chem.* **2024**, *133*, 454–463. [CrossRef]

MDPI AG
Grosspeteranlage 5
4052 Basel
Switzerland
Tel.: +41 61 683 77 34

Molecules Editorial Office
E-mail: molecules@mdpi.com
www.mdpi.com/journal/molecules

www.ingramcontent.com/pod-product-compliance
Lightning Source LLC
Chambersburg PA
CBHW061601210926
51850CB00009B/2172